Management Information and Optoelectronic Engineering

Management Information and Optoelectronic Engineering

Proceedings of the 2016 International Conference on Management, Information and Communication (ICMIC2016) and the 2016 International Conference on Optics and Electronics Engineering (ICOEE2016)

Guilin, China 28 – 29 May 2016

Editor

Yongsheng Gao

Hebei University of Science and Technology, China

World Scientific

NEW JERSEY • LONDON • SINGAPORE • BEIJING • SHANGHAI • HONG KONG • TAIPEI • CHENNAI • TOKYO

Published by

World Scientific Publishing Co. Pte. Ltd.
5 Toh Tuck Link, Singapore 596224
USA office: 27 Warren Street, Suite 401-402, Hackensack, NJ 07601
UK office: 57 Shelton Street, Covent Garden, London WC2H 9HE

British Library Cataloguing-in-Publication Data
A catalogue record for this book is available from the British Library.

MANAGEMENT INFORMATION AND OPTOELECTRONIC ENGINEERING
Proceedings of the 2016 International Conference on Management, Information and Communication (ICMIC2016) and the 2016 International Conference on Optics and Electronics Engineering (ICOEE2016)

ISBN 978-981-3202-67-2

Printed in Singapore

Preface

2016 the International Conference on Management, Information and Communication (ICMIC 2016) and 2016 the International Conference on Optics and Electronics Engineering (ICOEE 2016) have been held by The Scientific Research and Technical Service Center (SRTSC) and Hebei University of Science and Technology. The goal of two conferences are to provide a platform for researchers, engineers, academicians as well as industrial professionals from all over the world to present their new research results and development activities in respective fields. They also allow the attendees both in industry and academia to share their state-of-art results, to explore new areas of research and development.

ICMIC 2016 is focus on Management Science, Information and Communication. However, ICOEE 2016 is focused on Optics Engineering and Electronics Engineering. After receiving papers reviewed by two famous specialist in this field, we collected 80 papers into the conference proceedings.

There is a long list of individuals that we would like to thank for their support in organizing these conferences. We would like to take this opportunity to express our thanks to the individuals and organizations for their efforts to serve the conference. We would also like to extend our thanks to the conference academic committee, technical committee and organizing committee for their hard work. Lastly, we would like to express our appreciation to each participant.

The Organizing Committee of ICMIC 2016 and ICOEE 2016
Shijiazhuang, Hebei Province, China, 2016.

Table of Contents

Preface v

Section 1 Management, Information and Communication **1**

The Analysis of Entrepreneurs' Innovation Cognition in Hebei Province 3
Rong-Ping Li, Si-Ying Wang and Sen Zhao

Reviews of International Maritime Safety Management Philosophies and Approaches 9
Wei, RUAN

Efficient FEC Decoding Algorithm for 100Gbps Optical Transport Networks in FPGA 19
Jiang-Yi Qin, Zhi-Ping Huang, Jun-Yu Wei, Zhi Dong and Zhen Zuo

An Improved Line Non-symmetry and Anti-packing Model For Gray Images Representation.. 28
Fei Xiao, Jinwen Tian and Guangwei Wang

The Study of Competency Model about Endurance Ltd.'s Salesman 35
Li-Hong Han and Shao-Fan Zhang

Short-Term Electricity Price Forecasting Based on Nonparametric GARCH Residuals Correction-Least Square Support Vector Machine 42
Yuan-Sheng Huang,Shen-Hai Huang and Yin-Ping Hu

College Students' Work Innovation Model Research under the Background of Internet 49

Shu-Yang Zhu

An Empirical Research of the Input-Output Model on Demand Elasticity Based on the Energy Consumption 56
Yan Zhang and Xin Liu

Research on Network Public Opinion Evolution Model Based on Mobile Cellular Automaton 63
Chao-Bin Wang and Qian-Ren Mao

Rule and Maximum Entropy Model Based Method of Temporal Expression Recognition.. 73
Jie-Jun Li, Qing-Qiu Cai and Hao Tang

Feature Mining and Sentiment Orientation Analysis on Product Review 79
Jie-Jun Li, Han Yang and Hao Tang

Design of Raptor Code for Space-to-Ground High-speed Communication 85
Bao-Li Hong, Zhang-Wei and Li-Hong Bao

Fund Risk Early Warning System Design of Power Grid Enterprise Group 94
Lin-lin Wang

Layer Aggregation and Structure Detection of Interconnected Networks 105
Quantong Guo, Xin Jiang, Yanjun Lei and Zhiming Zheng

Design of Network Security Distributed Pre-Warning System 113
Ya-Lin Ye, Wei-Feng Sun, Kun Han, Quan Yuan, Ke-Li Yang and Ning Shan

Performance Evaluation of Different Grade INS Aiding GNSS Receiver Acquisition in Hot Start State 119
Zhou Qin, Wen-Xiang Liu, Ying-Xue Su and Fei-Xue Wang

Comprehensive Evaluation Index Model Construction and Application in the Transnational Power Transmission Project 129
Huang Qing

The Construction and Use Of Implicit Relationship In Knowledge 137
Jing Chen, Yun-Qing Li and Hu Kuang

The Efficiency Analysis of the Construction Industry in Linyi City Based on DEA Model 146
Xiang Liu and Zong-Ying Li

Loss Effect on the Dynamical Evolution of Rigid Holographic Bright Spatial Solitons in a Dissipative System 152
Guang-Yong Zhang, Mi-Zhuo Zhao, Guo-Zhen Chen and Xiao-Qing Zhang

The Efficiency Analysis of Logistics Industry in Shandong Province Based on VRS DEA Model 158
Zong-Ying Li and Xiang Liu

Study on International Electricity Market Regulation Models and China's Practice 164
Huang Qing

Building and Application of Project Management Maturity Model for Military Equipment Medium-Maintenance 171
Yong-Dong Zhang, Zheng-Ping Shu and Zhong-Guang Li

The Algorithm of the Regional Centroid Indoor Positioning Based on RSSI Filter 180
Ji-Hua Ye, Qing-Ping Lan and Jun Chen

Acquisition and Research of the Soil Building Environmental Data Based on the Sensor Network 190
Yuan Zhang, Chao-Ming Wang, Fan Zhang, Zhi-Yong Zhang and Xue-Wen Wang

Research on Influencing Factors of Employees' Acceptance of MOOC Training 197
Shuai Wang, Pin-Hua Dong and Shuai Shao

Decomposition and Modeling of Equipment Support Mission 204
Tian Zhao, Wei Zhang, Yao-Ping Lv and Xian-Jun Pan

Motion Simulation of Tracked Robot Under the Virtual Reality Environment 210
Ma Jiu-he, Chen Zhi-wei, Sun Hai-tao and Deng Hui-yong

Research on Social Media Marketing Strategy of Small and Micro Enterprises under the Background of "New Normal" of Economy 216
Fang Gao and Xin Zhao

Section 2 Optics and Electronics Engineering 223

A Novel Optical Fiber Transmission System Based on Dynamic Payload Clock Recovery Technology 225
Jiang-Yi Qin, Zhi-Ping Huang, Jun-Yu Wei, Zhi Dong and Zhen Zuo

The Research of Signal Power Quality Disturbance Identification and Classification Based on the GA-ELM 233
Bo Sang and Hong-Wen Liu

Research on Protection Configuration Optimization of Power System 239
Xian Li, Yubin Mao, Xinju Guo and Jinli Yu

Muiltidimensional Collaborative Constellation Design and Implement for MIMO Indoor Visible Light Communications 247
Wang-Feng Liang, Yi-Jun Zhu and Heng-Bin Cai

Motion of Transceivers in MCNS for Underwater Optical Wireless Communications 254
Hao Xu and Yi-Jun Zhu

An Instantaneous Frequency Extraction Method in the Application of Laser Doppler Velocimeter 260
Guo-Hua Wang, Si-Chen Xu and Wei-Shuo Zhang

The Improved Coaxial Bragg Structure 270
Xue-Yong Ding, Lian-Sheng Wang and Yuan Wang

The Influence of Return Stroke Current Waveforms to Approximation of Spatial Lightning Electromagnetic Field at Near Ranges 278
Li-peng Wang, Ya-zhou Chen, Hao-jiang Wan and Xiaojia Wang

Sorting Radar Signal based on the Resemblance Coefficient of Bispectrum Two Dimensions Characteristic 286
Chun-Lai Yu, Jun Han and Fang Wan

Research on Software Testing of FPGA Embedded System 295
Hong-Feng Wang, Hui Zhang and Qiang Li

A Judgmental Method of Illegal Electricity Based on BP Neural Network and Errors Theory 301
Jian-hao Lu, Sheng-ke Wen, Yi-xiong Nie and Yue-hui Yang

A Simple Frequency Domain Equalization Method for Atmospheric Laser Communication 310
Hui-Qin Wang, He Wang, Jing-wei Zhu, Ke-jun Jia and Ming-hua Cao

The Prediction of Time Domain Electromagnetic Pulse Waveform Based on the Minimum Phase Method 317
Qiang Li and Ming Wei

Design and Simulation of Neural Network Unit Based on FPGA 323
Zhu Fang and Ding Lei

Research on the Harmonic Governance Program for the Enterprise Distribution System with Photo-Voltaic 332
Yan-Qing Pan and Hui-Bin Sui

Design of Imaging System for Infrared Tracker Simulator 340
Ming-Chang Wang and Yang-Yu Fan

A Simple Implementation of Long Distance Visible Light Communication System 349
Ziyu Song and Shengliang Peng

The Influence of Magnetic Field on the Spatial Resolution of Pulse-Dilation Framing Tube 356
Yan-Li Bai

Research on Condition Evaluation Technology of Energy Storage Equipment in Photovoltaic System 364
Hai-Feng Bian, Rong-Jiang Jin, Chao Zhang, Tian-Chun Xiang and Xu-Dong Wang

Research on Laser Eavesdropping of Interference Method Based on Linear CCD 377
Jian An, Jia-Hao Song and Zhi-Yong Dai

Quantum Chemistry Calculation of the SOF_4 Product Formation Mechanism in SF_6 Discharge Gas 385
Zhen Li, Shao-yi Chen, Shang-kun Gong, Bing Feng and Zhou Zhou

A Power Estimation Method of Coarse Grained Reconfigurable Pipelined Processor Based on Macro-model 391
Wang Qi and Si-Kun Li

The Analytical Solutions of Two-Photon Spatial Solitons in Biased Photovoltaic Photorefractive Crystal 401
Guang-Yong Zhang, Guo-Zhen Chen, Mi-Zhuo Zhao and Xiao-Qing Zhang

Intellectualized Large Diameter Aspherical Reflector Manufacture Technology 407
Bao-Wei Qiu, Wen Guo and Ting-Ting Hao

Research on Gas Chromatography-Mass Spectrometer Method Measuring Corrosive Sulfur in Mineral Insulating Oil 413
Tao Wan, Jun Hu, Bing Feng, Zhou Zhou, Hui Qian and Ying-Jun Qiu

Lens's Design and Verification Used for Terahertz Beam Collimation 420
Qiu-Jie Yang, Jing-Guo Huang, Zhi-ping He, Zhi-Ming Huang and Rong Shu

Severity Diagnosis of Surface Partial Discharge Provoked by Immobilized Metal Wire on GIS Insulators 429
Xie Yaoheng, Zhou Weihua, Ye Huisheng, Liu Yun, Zhang Guoqi and Ding Yuzhu

Design of a Vertically Stacked Substrate Integrated Circular Cavity (SICC) Filter in LTCC Technology 435
Huan Wu, Jian-Rong Chen, Wen-Qiang Jia and Xiao-Yang Zhang

Research on the Impact of Meteorological Factors on Electric Power Accidents 443
Chen Jia, Chengrui Wang, Chao Zhang and Jing Zhou

Optical Analysis and Simulation of Spherical Vapor Cell 456
Yu-Ting Zhao, Gang Liu, Wei Quan and Ming Ding

Author Index 465

Section 1

Management, Information and Communication

The Analysis of Entrepreneurs' Innovation Cognition in Hebei Province

Rong-Ping Li†, Si-Ying Wang and Sen Zhao

Department of Economics and Management, Hebei University of Science and Technology, Shijiazhuang, 050018, China

†E-mail: 464157044@qq.com

According to the *entrepreneur innovation survey questionnaire* compiled by the National Bureau of Statistics, the paper makes the analysis from six aspects as follows: the affect of innovation on the survival and development of enterprises, the situation of enterprises to establish innovation strategy, the degree of influence of innovation types on enterprises innovation, the factors of the success of enterprises innovation, the effect of the incentive measures and the view of entrepreneurs on relevant policies of innovation, and then we find that the technological innovation of enterprises in Hebei Province has the major problems: the strategic goal of innovation is obscure, the innovative consciousness is weak, the enterprises put too much emphasis on material rewards, policy support is inadequate and so on, finally put forward appropriate countermeasures to solve these problems, hoping to provide references for the improvement of technological innovation capability of enterprises in Hebei Province.

Keywords: Hebei Province; Entrepreneurs; Innovation; Cognitive Analysis; Countermeasures and suggestions.

1. Introduction

Entrepreneurs play an important role in the process of carrying out technological innovation in the enterprises, they occupy the core position of technological innovation of enterprises[1–2]. To understand the comprehension of entrepreneurs to innovation and relevant policies, Hebei Provincial Bureau of Statistics has surveyed 21,142 entrepreneurs throughout the province in 2013-2014. By analyzing the results of the survey, we can identify the weak links of enterprises innovation and propose reasonable strategies and measures, thereby helping enterprises to develop right technological innovation strategy, enhancing the overall level of technological innovation capability of enterprises in the whole province, ultimately promoting the healthy and rapid development of enterprises in Hebei Province.

2. The analysis of entrepreneurs' awareness for innovation activity

2.1. *The effect of innovation on the survival and development of enterprises*

According to the survey results, 25.82% entrepreneurs believe that innovation has an important effect on the survival and development of enterprises, 60.66% entrepreneurs believe that innovation plays a role in the survival and development of enterprises, while 13.52% entrepreneurs believe that innovation makes no difference. It shows that although the majority of entrepreneurs recognize the importance of innovation in Hebei Province, only a quarter of entrepreneurs understand completely, there are more than one-tenth of entrepreneurs do not realize that innovation is the foundation of the survival and development of enterprises.

2.2. *The situation of enterprises to establish innovation strategy*

In the survey, 52.61% enterprises say that they have developed the objectives of innovation strategy for the next few years. There are 11,144 enterprises that have developed the objectives of innovation strategy, 52.9% enterprises have chosen "increase the investment in innovation, enhance the competitiveness of enterprises"; 20.8% enterprises have chosen "catch up with the domestic leading enterprises in the same industry"; 16.38% enterprises have chosen "maintain current level of technology and the state of production and management." This shows that there are few enterprises setting the goal of reaching the international standards, and even 47.39% enterprises do not develop the objectives of innovation strategy, innovation mentality is unclear.

3. The analysis of factors and their influences on the success of enterprises innovation

3.1. *The analysis of influencing factors*

In the survey of different factors on the influence of the success of enterprises innovation, 10,464 entrepreneurs have various perceptions. 43.31% entrepreneurs believe that "high-quality personnel" has a high degree of influence, 41.5% entrepreneurs believe that "innovative entrepreneurs" has a high degree of influence. Both of these factors are closely bound up with "talent," thus it can be seen sufficient talent is essential for the success of enterprises innovation. Besides, more than 35% entrepreneurs believe that "employees' sense of identity to enterprises," "incentive measures within the enterprises" and "unimpeded information channels" all have a high degree of influence on the success of enterprises innovation.

3.2. *Incentives and effect analysis*

In Table 1, it is shown that material rewards and promotion are the most effective means, acceptability of new incentives such as stocks or stock options is still relatively low.

Table 1. The measures and their effect of employee motivation in Hebei Province in 2013-2014

Incentive measures	The entrepreneurs' judgment of incentive effects							
	Good		General		Ineffective		Unused	
	Number (person)	Proportion (percent)	Number (person)	Proportion (percent)	Number (person)	Proportion (percent)	Number (person)	Proportion (percent)
Increase pay and bonus	5032	48.09	4391	41.96	209	2.00	824	7.88
Position adjustment or promotion	4066	38.86	4474	42.76	269	2.57	1644	15.71
Training or further education	3343	31.95	4410	42.14	293	2.80	2405	22.98
Material rewards such as cars and housing	1302	12.44	2570	24.56	273	2.61	6300	60.21
Stocks or stock options	1009	9.64	2585	24.70	299	2.86	6557	62.66

4. The analysis of the views of entrepreneurs on innovation policies

In the survey of effect of nine policies on enterprises' innovation activities, 30.2% entrepreneurs believe that "exemption of import taxes on technology development supplies" has no or a low impact on enterprises' innovation

Table 2. The effect of relevant policies on Hebei enterprises' innovation activities in 2013-2014

The types of policies	Number (person)	The influence is "low" or "no" (percent)
Exemption of import taxes on technology development supplies	6410	30.32
The incomes of technology transfer and development are exempted from value-added tax and the incomes of technology transfer enjoy income tax reduction	6129	28.99
High and new tech enterprises enjoy the reduction or exemption of enterprise income tax	5840	27.62
Special instruments and equipments of enterprises' R& D activities enjoy accelerated depreciation	5666	26.80
The enterprises can enjoy the policy of R&D expenses deduction	5561	26.30
Support policies of preferential developmental industries	4672	22.10
financial support policies	4664	22.06
The relevant policies of the creation and protection of intellectual property	4599	21.75
Encourage enterprises to attract and foster the talents	4244	20.07

activities, 80.41% entrepreneurs think the effect of "encourage enterprises to attract and foster the talents" is relatively high.

5. The main problems of enterprises technology innovation in Hebei Province

(1) The strategic goal of innovation is obscure; more enterprises do not formulate long-term development plans.
(2) The innovative consciousness is weak; entrepreneurs themselves have not yet formed strong consciousness of innovation.
(3) The enterprises put too much emphasis on material rewards; to provide learning and training opportunities is more effective than giving material rewards and promotion for scientific researchers[3]–[6].
(4) Policy support is inadequate; it results in a great reduction in the conversion rate of technological innovation product and application.

6. The countermeasures and suggestions on improving the technological innovation capabilities of enterprises

6.1. *Strengthen the concept of innovation, improve the consciousness of employees on innovation*

Enterprises need to pay attention to the importance of concept and consciousness of innovation, enhance the consciousness of entire personnel on innovation, the entrepreneur spirit should be cultivated at first. Entrepreneurs only have the innovation spirit is far from enough, they should also make the staff understand and pay attention to the code of conduct on innovation, and promote the entire personnel to participates in and support innovation activities.

6.2. *Emphasize the cultivation of talents, establish efficient staff incentive system*

The enterprises should not only attract more talents through various channels and methods, but also retain outstanding talents and make good use of talents[7]. Firstly, the enterprises should select professional talents preferentially; secondly, the enterprises and universities can cooperate to cultivate talents. Meanwhile, the enterprises also need to perfect the incentive measures and pay attention to the incentives such as stocks or stock options.

6.3. *Clearly strengthen enterprise management innovation, technological innovation to develop a reasonable strategy*

In the selection of innovation strategies, the enterprises must be clear about the importance of enhancing the capacity for independent innovation; moreover, the enterprises should pay attention to the secondary innovation of imported technology[8]. If the enterprises are lack of the manpower, material resources and financial resources, then we need to introduce advanced technologies and to digest, absorb and recreate the technologies introduced, which is the most effective way of conducting technological innovation for the enterprises.

6.4. *Perfect the fiscal taxation policies and financial support policies, build the policy environment of promoting enterprises' innovation*

The government should implement appropriate fiscal taxation support policies to encourage enterprises to increase the investment in science and technology[9][10]. In addition, the government also needs to implement financial support policies, give more support in enterprises technological innovation loans, and adjust the structure of science and technology loans appropriately to encourage financial institutions to support the enterprises in carrying out technological innovation.

6.5. *Improve the external environment of innovation, create a good cultural atmosphere of innovation*

The government can propaganda the important ideology about the innovation through various mass media to make everybody have a spirit of adventure, an attitude of grasping the opportunity and a mentality of aggressive innovation. Furthermore, the government can also create a social and cultural atmosphere conducive to enterprises technological innovation by taking entrepreneurs and scientific researchers who have achieved innovation success as examples for encouraging more people to innovate and to pursue the success.

Acknowledgment

This work is supported by the Natural Science Foundation of Hebei Province (No. G2012208057).

References

1. Ying Lv, Industrial Entrepreneurs' Cognition of Innovation Activities in Ningxia (Science and Economy, 2008) pp.56–57.
2. Qun-Lin Fan, Hua-Ping Wu, Yun-Fei Shao, The Intermediary Role of Entrepreneurship and Technological Innovation Ability in Cultural and Technological Industry (Finance and Accounting Monthly,2015) pp.29–33.
3. Zhong-Ju Liao, Hua Cheng, Research on The Influencing Factors of Incentive Measures and Performance of Enterprises Technological Innovation (Science Research Management, 2014) pp.60–66.
4. Zhong-Ju Liao, Dao-You Wu, Hua Cheng, Analysis on The Preference for Innovation Incentive Measures of Scientific and Technological Talents and Countermeasures: Based on The Statistical Research of the 2196 questionnaires in Zhejiang Province (Science and Technology Progress and Policy, 2013) pp.145–149.
5. Shi-Min Wu, The Survey and Analysis Report on Entrepreneurs Innovation Survey Questionnaire in Shanghai (Statistical Science and Practice, 2015) pp.40–42.
6. Jian-Xiang Sun, Entrepreneurs' Knowledge of Science and Technology Innovation: Analysis on The Results of Questionnaire Survey of Industrial Entrepreneurs in Jiangsu Province (Jiangsu Science and Technology Information, 2007) pp.8–12.
7. Jue-Liang Bu, Exploration on Integration of Enterprise with University to Train Talents (Management and Technology of SME, 2013) pp.42-45.
8. Yue-Ya Hu, Investigation and Research on the Innovation Activities of Industrial Enterprises in Wuxi (Jiangnan Forum, 2008) pp.33–35.
9. Zhi-Run Xiao, Li-Juan Zhang, Wei Chen, Research on The Policy Environment for Promoting Independent Innovation of Enterprises (Jiangsu Commercial Forum, 2010) pp.1441–146.
10. Sai-Qin Zhou, Xiao-Yuan Lan, Research on The System of Independent Innovation Policy (Development Research, 2011) pp.13–20.

Reviews of International Maritime Safety Management Philosophies and Approaches

Wei, RUAN†

Merchant Marine College, Shanghai Maritime University,
Shanghai, Shanghai 201306, China
†*E-mail: weiruan@shmtu.edu.cn*

The international maritime safety has been always the interest of the international economy due to the fact that over 90% world trade volume is shipped by sea. In reality, the international maritime safety is maintained by executing a maritime standard system consisting of maritime conventions, rules, regulations and guidance, all made by International Maritime Organization (IMO). The value of the paper lies in its distillation, summarization and remarks on the key philosophies and approaches among major international maritime instruments and maritime management practices, focusing on general system management methodologies, quality management system, Formal Safety Assessment (FSA) and Goal-Based Standards (GBS), with an aim to discover the paths of achieving modern maritime safety and to remind the interested parties to work in light of the trends of international maritime safety management.

Keywords: Maritime system; safety management engineering; system management; maritime legislation

1. Introduction

The safety system is a complicate "man-machine-environment" system. The relationships between structures, functions and behaviors of the system, the environment and the system, are dynamic, dialectical and antilogous. The safety system engineering is a theory to identify, diagnose, assess, predict, remove and control various risks and hazards in the operations of a system, by applying the axioms and approaches of system engineering, and to analyze, evaluate and handle the processes of manufacturing, equipment, circle of production and funds, so that best performances of the system can be achieved. The main tasks in the following figure should be accomplished in other words to obtain the system safety [1].

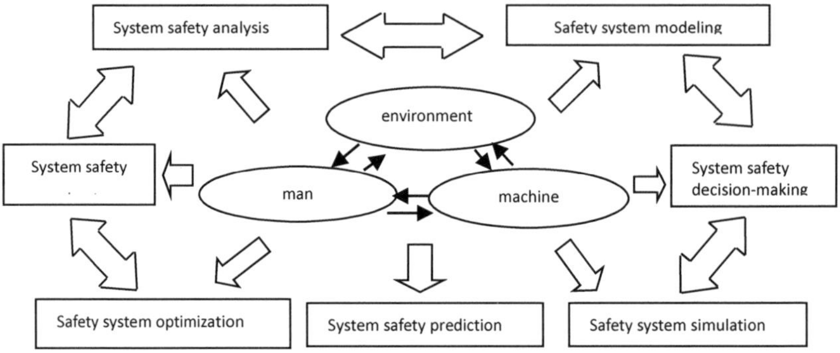

Fig 1. Theoretic framework of safety system engineering.

The international maritime industry owns a great variety of operations, constituting thus a giant system. The great challenge faced by the International Maritime Organization (IMO) confronting the system is to manage such a huge "man-machine-environment" system to be "safe, secure, environmentally sound, efficient and sustainable" [2]. Safety system engineering is an important approach which enables the organization to fulfill the tasks. Based on Fig 1, the Safety System Optimization, for the purpose of international maritime safety management, can be realized by introducing a standardized quality management system, and System Safety Analysis, System Safety Evaluation, System Safety Prediction, and System Safety Decision-making can be done through Formal Safety Assessment (FSA) and Goal-Based Standards (GBS). Lastly Safety System Modeling and Safety System Simulation are widely used in maritime infrastructure designing and operations.

Digging into the essences of the works of IMO, it can be observed that IMO sticks to the strategies that a sound maritime standards internationally and domestically is the prerequisite for maritime safety. This means the international and also national maritime safety standards should be well devised. The next step is then to fully implement the standard system. The following philosophies and approaches are therefore well applied.

2. The system safety engineering philosophies and approaches

2.1. *Systematic maritime legislations*

In the maritime "man-machine-environment" system, "man" is referred to those personnel such as crewmembers onboard, pilots, surveyors, maritime administrators, PSCO and Flag State officers, etc. who are regulated by STCW, MOUs, III Code, etc; and the "machine" refers to the ships and all installations onboard, for which the SOLAS, MARPOL and other instruments apply; while the "environment" means the spaces where "man" and "machine" locates. The MARPOL is the leading instrument for the purpose, together with Anti-fouling Convention, Ballast Water Management Convention, Ship-Recycling and Codes for dangerous cargo transportation (INF, IMSBC, IMDG, IBC, IGC, etc,). IMO has therefore built up an international maritime standard system.

The safety management exerted by the IMO can also be evidenced by the tracks that more system units and more maritime activities are being involved into the system, from traditionally the maritime safety and pollution to security and efficiency of transportation, from general cargo to specialized cargo transportation, from the discharge of basic ship-borne pollutants to the control of vibration and noises from the ships, from international navigable waters to polar zones, and from the discharge of SOx and NOx to black carbon. Simultaneously, the safety management becomes more thorough – all the products supplied onto new ships under construction should be reported and recorded for their content of hazardous substances, according to the 2009 Ship-Recycling Convention [3].

It can be also observed that the maritime legislation system is following a time sequence and is forming an enclosed circuit of a maritime operation (Fig 2).

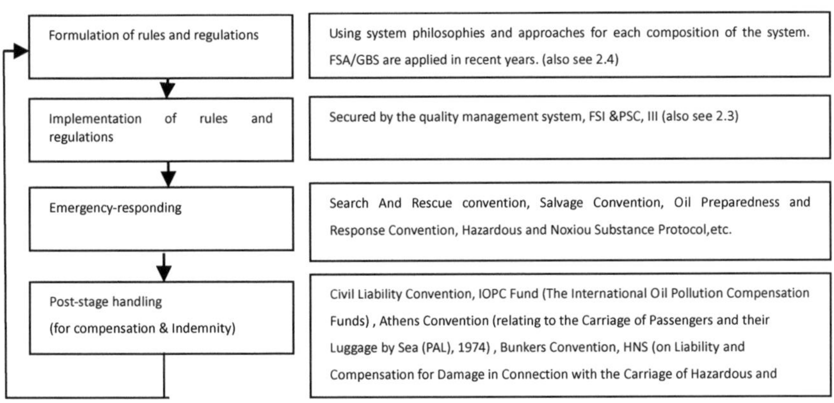

Fig 2. The enclosed maritime legislation system.

2.2. *Systematic implementation of maritime standards*

A systematic legislation should be supplemented by a systematic ways of enforcement. IMO developed therefore a global implementation system and the core thereof is that it is guaranteed by an internationally-networked maritime administrations and organizations, where maritime safety standards are able to be executed from maritime nations down to individual ships and where international ships are seamlessly and continuously monitored, regardless of their locations of being in ports or open seas (Fig 3). Some management approaches such as risk management, as well as information technologies are widely used in such a system.

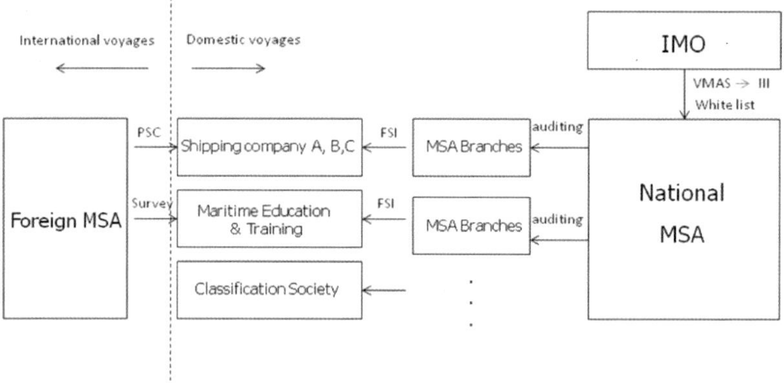

Fig 3. Global maritime safety supervision system

It is worthwhile to pay more attention to more details of the supervision and monitoring system as illustrated by Fig 3, to see more trends in maritime safety management. First, some safety management practices are interlinked, e.g, the results of inspection initiated by III Code would be cross-referenced in evaluating and awarding "low risk profile" ships as defined in the Paris MOU and Tokyo MOU[4]. Secondly, a "credit-based" PSC scheme (similar to the "21st quality shipping" as practiced by USCG) is promoted in which ships with low risks are awarded with a favorite policy of extending inspection intervals to be up to 36 months. Lastly, quantified and informatized management approaches for maritime safety are applied in the system. For example, both FSI and PSC are operating a database which helps the inspectors to target substandard vessels, evaluate the safety level of the ships and record the performance of the ships.

2.3. *Quality management system approach*

The quality management system is first of all an approach to secure the performance of a system. It accentuates the use of procedural approach to induce an organization to manage and operate systematically over various processes and their interactions in light of policy and strategic direction. Derived from the general industrial practices in the use of ISO standards, ISM is dedicated for the safety of maritime systems. Since 1998, the Code has been promoting effectively the management policies and regulations, the input of various resources for the maintenance of maritime safety, the capacity of emergency-responding of individual shipping companies and the reduction of human errors at sea. The latest development of ISO standards, according to ISO9001:2015, is forging the combination of Plan-Do-Check-Action (PDCA) and risk-based thinking into the quality management systems, the former being able to secure proper management for the processes, to provide adequate resources and to ascertain opportunities to improve and action, and the latter being helpful for fixing out various factors leading the processes and system to deviate from the planned outcomes, to take preliminary controls to minimize adverse impacts, and to harness all potential opportunities[5]. The development of general ISO standards will improve the ISM correspondingly.

More importantly, IMO has extended such an approach to the implementation of international maritime standards through the "White-list" scheme of STCW and III Code. The III code, in particular, which has been entered into force since 1st, Jan, 2016, is essentially an application of quality management system, where maritime countries applicable are regarded as an "individual entity" (like an individual shipping company) and required to establish a quality management system internally. Procedural approach, records and evidences, PDCA circles are requested to be established and obtained to show full implementation of SOLAS, MARPOL, STCW, Loadline Convention, Tonnage Convention and COLREG.

2.4. *Formal Safety Assessment (FSA) approach*

The assessment and decision-making based on the risk analysis is a trend for maritime safety management because of their functions in predicting the potential risks and hazards [6]. That is, risks existing in the maritime systems are to be identified and evaluated, and then reasonable risk management tools are adopted to make correct decisions in order to control the risks in the principle of As Low As Reasonably Practicable (ALARP). FSA, the maritime risk assessment approach formally recognized by IMO, is a structured, quantitative and systematic approach for comprehensive safety of the system. At present, some FSA research reports have been submitted to IMO, which carried out high-level analysis for relevant ships like container ships, bulk carriers, tankers, Roll-on & Roll-off ships, etc. Those analysis reports accomplished completely the five steps of the FSA process and quantitative analysis [7].

In recent years, IMO has realized ulteriorly the potential application of risk management on its legislative activities. FSA can be used to ascertain the safety level determined by the IMO standards, to compare and assess such a level with other industries, i.e, making risk analysis based on statistical analysis to determine the safety level, and then decide whether it is necessary to amend the existing legislation so as to adjust the safety levels (some researchers have been done[8]), and to it combines cost – effectiveness analysis into the decision-making process[9].

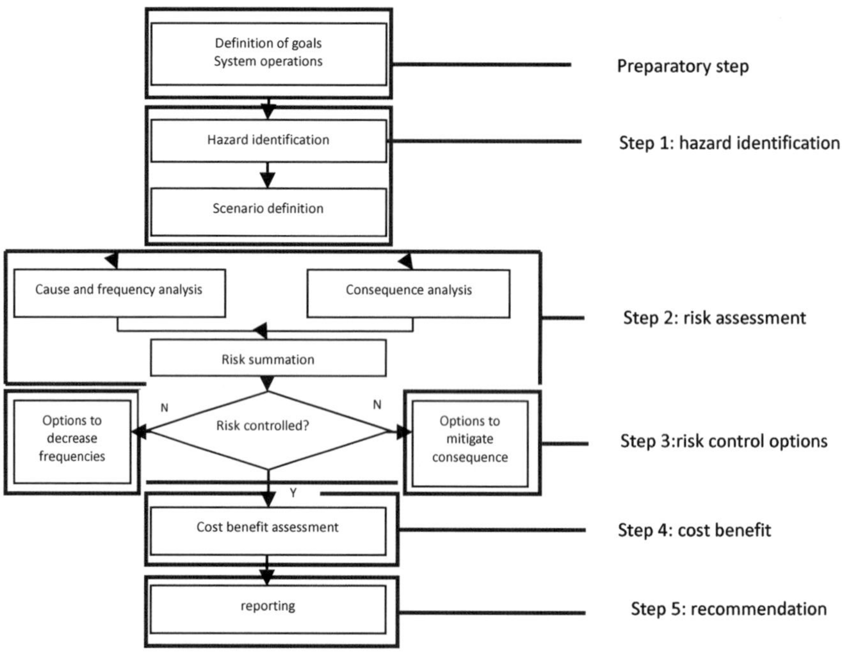

Fig 4: Rule-making in IMO and FSA [10]

2.5. *Goal-Based Standard（GBS）approach*

GBS aims to initially make ship-building standards of classification societies to be consistent with safety goals and functional requirements set by IMO. GBS is derived from ship-building, but it eventually becomes an approach for safety management due to its influences upon the formulation and implementation of international maritime standards. The greatest value of GBS for international maritime safety is to secure the dynamic conformities between "objectives &

functions" and concrete "criteria and norms" (see the Fig 5 & 6), and the other being the introduction of risk management approaches.

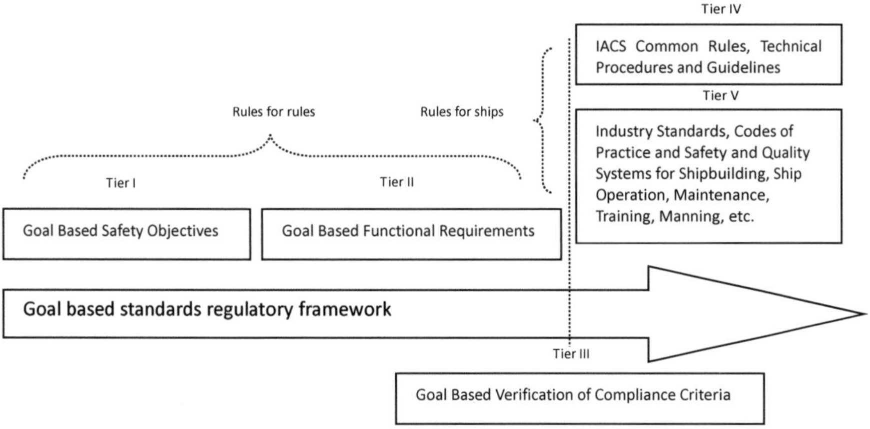

Fig 5. the GBS structure

GBS has thus become a substantial approach for future maritime safety management. The Maritime Safety Committee (MSC) of IMO considers even, with the application of GBS/FSA, whether all IMO instruments under development and classification rules and other compulsory standards should clearly describe the functional requirements and the significant reasoning for such legislation by following a standardized assessment formats [11].

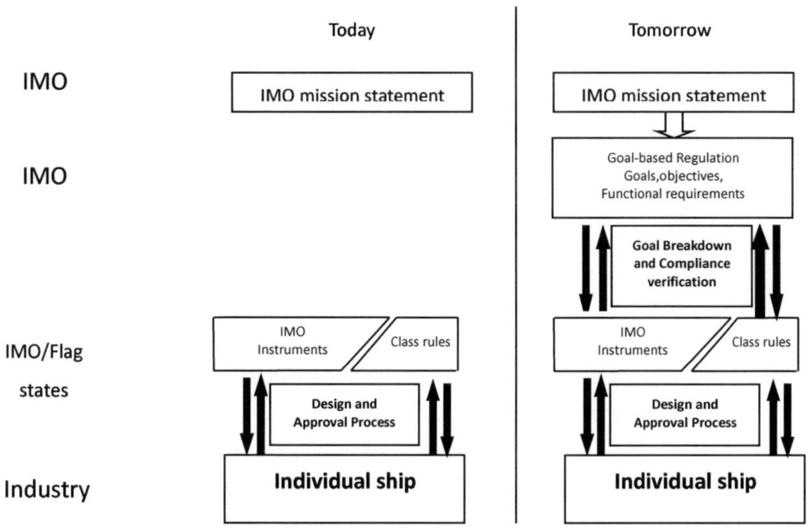

Fig 6. Comparison between the framework of GBS[12]

GBS can be achieved by “Prescriptive Approach (PA)” and “Safety Level Approach(SLA)”. The former refers to descriptive and mandatory norms however the approach uses “objective and functional requirements” and requires a five-tier system. PA is first stage to explore GBS and IMO has been applied in new shipbuilding standards for bulk carriers and tankers, for instance, “Chapter II-1 Amendment of SOLAS Convention”(Res.MSC 290(87), “Goal-Based Standards of international bulk carriers and tankers (Res.MSC 287(87)”, “Guidelines for the verification of the Goal-based Standards (Res. MSC 296(87)”, etc”. PA currently done is based on definite functional requirements derived from long-term practical experience in the field of bulk carrier and tanker operations. However, the approach is on the basis of historical experience rather than the risk level on the whole.

SLA is a risk-based approach and aims to set up a unified safety level for ships in order to describe the ship safety goals and objectives by risk levels. When the SLA is applied, the goal-based standards will use IMO risk acceptance approach to define different acceptable and reliable levels, which can promote the development of modern maritime legislation in a consistent, transparent and reliable way. In accordance with the high-level definition of the SAFEDOR project of Europe Union: the risk-based designing is a comprehensive approach which can involve risk assessment into the designing process, and integrate the goals of risk prevention and alleviation with traditional designing goals. The required safety level is the maximum acceptable risk level in making the rules. The risk-based ship-designing brings the risk analysis into the traditional designing process. It integrates safety performance assessment into the ship designing, and risks are used for the measurement of safety performance. Figure 7 illustrates the concept and process to adjust the international maritime instruments according to safety levels, by turning the “safety knob”.

SLA sets objectives only but never forcibly stipulates the ways of compliance, which hints that it allows to achieve the objectives with “equivalent and alternative” ways approved by competent administrations. A few works have been done based on the approach such as “the alternative design and layout of fire safety” in the chapter II-2 of article 17 of the SOLAS Convention, the alternative tanker design and special boats in the appendix I-4/19 of the MARPOL convention, etc.

So the SLA is a potential approach for formulating IMO legislation and FSA plays importantly therein[14]. The GBS/SLA makes risk assessment for compulsory regulations affecting ship safety, evaluates the safety levels of relevant ship types by FSA, and decides upon the future risk acceptance criteria.

Meanwhile, it applies the principle of ALARP in FSA to make comprehensive assessment on the cost effectiveness of the risk-aversion measures. FSA is also applicable in evaluating the criteria and norms of shipbuilding to be in compliance of the objectives and functions in a rigorous quantitative reasoning process.

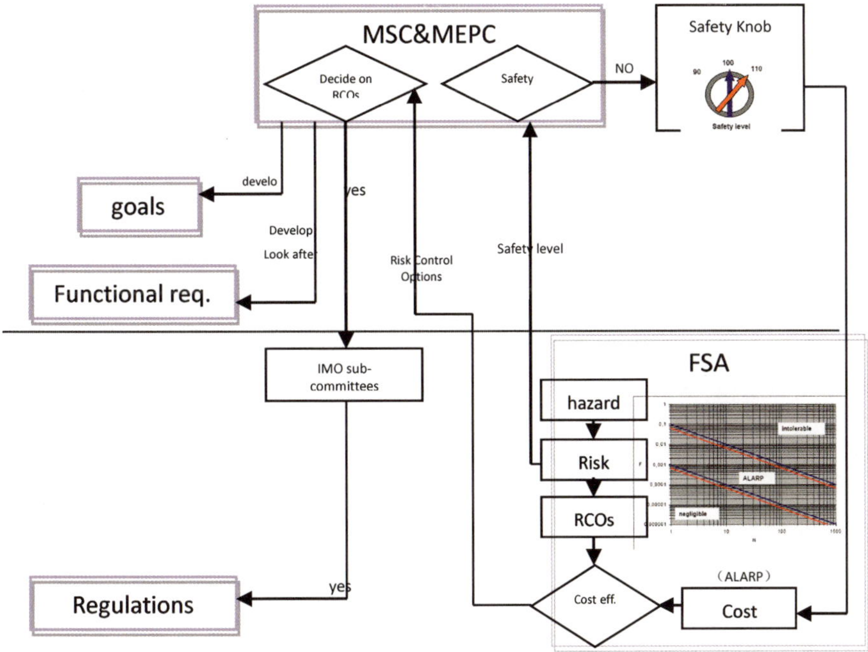

Fig 7. Concept of GBS/SLA[13]

GBS is still evolving. Since the MSC91 meeting, documents such as the Guidance for Approval of Risk-based Ship Design and the general guideline of risk-based equivalent alternative design have been discussed and determined[15].

3. Trend analysis

The international maritime system safety relies on a systematic maritime standard system and a full implementation thereof. Driven by the said philosophies and approaches, the safety system will more subject of matter into the system. Based on FSA and GBS, a main stream of "risk management" in maritime safety management can be concluded clearly and quantitative analysis processes will be more important for the accuracy of management. Therefore digitization, informatization and networking become the general trends in the industry.

Yet, the challenges exist in the fact that the maritime system can be endlessly sub-divided, through it is huge enough, that to record and evaluate fully and correctly the performance of a system is hard. Meanwhile, the evaluations for the conformity between static maritime legislation and dynamic maritime operations, for the accuracy and effectiveness of actions taken against the system, and for determination of safety level of the industry as required by GBS are also difficult, because there is an acute of shortage of safety data and information.

Acknowledgments

Work partly supported by the Chinese Maritime Safety Administration.

References

[1] Wang Dehong (eds), *Safety System Engineering* (National Defence Publication, China), P13, 2013.

[2] IMO A 28/Res.1060, *Strategic plan for the organization (for the six-year period 2014 to 2019)*, London, UK, 2014.

[3] Ruan Wei, *Maritime safety management and new technologies* (Pu Jiang Press, Shanghai, China, 2012).

[4] *Tokyo memorandum of understanding* (China Communication Press, 2014).

[5] DNV.GL, *The revised standards of ISO9001: 2015*, DNV.GL, 2015.

[6] Zhang Shuang, *The status and influences of risk philosophy in the framework of IMO*, ship-building, Vol 1, 2015.

[7] MSC 83/INF.8, 2007, *FSA – container vessels, Details of the Formal Safety Assessment*, submitted by Denmark, IMO, London, 2007.

[8] Eliopoulou Eleftheria, et al, *Statistical analysis of ship accidents and review of safety level,* Safety Science, 85 (2016) 282–292, Elsevier.

[9] MSC-MEPC.2/Circ.12, *Revised guidelines for formal safety assessment for use in the IMO rule-making process*, IMO, London, 2013.

[10] MSC/Circ. 1023, *Guidelines for formal safety assessment for use in the IMO rule-making process*, IMO, London, 2002.

[11] MSC 88/5, *Outcome of MSC 87: Future work on goal-based standards,* IMO, London, 2010.

[12] MSC80/6/4, *Introducing goal-based standards (GBS) into the existing regulatory framework,* submitted by Germany, IMO, London, 2005.

[13] MSC91/5/1, *Future safety-level based standards,* Submitted by Germany and Poland, submitted by Germany and Poland, IMO, London, 2012.

[14] MSC 80/WP.9, *Formal Safety Assessment – Report of the Group*, IMO, London, 2005.

[15] MSC95/22, *Report of the maritime safety committee on its ninety-fifth session,* IMO, London, 2015.

Efficient FEC Decoding Algorithm for 100Gbps Optical Transport Networks in FPGA

Jiang-Yi Qin†, Zhi-Ping Huang, Jun-Yu Wei, Zhi Dong and Zhen Zuo,

College of Mechatronic Engineering and Automation, National University of Defense Technology, Changsha 410073, Hunan Province, P.R. China
†E-mail: qjyacmilan@163.com

In 100Gbps optical transport networks (OTN), the forward error-correction (FEC) techniques are used to reduce the errors which occur in transmissions. Normally, field-programmable gate array (FPGA) is used to decode such high speed signal. Staircase codes combining with Bose-Chaudhuri-Hocquenghem (BCH) codes are widely used in FEC OTN for their powerful error-correction capabilities based on the iterative decoding procedure. However, the structure of BCH staircase codes is complex so the decoding procedure will cost lots of field-programmable gate array (FPGA) logic resource. In this paper, an efficient FEC decoding algorithm in 100Gbps OTN is proposed to realize FEC decoding with less FPGA logic resource. The normal FEC decoding algorithm of BCH staircase codes based on matrix transposition, Berlekamp iterative algorithm and Chien-search cost lots of FPGA logic resource. To improve the algorithm, an efficient BCH codeword acquisition method and an improve BCH error correction algorithm based on look-up table is deduced in detail. Combining with the characteristics of staircase and BCH codes, an improved FEC decoding algorithm is proposed to save FPGA logic resource. Meanwhile, this improved algorithm can be extended to other FEC decoding systems in OTN. Finally, the simulation results show that the improved FEC decoding algorithm use the least logic resource comparing with the normal one.

Keywords: Optical transport networks; forward error-correction; BCH code; staircase code; field-programmable gate array.

1. Introduction

As the development of OTN, optical signal has faster transmission rate and more complicated modulation mode. As these technologies mature, system designers have increasingly focused on FEC techniques and theory to increase transmission capacity and decrease transmission costs at the same time [1–3]. Currently, low-density parity-check (LDPC) codes, large scale BCH staircase codes and large scale Reed-Solomon (RS) staircase codes have been investigated as the FEC codes in OTN. What is more, these codes usually have multilevel coding and interleaving method. In order to reduce transmission bit error rate (BER), a great of compute resource will be used to decode and encode these complex transmission codes [4–5]. For high speed real-time processing, FPGA has been the best compute circuit. However, how to conserve logic resource achieving decoding is the most difficult problem. In this paper, an efficient decoding algorithm of BCH staircase codes in 100Gbps OTN has been presented to improve the normal FEC decoding algorithm [6–7].

2. BCH and staircase code

According to the linear encoding theories, the linear mapping between information polynomial and codeword polynomial can be expressed as follows.

$$C(x) = M(x) \times G(x) \tag{1}$$

In Equation (1), $M(x)$ is the information polynomial, each line in the polynomial represents a packet of information. It can be expressed as

$$M = \begin{bmatrix} m_1 & m_2 & \cdots & m_k \\ m_{k+1} & m_{k+2} & \cdots & m_{2k} \\ m_{2k+1} & m_{2k+2} & \cdots & m_{3k} \\ \vdots & \vdots & \vdots & \vdots \end{bmatrix} \tag{2}$$

C(x) is the codeword polynomial, which are all defined in the extension field $GF(2^m)$ $(1 \le m)$. It can be expressed as

$$C = \begin{bmatrix} c_1 & c_2 & \cdots & c_n \\ c_{n+1} & c_{n+2} & \cdots & c_{2n} \\ c_{2n+1} & c_{2n+2} & \cdots & c_{3n} \\ \vdots & \vdots & \vdots & \vdots \end{bmatrix} \tag{3}$$

Here, $G(x)$ is the generator polynomial and it is a row full rank matrix. The linear mapping processing is shown in Fig .1.

The binary BCH code is one of the most important linear coding and it can correct multiple random errors. They are widely used for their powerful error-correcting capability, convenient design of encoders, and efficiency of decoding algorithms.

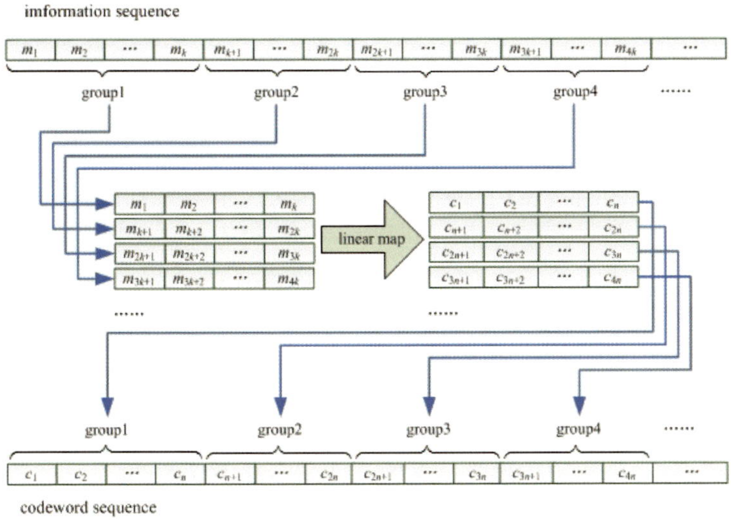

Fig.1. The linear mapping process between information polynomial and codeword polynomial

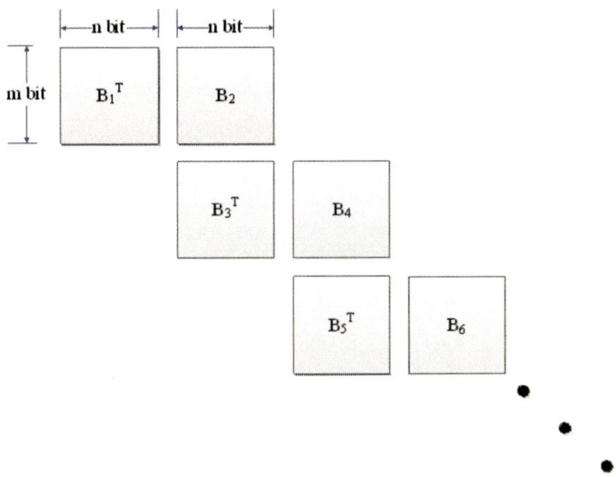

Fig .2. The basic structure of staircase codes

The staircase codes are similar to but different from two-orthogonal-concatenated (TOC) codes [8]. In the transmitted frames coding with staircase codes, each bit is protected by a row and a column component codes. But the number of row and column is certain. This feature is different from TOC. The basic staircase codes mode is shown in Fig.2. In this paper, one row in B_1 and the corresponding column in B_2^T can constitute a BCH codeword.

And one column in B_2^T and the corresponding column in B_3 can constitute another BCH codeword. Therefore, by decoding the each BCH codeword alternately, staircase codes can be iteratively decoded [9].

3. Efficient BCH codeword acquisition method

In this paper, a usual staircase code is utilized to describe the efficient BCH codeword acquisition method. The staircase codes is consist of a 510 lines 512 columns matrices, and the BCH (1022, 990) code is fill in the matrices. Its generator polynomial can be expressed as follow.

$$G(x)=\prod_{i=0}^{3} m_i(x) \tag{4}$$

Here, the minimal polynomial can be expressed as follow.

$$\begin{cases} m_0(x)=x+1 \\ m_1(x)=x^{10}+x^3+1 \\ m_2(x)=x^{10}+x^3+x^2+x+1 \\ m_3(x)=x^{10}+x^8+x^3+x^2+1 \end{cases} \tag{5}$$

The maximum error correcting capability of this BCH code is triple-error. Therefore, the highest degree of error location polynomial is three.

According to the Fig .2, the m is equal to 510, n is equal to 512 and k is equal to 32. However, realizing the large-scale matrix transposition in FPGA is impossible. In order to decode the staircase codes, an efficient BCH codeword acquisition method is proposed in this paper.

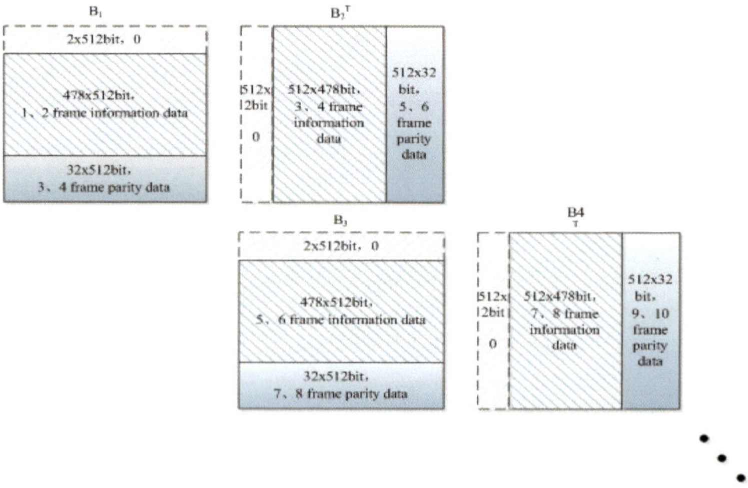

Fig .3. The structure of staircase codes for FEC decoding

Firstly, according to Fig.3, the matrix will be added 2 lines which are all 0 data and constitute a new matrix which has 512 lines and 512 columns.

Secondly, the syndrome will be cached in a 31 lines 512 columns bit buffer. In each FPGA clock cycle, a line of 512 bits in B_1 will be send to the error correction model. The first clock cycle of error correction model will calculate and cache the temporary syndrome of the first line in B_1. Next clock cycle, it will calculate and cache the temporary syndrome of the second line in B_1, until all the data in B_1 have been processed. Because, it will cost two FPGA clock cycles for adding two zero lines into the B_2. Exactly, utilizing the two FPGA clock cycles, error correction model can complete calculate and cache the temporary syndrome of the last line in B_1. At the same time, B_2 data will be parallel processed in FPGA. When the third line of B_2 is send to the error correction model, the model simultaneously calculates and caches the 512 codewords temporary syndromes. When the fourth line of B_2 is send to the error correction model, the model simultaneously calculates temporary syndromes and xor the temporary syndromes with the last one. Successively, when all the lines in B_2 is processed by error correction model. The whole process is shown in Fig.4.

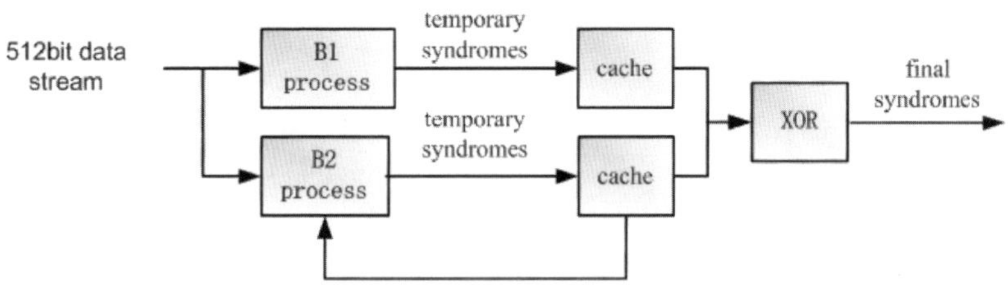

Fig .4. The improved syndrome calculation process

Finally, another error correction model will xor the temporary syndrome of B_1 with the one of B_2 and the result is the final syndrome of the 512 codewords. This method can avoid the large-scale matrix transposition and significantly reduce the logic resource of FPGA. This process is the base of decoding staircase codes in FPGA.

4. The improve BCH error correction algorithm based on look-up table

To the traditional BCH decoding method, calculating syndrome is the first step. The second step is calculating error location polynomial based on the Berlekamp iterative algorithm. Then, Chien-search is used to calculate the root of error location polynomial. Finally, the final error location can be calculated by Galois

Field logarithm operation. However, the Berlekamp iterative algorithm and Chien-search will cost lots of FPGA logic resource. In this paper, a novel BCH decoding algorithm which is taking advantage of BCH encode characteristic will be used to save FPGA resource [10].

Firstly, estimating the number of error bit. Supposing S_1, S_2 and S_3 have been calculated. Calculating the parameters $D_3 = S_1^3 + S_3$ and $D_5 = S_1^5 + S_5$. For the triple-error correcting decoder, there will be four cases.

(1)When $S_1 = S_2 = S_3$, then the number of error bits is 0;

(2)When $S_1 \neq 0$ and $D_3 = D_5 = 0$, then the number of error bits is 1;

(3)When $S_1 \neq 0$, $D_3 \neq 0$ and $S_1 D_5 = S_3 D_3$, then the number of error bits is 2;

(4)When $S_1 \neq 0$ and $v \neq 2$, then the number of error bits is 3.

For the case (1), there is no error in the received data, put out data directly.

For the case (2), there is one error in the received data and the error location is only connected with S_1 which is a vector of 10 bits wide. Therefore, a look-up table can be stored in FPGA with little on-chip memory to get the error location easily and the address number of look-up table is 1024.

For the case (3), the error location equation can be expressed as

$$x^2 + s_1 x + \frac{s_1^3 + s_3}{s_1} = 0 \tag{6}$$

The equation root is the error location. In order to get the root, x can be expressed as $x = s_1 y$, then the error location equation can be expressed as

$$s_1^2 \left(y^2 + y + \frac{s_1^3 + s_3}{s_1^3} \right) = 0 \tag{7}$$

The error location is only connected with $(s_1^3 + s_3)/s_1^3$ and $(s_1^3 + s_3)/s_1^3$ is a binary number of 10 bits wide. Therefore, the root can be got by a look-up table which has 1024 address numbers.

For the case (3), the error location equation can be expressed as

$$x^3 + s_1 x^2 + \frac{s_1^2 s_3 + s_5}{D_3} x + s_1 \frac{s_1^2 s_3 + s_5}{D_3} + D_3 = 0 \tag{8}$$

x can be expressed as $x = s_1 + y$, then the equation can be expressed as

$$y^3 + \frac{D_5}{D_3} y + D_3 = 0 \tag{9}$$

When $D_5 = 0$, the cube root of D_3 can be got by the look-up table. And the sun of S_1 and the root is the error location is also a vector of 10 bits wide, therefore, the look-up table address number is 1024, too. When $D_5 \neq 0$, the equation can be expressed as follow.

$$\left(\frac{D_5}{D_3}\right)^{3/2}\left[z^3+z+\left(\frac{D_3^{\ 5}}{D_5^{\ 3}}\right)^{1/2}\right]=0 \tag{10}$$

The multiplication result between the root of equation (8) and (D_5 / D_3) is equal to y. And x is the sum of y and S_1. The logarithmic result of x is the error location. Also, the error location is connected with the binary number D_5 / D_3.

In conclusion, no matter how many errors exist in the received data, the error location can be got by a look-up table with 1024 address numbers [11–12].

Furthermore, if FPGA resource allowed, multilevel iteration can gain very excellent error-correcting capability. The multilevel iteration decoding process is a parallel work in the form of pipeline. The data flow is shown in Fig.5.

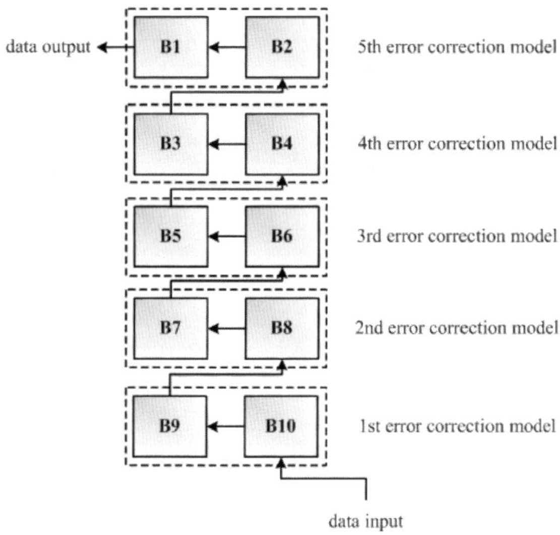

Fig .5. Five-steps iterative decoding data stream

5. Results and Discussion

In this section, simulations are used to show the efficiency of the improved FEC decoding algorithm proposed in section 2 and section 3, and the corresponding decoding performance is tested and given. The simulations use the 127Gpbs DP-QPSK signal communication system and transmission channel is AWGN channel whose signal to noise ratio (SNR) is 3dB. The encoding form is declared in section 2 and section 3 and utilizing five steps iterations to realize excellent error correction. In the simulations, five steps iterative decoding is implemented in a Stratix V FPGA device, which is 5SGXEA7N2F45I3N. This FPGA has 622K logic memory bit resource and 52M memory bit resource.

Three decoding methods are given in the simulations. The efficiency can be seen by the cost of the FPGA logic resource. The first method use normal matrix transposition, Berlekamp iterative algorithm and Chien-search to realized triple-error correction. The second method use efficient BCH codeword acquisition method, Berlekamp iterative algorithm and Chien-search to realized triple-error correction. The third method use efficient BCH codeword acquisition method and improve BCH error correction algorithm based on look-up table to realized triple-error correction. The simulation results are shown in Fig.6. From the Fig.6, it is obvious that the third method cost the least FPGA logic resource and the first one cost the most at each iteration. The resource used in five steps iterations is shown in Fig.7. Therefore, the simulation results verify that the improved FEC decoding algorithm proposed in this paper can save the FPGA logic resource obviously.

FEC decoding method	Logic bit	Memory bit
Matrix transposition	524288/662000	41943040/52428800
Efficient BCH codeword acquisition method	16384/662000	5242880/52428800
Berlekamp iterative and Chien-search algorithm	33100/662000	No use
Improve BCH error correction algorithm based on look-up table	26480/662000	No use

Fig.6. The cost of FPGA logic resource in each iteration

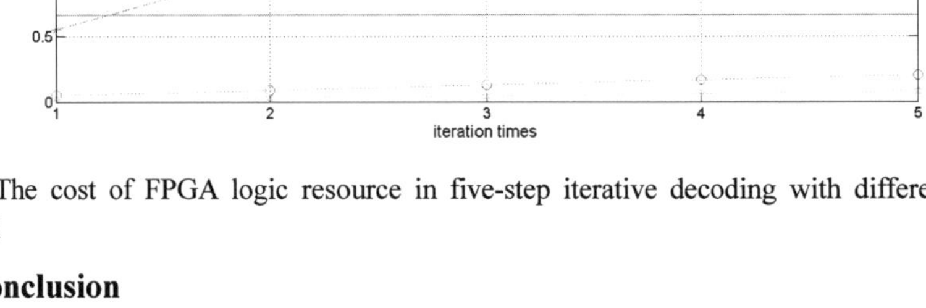

Fig.7. The cost of FPGA logic resource in five-step iterative decoding with different method

6. Conclusion

In this paper, an efficient FEC decoding algorithm in 100Gbps optical transport networks is proposed. Comparing with the algorithm based on normal matrix transposition, Berlekamp iterative algorithm and Chien-search, the one based on efficient BCH codeword acquisition and look-up table is developed to improve the FEC decoding performance with less FPGA logic resource. 127Gbps DP-QPSK signals which is encoded by BCH and staircase code is used to test the performance of these algorithms. The simulation results mean that the improved algorithms can save lots of FPGA logic resource obviously. What is more, these algorithms can be used by other FEC decoding system in OTN.

References

1. M. Carroll, J. Roese, T. Ohara, The operator's view of OTN evolution, IEEE Commun. Mag. 48 (2010) 46–52.
2. W.D. Grover, Forward error correction in dispersion-limited light-wave systems, J. Lightwave. Technol. 6 (1988) 643–645. ITU-T G.975, Forward error correction for submarine systems, 2000.
3. ITU-T G.975.1, Forward error correction for high bit rate DWDM submarine systems, 2004.
4. A. Darabiha, A. C. Carusone, and F. R. Kschischang, Power reduction techniques for LDPC decoders, IEEE J. Solid-State Circuits, vol. 43, no. 8, pp. 1835–1845, Aug. 2008.
5. K. Onohara, T. Sugihara, Y. Miyata, et al., Soft-decision forward error correction for 100 Gb/s digital coherent systems, Opt. Fiber Technol. 17 (2011) 452–455.
6. J. Yuan, W. Ye, Z. Jiang, A novel super-FEC code based on concatenated code for high-speed long-haul optical communication systems, Opt. Commun. 273 (2007) 421–427.
7. T. Mizuochi, Y. Konishi, Y. Miyata, et al., Experimental demonstration of concatenated LDPC and RS codes by FPGAs emulation, IEEE Photon. Technol. Lett. 21 (2009) 1302–1304.
8. Zhou Jing, Huang Zhiping, Liu Chunwu, Su Shaojing, Zhang Yimeng, A resource-saving decoding scheme of the orthogonally concatenated codes for optical transport networks in FPGA, Optik 125 (2014) 1174–1177.
9. B.P. Smith, A. Farhood, A. Hunt, F.R. Kschischang, J. Lodge, Staircase codes: FEC for 100 Gbps OTN, J. Lightwave. Technol. 30 (2012) 110-117.
10. I. S. Reed and X. Chen, Error-Control Coding for Data Networks, Boston, MA: Kluwer, 1999.
11. R. T. Chien, B. D. Cunningham, and I. B. Oldham, Hybrid methods forfinding roots of a polynomial with application to BCH decoding, IEEE Trans. Inf. Theory, vol. 15, no. 2, pp. 329–335, Mar. 1969.
12. E. R. Berlekamp, H. Rumsey, and G. Solomon, On the solution of algebraic equations over finite fields, Inf. Control, vol. 10, pp. 553–564, 1967.

An Improved Line Non-symmetry and Anti-packing Model For Gray Images Representation

Fei Xiao, Jinwen Tian

School of Automation, Huazhong University of Science and Technology, Wuhan, China

Guangwei Wang†

School of Computer and Information Science, Hubei Engineering University Xiaogan, Hubei Province, China

†*E-mail: wgw1949@gmail.com*

In this paper, inspired by the idea of Non-symmetry and Anti-packing Model (NAM) representation method, we proposed an improved algorithm for the gray image representation by using the Line Non-symmetry and Anti-packing Model (LNAM), which is called the LNAM image representation method. Also, we present a LNAM representation algorithm of gray images, Storage structure and data analysis. By comparing our proposed LNAM method with the conventional linear quadtree (LQT) method, the experimental results presented in this paper show that the former can significantly reduce the lower bit rate and the number of storing node than the latter whereas remaining the satisfactory image quality.

Keywords: Image representation; quadtree image; NAM model; parking; encoding.

1. Introduction

Designing efficient image representations is an important problem in computer graphics, computer vision, image processing, and pattern recognition [1]. An efficient image representation can save space and facilitate the manipulation of the acquired images [2]. There are many binary image representations, such as strings and sets of codes, and many efficient algorithms on these spatial data structures (SDS) have been developed [3].

Klinger proposed a quadtree representation method for binary images. However, the earlier quadtree representation is based on the pointers. Later, in order to reduce the storage room further, Gargantini removed the pointers and put forward a linear quadtree (LQT) representation method [4]. Generally speaking, the LQT can reduce the storage space by 66% and even in some special cases up to 90% [5].

Although the LQT representation stated above have many merits, they put too much emphasis upon the symmetry of segmentation. Therefore, they are not the optimal representation methods. Inspired by the concept of the rectangle packing problem, Chen et al. proposed a novel representation algorithm based on the non-symmetry and anti-packing model (NAM) by using the binary bit-plane decomposition (BPD) method.

Therefore, although the NAM representation for gray images is an effective image representation method, there is still much space left for optimization. In this paper, we proposed an improved line non-symmetry and anti-packing model (LNAM) for gray image representation in order to represent gray images more effectively. By comparing the representation algorithm of the LNAM with those of the latest NAM and the popular LQT, the theoretical and experimental results show that the former can greatly reduce the numbers of sub-patterns or nodes and simultaneously save the data storage much more effectively than the latter.

2. Description of non-symmetry and anti-packing pattern representation model (NAM)

2.1. *Abstract description of NAM*

The Non-symmetry and Anti-packing pattern representation Model (NAM) is an anti-packing problem. The idea of the NAM can be described as follows: Giving a packed pattern (container) and n predefined sub-patterns (objects) with different shape, pick up these sub-patterns (objects) from the packed pattern (packed container) and then represent the packed pattern (container) with the combination of these sub-patterns (objects) [6, 7, 8].

Suppose an original pattern is Γ and two reconstruction non-distortion and distortion patterns Γ' and Γ'', respectively. Then, the NAM is either a non-distortion transform model from Γ to Γ' or a distortion one from Γ to Γ''. The procedure of the transform can be written as Eq. (1).

where T(▪) is a transform or a coding function.

The procedure of the non-distortion coding can be obtained by the Eq.

where is the reconstruction pattern; p is a set of some predefined sub-pat k is the total number of the sub-pattern instances; p_j is the jth sub-p instance; v is the value of p_j; A is a parameters set of the sub-pattern p_j; residue pattern.

2.2. *Description of LNAM*

The main improvement of the LNAM is embodied by simplifying stru of sub-pattern. In the LNAM algorithm, line is selected as a sub-pattern following Fig. 2 illustrates the feasibility of line sub-pattern in the LNA

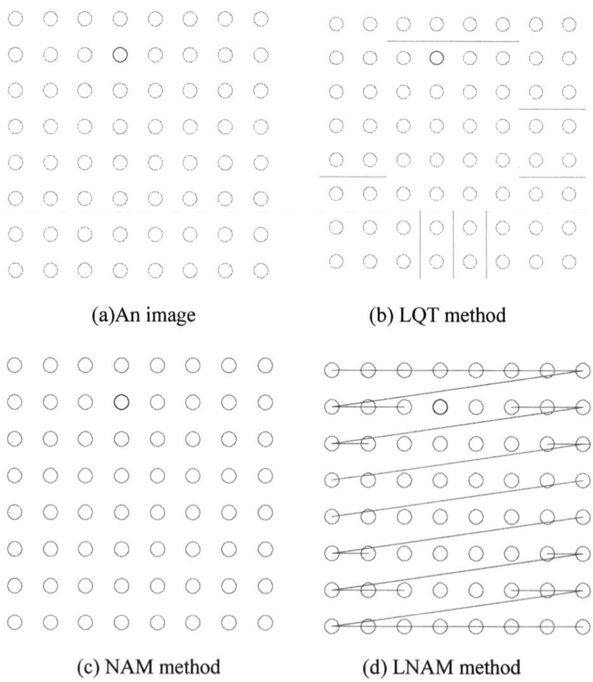

Fig. 2. A binary image and its different representation methods

Fig. 2 (a) is a binary image of size $2^n \times 2^n$ (n =3), where objects are m with black pixels. Since only the black pixels are anti-packed fo representation methods of the LQT, the NAM, and the LNAM, we need not store the bit depth m for the binary image.

Fig. 2 (b), (c), and (d) show the representation results of Fig. 2 (a) by using the LQT, NAM and LNAM representation methods. The total storage space for the LQT, the NAM, and the LNAM are 224 bits, 59 bits, and 31 bits, respectively. Therefore, the total data amount of the LQT is 4 and 7 times of those of the NAM and the LNAM, respectively. What's more, the total data amount of the NAM is 1.7 times of that of the LNAM, which shows that the LNAM can save the storage space more effectively than that of the NAM.

3. Description of the ALGORITHM OF THE IITNAM FOR GRAY IMAGES

3.1. *Description of the LNAM encoding algorithm*

For a given gray image I, an encoding result of I is stored into a queue $Q=\{p_1, p_2, \ldots, p_k\}$, where $p_i(1<=i<=k)$ denotes the ith instances of rectangle sub-pattern and k is the total numbers of instances in Q. The encoding part of our algorithm has the following steps:

Input: A given gray image I with a bit depth m.

Output: An encoding queue set Q for I, where $Q=\{p_1, p_2, \ldots, p_k\}$, $p_i(1<=i<=k)$ denotes the ith instances of **line** sub-pattern and k is the total numbers of instances in Q.

•***Step** 1*. Decompose a gray image I to m binary-bit plane images BP_i $(1\leq i\leq m)$ by the BPD method, which was presented in our previous work [7, 9], and let i=1.

•***Step** 2*. Set every pixel in image BP_i unmarked.

•***Step** 3*. Establish an unmarked start point $sp(x, y)$ from the first entrance of the image pattern BP_i according to the raster scanning order, trace the corresponding **line** sub-patterns.

•***Step** 4*. Find out the longest sub-pattern p in term of the areas of the sub-pattern line and mark the line in the image BP_i.

•***Step** 5*. Record the parameters of the found sub-pattern, i.e., $Q\leftarrow(sp, length)$.

•***Step** 6*. Repeat step 3 to step 5 until there are no unmarked pixel in the image BP_i.

•***Step** 7*. Output the encoding result Q.

3.2. *Storage structure and data analysis*

In this section, the storage structure of line sub-pattern and the data amount are analyzed. Suppose the size of binary image is $2^n \times 2^n$. The storage structure of line can be denoted by the expression $p=\{r|r=(sp, length)\}$. The start point *sp* is stored as a relative value Δsp, i.e., $\Delta sp=sp_i-sp_{i-1}$. Therefore, the length of Δsp is n bits in the sight of statistical concept. In fact, if the length of Δsp is greater than n, then one record can be divided into two records. Also, the maximal length of length is $n/2$ bits. Therefore, storing a line takes up $3n/2$ bits. For the method of NAM, storing a rectangle node takes up $2n$ bits.

Let the total data amount of the binary image BP_i be H_{LNAM} and H_{NAM} when BP_i is represented by LNAM and NAM encoding algorithm. We can write H_{LNAM} and H_{NAM} as in Eq. (3) and (4).

$$H_{LNAM} = 3n/2N_{LNAM} \quad (3)$$

$$H_{NAM} = 2nN_{NAM} \quad (4)$$

where N_{LNAM} and N_{NAM} are the total number of the sub-pattern instances.

For the method of the linear quadtree, storing a record of a node takes up $2n+\lg n$ bits for a binary image [4]. Suppose the total data amount of the image BP_i is H_{LQT} when BP_i is represented by the linear quadtree. We can write H_{LQT} as n Eq. (5).

$$H_{LQT} = (2n+\lg n)N_{LQT} \quad (5)$$

where N_{LQT} is the total nodes number of the linear quadtree.

Let φ_{LQT_LNAM} denote the ratio of H_{LQT} to H_{LNAM}, and φ_{NAM_LNAM} denote the ratio of H_{NAM} to H_{LNAM}. Then, H_{NAM} and H_{LNAM} can be expressed by Eq. (6) and (7).

$$\varphi_{LQT\_LNAM} = \frac{(2n+\lg n)N_{LQT}}{3n/2N_{LNAM}} \quad (6)$$

$$\varphi_{NAM\_LNAM} = \frac{2nN_{NAM}}{3n/2N_{LNAM}} \quad (7)$$

Thus, we can easily deduce the expression. For example, $\varphi_{NAM\_LNAM} > (2n)/(3n/2)>1$, when both *m* and *n* equal 8, φ_{NAM_LNAM} is greater than 1.3333.

Therefore, the theoretical analyses in this section prove that the representation method of the LNAM for gray images can reduce the data storage much more effectively than that of the popular LQT and NAM method, and it is a better method to represent gray images.

4. Experiment Result

We have implemented our algorithm of the LNAM for gray images with the Matlab 7.0 software and made a comparison with those of the NAM [9, 10] and the LQT [4, 5]. Four test gray images of size $2^8 \times 2^8$, i.e., $n = 8$ are selected. The bit depth of these images is 8, i.e., $m = 8$. The following Table 1 shows the comparison of the performance among them.

In Table 1, we know that the sub-patterns number of the LNAM is much less than the nodes number of LQT and NAM. Since the LQT segmentation is symmetrical, the segmentation methods suffer from great confine. However, since the LNAM segmentation is asymmetrical, the segmentation methods are unrestricted.

By expression (6) and (7), we can judge which method is better to represent the binary image. From the value of φ_{LQT_LNAM} and φ_{NAM_LNAM} in Table 1, the total data amount of LNAM is less than that of NAM and LQT. The results show that the LNAM can save the storage space more effectively than that of the NAM and the LQT.

Table 1. Comparison of the performance between the LNAM, NAM and LQT

Image	N LQT N_{LQT}	NAM N_{NAM}	LNAM N_{LNAM}	φ_{LQT_LNAM}	φ_{NAM_LNAM}
Lena	17889	5199	5180	4.9686	1.3324
Peppers	10822	2458	2449	6.3576	1.3679
Boat	16137	4324	4297	5.3890	1.3120
Flight	13435	3465	3430	5.5989	1.3512

5. Conclusions

This paper presented an improved image representation method, called line non-symmetry and anti-packing pattern representation model (LNAM). Based on the idea of NAM, a binary image representation algorithm based on line sub-pattern was proposed. By comparing the representation algorithm of the LNAM with those of the latest NAM and the popular LQT, the theoretical and experimental results show that the former can greatly reduce the numbers of sub-patterns or nodes and simultaneously save the data storage much more effectively than the latter, and therefore it is a better method to represent gray images.

Acknowledgment

This work is supported by grant No. 61370092 of the National Natural Science Funds of China.

References

1. K.S. Gurumoorthy, A. Rajwade, A. Banerjee, and A. Rangarajan, A method for compact image representation using sparse matrix and tensor projections onto exemplar orthonormal bases, *IEEE Transactions on Image Processing*, vol. 19 (2), 2010, pp. 322–334.
2. P. Yap, X. Jiang, and A.C. Kot, Two-dimensional polar harmonic transforms for invariant image representation, *IEEE Transactions on Pattern Analysis and Machine Intelligence*, vol. 32 (7), 2010, pp. 1259–1270.
3. S. Chen, An exact closed-form formula for d-dimensional quadtree decomposition of arbitrary hyperrectangles, *IEEE Transactions on Knowledge and Data Engineering*, vol. 18 (6), 2006, pp. 784–798.
4. I. Gargantini, An effective way to represent quadtrees, *Comm. ACM*, vol. 25, no. 12, 1982, pp. 905-910.
5. G. Minglun, and Y. Yee-Hong, Quadtree based genetic algorithm and its applications to computer vision, *Patter Recognition*, vol. 37, no. 8, 2004, pp. 1723–1733.
6. Yunping Zheng, and Xi Guo, An Improved Indirect Triangle Non-symmetry and Anti-packing Model for Gray Image Representation, *2011 International Conference on Multimedia and Signal Processing*, 2011, pp. 117–121.
7. Guangwei Wnag, Xinggang Zhang, Cenggang Xiong, and Yihua Lan, A Novel Image Representation Model for Color Image Compression, *Journal of Application Science*, vol. 13(13), 2013, pp. 1589-2593.
8. Yunping Zheng, Chuanbo Chen and M. Sarem, Novel algorithm for triangle non-symmetry and anti-packing pattern representation model of gray images, *Proceedings Third International Conference on Intelligent Computing*, 2007, pp. 832–841.
9. Yunping Zheng, Chuanbo Chen, Wei Huang, and Shaohong Fang, A new non-symmetry and anti-packing model with k-lines and its application to color image representation, *Proceedings of the 2007 International Conference on Image Processing, Computer Vision*, 2007, pp. 526–532.
10. Wei Huang, Fan Xiong, Chuanbo Chen, and M. Sarem, Accurate computation of zernike moments using non-symmetry anti-packing image representation, *2009 Asia-Pacific Conference on Information Processing*, vol. 2, 2009, pp. 156–159.

The study of competency model about Endurance Ltd.'s salesman

Li-Hong Han†, Shao-Fan Zhang

Hebei University of Science and Technology

Shijiazhuang, Hebei, China

†*E-mail: 1909162307@qq.com*

Salesman is the main executor of product sales activities, playing an important role in the development of the enterprise. In this paper, we mainly studied competency model about the salesperson of Endurance Ltd. We adopted Behavioral Event Interview and questionnaire to analyze the salesman of the Endurance Ltd. And then defined the competency features and competency model of the salesman. In the article, the competency model consists of 27 competency features which was divided into 6 modules, including basic quality, customer-oriented, professional knowledge and skills, communication skills, self-learning ability and cooperation and solidarity. The competency model can provide reference for its employee selection and training and performance evaluation.

Keywords: Endurance Ltd.; salesman; competency model; behavioral event interview; questionnaire.

1. Guidelines

Competency was first put forward in foreign. In 1993, the concept of competency that was proposed by Spencer is widely recognized. He thought that competency was some potential features that can be measured, including motivation, personal characteristics and values, and these features can separate outstanding person from ordinary people[1]. Competency model is the sum of some different competency features that were required for a particular task[2]. Constructing competency model plays a fundamental role in a company's employee selection and training, performance evaluation.

Endurance Ltd., a manufacturing enterprise which is mainly engaged in the manufacture and sale of air compressor. Enterprise now is in the development stage, and its product is not only for the domestic market, but also gradually moves to overseas markets. Salesman plays an important role in the development of the enterprise. How to select and retain a number of high-quality and high-capacity sales staff is the key that enterprise value will be recognized by the market. This paper mainly studied the salesman of Endurance Ltd., and combined with the actual situation of the enterprise to construct competency model for its salesman. The competency model can provide reference for its employee selection and training and performance evaluation.

2. Research methods

Now, the methods about constructing competency model are generally divided into five categories: Behavioral event interview, Functional analysis, Situation method, Performance method and Multi-dimensional method[3]. This study used Behavioral Event Interview to preliminary get competency features, and then used questionnaires to get and analyze data, and finally drew conclusions.

2.1. *Behavioral Event Interview*

Behavioral Event Interview was put forwarded by American psychologist McClelland[4]. Behavioral event interview is based on the past behavior, letting interviewer describe three the most successful and three the most unsuccessful things they have encountered in their work, including origin, process, results and their attitude and solutions to the things[5]. And then analyzing the content of the interview to select competency characters. Finally giving the contrast analysis of the difference of characters between Outstanding persons and general persons to determine the competency characters of the task.[6].

2.1.1. *Sample Selection*

According to the Performance Evaluation's Result the enterprise provided, we interviewed 26 salesperson, and the 26 salesperson is divided into the general group and outstanding group. Both groups include 13 salesman. The specific circumstances of the samples is shown in Table 2-1:

Table 2-1 The specific information of the samples

Attribute	Category	Number of people
Sex	Male	18
	Female	5
Age	Under 20 (20 included) years old	4
	21 to 30 years (30 included)	12
	31-40 years (40 included)	5
	Above 40 years old	2
Educational background	Master degree or above	3
	Undergraduate course	16
	College degree or under	4
Position	Department manager	2
	General sales staff	21

2.1.2. *Depth Interviews*

We interviewed these 26 salesman, and recorded the interview. Next we transformed the recording into text, and we got 23 interview transcriptions (not including 3 unsuccessful samples), which contains more than 60000 words.

2.1.3. *Data Encoding*

We next coded interviews text and divided the text into different coders, when the result of the code achieves highly consistent, we began to code formally. Examples are as follows:

Interview	Analysis process	Competency
Since many customers do not understand or believe our products, so they will ask a lot of questions, ①at this point we should patiently explain our products to customers. ②Meanwhile, we should have some knowledge about own products and similar products, So that we can analyze the differences between two product.....	①Be Patient to the questions customers raised ② Know about our own products and pay attention to similar products	① Patience ②Basic knowledge and Market sensitivity

After arranging the interview report, we generalized and refined competence characters from the successful things that each salesperson has described before. And then we selected the competency characters and the final result showed there were 13 competency characters that had been mentioned repeatedly, which were shown in Table 2-2. Since the technology was limited, we only counted their frequency that had appeared in two groups.

Table 2-2 Comparison table of frequency between two different groups

Competence Characters	Outstanding Group	General Group
customer-oriented	9	6
responsibility	10	7
patience	9	5
confident and optimistic	9	6
Integrity	11	8
desire for achievement	9	4
product knowledge	11	9
communication skills	9	6
self-learning ability	6	4
unity and cooperation ability	7	4
market sensitivity	9	5
marketing knowledge and skills	9	4
Observation	8	5
total	116	73

The frequency of 13 competence characters that appeared in two different groups is significantly different, so they can be used to build competency model[7].

2.2. *Questionnaire*

According to the above competency characters, we designed questionnaire. Questionnaire consists of two parts. The first part is basic information about the respondents. The second part of the questionnaire is multiple-choice. We adopted Likert Scale, and each question contains five options, where 1 means completely unimportant, 5 means very important[8]. Some questions of the questionnaire are as follows:

Question	Importance
1.love your job	1 2 3 4 5
2.Be good at observing customers and can meet their requirement	1 2 3 4 5
3.Be passionate about your job and pursuit for excellence	1 2 3 4 5
……	1 2 3 4 5

We issued 80 questionnaires to salesman, and recovered 73 questionnaires(the recovery is 89%), in which 68 questionnaires is valid, accounting for 85% of total questionnaire. We used SPSS19.0 to process data.

3. Exploratory Factor Analysis

3.1. *Reliability Analysis*

Using the Cronbach'α to test the reliability of the whole scale, and can be seen from Table 3-1 that the overall homogeneity reliability of the questionnaire was 0.809, which proved the consistency reliability of the questionnaire is high and the test results is very reliable.

Table 3-1 The reliability statistics

Cronbach'α	Number of items
0.809	27

3.2. *KMO and Bartlett Test*

Using SPSS19.0 to test the feasibility of the factor analysis, as can be seen from Table 3-2, after the factor analysis, the KMO was 0.746, indicating that the result of factor analysis was good[9]. The X2 of Bartlett test was 2396.478, which indicates that there are common factors in the correlation matrix. So we can conduct factor analysis.

Table 3-2 KMO and Bartlett Test

K-M-O	0.746	
Bartlett test	Approx.Chi-square	2596.478
	df	478.000
	sig.	0.000

Under normal circumstances, when the cumulative variance of factors we got from principal component analysis is more than 60%, indicating that construct validity of the scale is good[10]. By analyzing data, we got the accumulated variance rate reached 64.87%, indicating that the questionnaire has good validity. Detailed results are shown in Table 3-3:

Table 3-3 Eigenvalues and variance explained

	Initial Eigenvalues			Rotation sums of squared loadings		
Component	Eigenvalue	% of variance	Cumulative %	Eigenvalue	% of variance	Cumulative %
1	8.244	26.121	26.121	5.779	18.389	18.389
2	6.321	19.783	45.904	3.987	12.562	30.951
3	2.346	7.248	53.152	3.453	10.851	41.802
4	1.658	5.245	58.397	3.154	9.839	51.641
5	1.623	5.089	63.486	2.156	6.698	58.339
6	1.075	4.061	67.547	2.086	6.534	64.873

Through orthogonal rotation, we extracted 6 factors of which the characteristic roots was greater than 1, and the total variance explained reached 64.87%, and all the factor's load level was more than 0.4. The results are shown in Table 3-4:

Table 3-4 Load level

Component	F1	F2	F3	F4	F5	F6
love your job	0.889					
Be patient	0.859					
Responsibility	0.795					
Confident and optimistic	0.789					
Be integrity	0.746					
Can calmly deal with unexpected matters	0.687					
When encounter problems, you can control your emotions	0.563					
Be proactive and contact past clients about what you can do for them		0.799				
Sincerely treat every customer		0.781				
Can offer customers information timely		0.770				
Help customers solve problems and give them advice		0.712				
Offer personalized services.		0.560				
Master the basic information of own product			0.801			
pay attention to similar product			0.741			
Market sensitivity			0.732			
Marketing knowledge and skills			0.725			
Have good language skills				0.819		
Be good at communication				0.775		
Can change the way of communication according to customer's reaction				0.491		
Have strong persuasion				0.484		
Have a strong thirst for achievement					0.793	
Constantly challenge and develop yourself					0.559	
Set higher goals for yourself					0.508	
Study actively and improve your ability					0.419	
Be good at Cooperation with others						0.661
Be good at solving problems together						0.658
A strong sense of solidarity						0.620

4. Factor classification

We named the 6 factors according to the meaning that these factors contained. The results are as follows: the first factor was named basic quality, the second factor was named customer-oriented, the third factor was named professional knowledge and skills, the forth was named communication skills, the fifth factor was named self-learning ability and the sixth factor was named cooperation and solidarity.

5. Conclusion

In this paper, we concluded that the competence character about salesman of Endurance Ltd. consists of 6 dimensions: basic quality, customer-oriented, professional knowledge and skills, communication skills, self-learning ability and cooperation and solidarity. The model can not only be used to evaluate its salesman, but also provides a reference for recruitment and selection, training and performance evaluation of the company's salesperson.

Our study also has certain limitations, which mainly reflect in the following aspects: the sample size was too small (just 80 salesperson participated in the questionnaire survey), and our interview techniques still should be improved. In this paper, we preliminarily analyzed the competency that the salesperson of Endurance Ltd. should have. We didn't compare the relationship between these competency characters in detail, so there are still some questions remain to be further studied.

References

1. Ming-Fei Li and Xiao-Jun Lu, The study about method to build competency model. Journal of Dalian University of Technology (2004).
2. Ming Feng and Ming-xin Yin, Method Review about building competency model [J]. Science and Technology Management Research (2007).
3. Ben-xin Huang, The study on competency model about automobile sales personnel[J]. Journal of Business Strategy (2009).
4. Boyatzis R E. Rendering into competence the things that are competent[J]. American Psychologist (1994).
5. Spencer L M and Spencer S M. Competence at work: Models for superior performance[J]. New York: John Wiley & Sons Inc (1993).
6. Yang-Hua Jin, Wei-Qi Chen and Chong-Ming Wang, Research on Management Competency and Job Performance [J]. Psychological Science(2004).
7. Yan-Ting Xia, The study about competency structure on the basis of customer managers in State-owned Commercial Bank[D]. Master thesis in Chengdu Univerisity of Technology (2012).
8. McClelland, D. C. Identifying competencies with Behavioral Event Interviews[J]. Psychological Science (1998).
9. Hai-Long Qing, The study on competency structure about salesman in Back of China[J]. Journal of Jiannan University (Philosophy and Social Sciences Edition), (2006).
10. Lei Li, The study on competency structure about personal customer managers in State-owned Commercial Bank-on the basis of A back[D]. Master thesis in Capital University of Economics(2014).

Short-Term Electricity Price Forecasting Based on Nonparametric GARCH Residuals Correction-Least Square Support Vector Machine

Yuan-Sheng Huang, Shen-Hai Huang†, Yin-Ping Hu

Department of Economic Management, North China Electric Power University, Baoding P. R. China

†E-mail: hshncepu@163.com,

www.ncepu.edu.cn

Accurate prediction of short-term electricity price provides effective decision-making guide for market participants. In this paper, least squares support vector machines (LSSVM) model is proposed and nonparametric generalized auto regressive conditional heteroscedasticity (NPGARCH) model is built for electricity price volatility of forecasting residuals to improve the prediction accuracy of price volatility, as well as precision of LSSVM model. This proposed model is applied to short-time price forecasting of the electricity market of Pennsylvania-New Jersey-Maryland (PJM), compared with traditional LSSVM and generalized regression neural network (GRNN), the results show that NPGARCH-LSSVM outperforms the two models mentioned before, which can improve the prediction accuracy and reduce the prediction of price fluctuation.

Keywords: Nonparametric GARCH Model; Residuals Correction; Least Squares Support Vector Machines; Short-time Electricity Price Forecasting

1. Introduction

In the electricity market, accuracy of short-term price forecasting directly affects the bidding decisions of market competitors and their economic benefits, and as a result, more and more attentions are focused on short-term electricity price forecasting. Most of the current short-term forecasting models stem from financial bond prediction problems, but there are some differences between electricity price prediction problems and financial bond prediction problems. Therefore, we need to optimize the models to reach a better forecasting result to the electricity price. The traditional methods of price forecasting are mainly based on time series analysis and statistical models, including linear regression model [1], and auto regressive moving average model [2][3]. However, with the development of artificial intelligence algorithms, the self-learning theory is widely used in the field of electricity price forecasting. The most common used algorithms are artificial neural network (ANN) [4] and genetic algorithm (GA) [5]. Besides, wavelet transform method is also proposed to decompose historical data and to establish a multivariate time series model for the randomness and volatility of short-term price, to some extent, to improve the prediction accuracy [6]. In order to solve the traditional statistical models in the problem of "curse of dimensionality", Zhang Jinliang proposed wavelet transform, by using PSO to search for the optimal decision parameters of least squares support vector machines (LSSVM), and predicted day-ahead electricity price. The prediction results showed that the ability of learning and generalization was greatly improved [7].

In the electricity market, day-ahead electricity price is influenced by many uncertain factors, including changes of natural environment, electricity demand, constraints of network operation and bidding strategies of buyers and sellers. The uncertain factors sometimes lead to strong price twisting which is performed as heteroscedasticity in the general time series. During the former electricity price forecasting processes, whether by using traditional or artificial intelligence models, the problem of heteroscedasticity appeared in time-series method is rarely analyzed or solved in the process of data preconditioning. The lack of data preconditioning will obviously cause the imprecise of logical and will affect the prediction accuracy.

In this paper, auto regressive conditional heteroscedasticity (ARCH) model is used to test the past time series and least squares support vector machines (LSSVM) is proposed to predict electricity price. Generalized auto regressive conditional heteroscedasticity (GARCH) model is normally used to eliminate the effects of heteroscedasticity of price time series. However, consider the volatility of electricity market, it is very difficult to determine the conditional mean and conditional variance of GARCH model. Nonparametric GARCH can successfully solve the parameter problem. Based on this, nonparametric GARCH is presented for residual correction, so that the standard conditional variance of the price time series will be stable. By using the method, we can increase the accuracy of prediction through getting a more stable residual sequence. In this paper, we present a novel LSSVM algorithm based on nonparametric GARCH residuals correction (NPCARCH-LSSVM). Compared with the traditional LSSVM and generalized regression neural network (GRNN), NPGARCH-LSSVM increased the forecasting accuracy. In addition, the results also shows that this model is more stable than the other two models mentioned before, and the results can verify the superiority of this model.

2. Least Square Support Vector Machine (LSSVM)

LSSVM is proposed as an improved algorithm based on support vector machine(SVM), given the training data set $\{x_i, y_i\}(i=1,2,\cdots,N)$ with the input $x_i \in R^n$ and the output $y_i \in R^n$ .the following regression model is constructed by using nonlinear mapping function $\phi(x_i)$ .Lssvm is to solve the problem of this nonlinear programming problems

$$f(x_i) = W^T \phi(x_i) + b \tag{1}$$

Given the training data set $\{x_i, y_i\}(i=1,2,\cdots,N)$, the optimization problem of LSSVM is defined as follows,

$$\begin{cases} \min \quad J(W,\xi_i) = \dfrac{1}{2} W^T W + \dfrac{C}{2}\sum_{i=1}^{M}\xi_i \\ s.t. \;\; y_i = W^{\mathrm{T}}\varphi(x_i) + b + \xi_i, \quad i = 1,2,\cdots,n \end{cases} \tag{2}$$

Where W is n-dimension weight vector; ξ_i is the training error; $C>0$ is the punish factor, which determines the trade off between minimizing the training error and minimizing model complexity.

3. Short-Term Price Forecasting Based on NPGARCH-LSSVM Optimization Algorithm

3.1. *Nonparametric GARCH (1, 1) residual correction*

The result of short-term electricity price forecasting is not only related with the model,[8][9] but also affected by the time series and other factors, including changes of natural environment, electricity demand, constraints of network operation and bidding strategies of buyers and sellers, etc.

This makes an error in a certain tendency to expand, which could generate the random and volatility [10][11]. Prediction error correction is a supplement for traditional and artificial intelligence forecasting models, which can improve the accuracy of the final prediction in a certain degree.

This paper presents a nonparametric model for the error fluctuation, which does not need a lot of assumptions.

3.2. *Pretreatment of time series data*

In order to meet the multidimensional input variable condition of least squares support vector machine, the time series of electricity price is dealt with sliding window. The time series $(x_1, x_2, \cdots, x_n)$ is divided into two portions, the first part of which is as a parameter estimation, the other one is a test for predicting. However this part of the time series is one-dimensional, in order to make a more effective prediction, the one-dimensional vector need to have a correlation conversion for obtaining more information. For example, in order to obtain prediction data $\{x_p\}$, then these data will be as linked data, the relationship between the sliding window $\{x_{p-1}, x_{p-2}, \cdots, x_{p-d}\}$ and $\{x_p\}$ is established.

Here *d*, represents the embedded dimension, if *d* is too small, it is not possible to express the complexity of the real structure, but *d* is too large ,it is difficult to explain the relationship between the data. Generally, if the dimension of the original system is *D* dimension, then $d \geq 2D+1$.

4. Illustrative Example

When the time series is dealt with sliding window, considering that the original data of time series is one-dimensional, therefore, according to the principle of $d \geq 2D+1$, *d* is set to 3.

Data which are from 10/9/2014 to 12/9/2014 are chosen from the US PJM electricity market to be as the original data, and the price changes in the hourly period is analyzed. Firstly the electricity price of 13/9/2014 is forecasted by using LSSVM.

In order to quantitatively determine whether the time series of forecasting price heteroscedasticity, ARCH test is needed. Based on the Eviews 6.0 platform, Lagrange Multiplier(LM) is equal to 5.736,the specific expression of which is as follow:

$$LM = T \times R^2 = 24 \times 0.239 = 5.736 > \chi^2_{0.05(1)} = 3.841$$

Where T is the number of data, R^2 is the coefficient of determination which describes the degree of fitting .

The results shows that the existence of autoregressive conditional heteroscedasticity, and higher order ARCH effect, namely GARCH effect. So nonparametric GARCH is applied for residual correction. M is set to 8, and the nonparametric GARCH (1,1) modified residual graph is as shown below

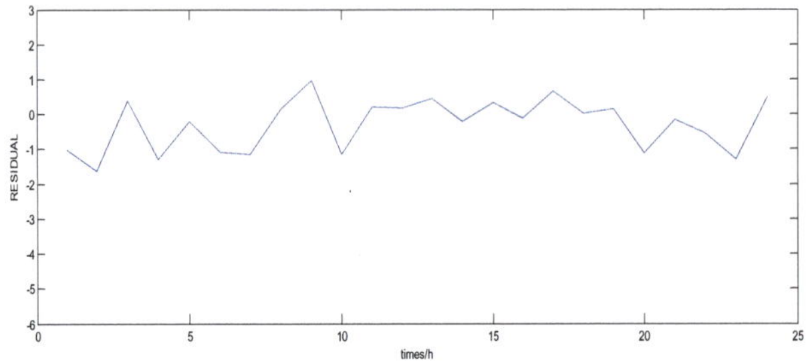

Fig.1 Resides modified by the proposed method

In Fig.1, Obviously, by using nonparametric GARCH (1,1) model to predict the results of the LSSVM, the predicted results of the residual data are more smoothly, and the peak appears less.

In order to verify the prediction effect of the algorithm, the historical data from10/9/2014 to 12/9/2014 are as samples, to forecast price of 13/9/2014, the results are shown as follows:

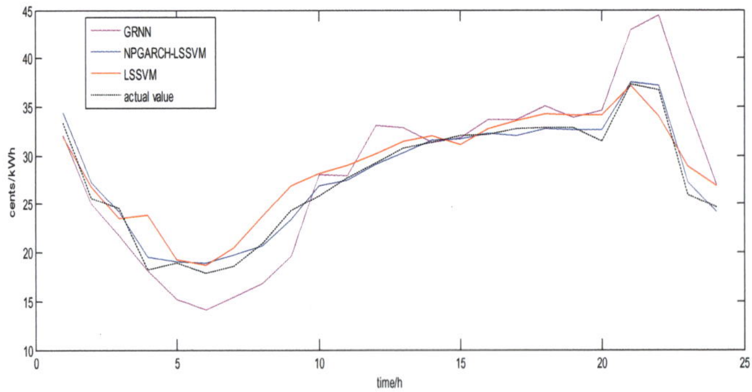

Fig.2 The comparison between the prediction results

According to Fig 2 and Tab 1, through three different models to train and fit the sample, in the electricity price forecasting results have big difference among different algorithms.

NPCARCH-LSSVM has the lowest MAPE, only 2.52%.GRNN has the highest MAPE, up to 11.44%.The MAPE of LS-LSVM is respectively 6.56% which is between the above two values.

And the number of relative error less than 5% of NPCARCH-LSSVM is twenty, the value of which is low. The number of relative error less than 5% of LSSVM is sixteen. But the prediction of GRNN has a large fluctuation, the vast majority of the relative error is more than 5%, the forecast is not satisfactory, a greater error is between the predicted and actual values. Therefore, it is proved that NPGARCH-LSSVM has a better prediction effect.

Table 1 The comparison between the prediction results

Time/h	Av/(cents/kWh)	LS-SVM Fv/(cents/kWh)	ER/%	GRNN Fv/(cents/kWh)	ER/%	NPCARCH-LSSVM Fv/(cents/kWh)	ER/%
01	33.38	31.92	4.38	32.07	-9.74	34.39	-3.02
02	25.62	26.79	-4.57	25.07	2.16	27.24	-6.34
03	24.63	23.54	4.42	21.72	11.82	24.24	1.59
04	18.25	23.87	-30.82	18.08	0.92	19.55	-7.10
05	18.9	19.29	-2.06	15.17	13.30	19.10	-1.06
06	17.86	18.70	-4.70	14.18	20.58	18.93	-6.00
07	18.64	20.53	-10.16	15.50	16.87	19.78	-6.11
08	20.92	23.73	-13.43	16.87	19.37	20.75	0.82
09	24.34	26.96	-10.77	19.63	19.35	23.38	3.96
10	25.81	28.17	-9.13	28.09	-8.82	26.94	-4.39
11	27.72	29.00	-4.60	27.97	-0.89	27.49	0.83
12	29.28	30.20	-3.13	33.10	-13.03	29.10	0.60
13	30.81	31.44	-2.05	32.87	-6.68	30.35	1.49
14	31.42	32.02	-1.91	31.39	0.10	31.62	-0.65
15	32.12	31.18	2.92	31.83	0.89	31.77	1.10
16	32.19	32.73	-1.67	33.68	-4.63	32.32	-0.39
17	32.8	33.54	-2.25	33.73	-2.84	32.12	2.06
18	32.86	34.30	-4.39	35.06	-6.69	32.81	0.14
19	32.86	34.21	-4.11	33.99	-3.43	32.69	0.50
20	31.50	34.12	-8.32	34.63	5.58	32.61	-3.53
21	37.37	37.24	0.36	42.93	-14.88	37.53	-0.42
22	36.70	34.03	7.28	44.47	-47.53	37.23	-1.45
23	26.03	28.88	-10.97	35.14	-35.01	27.31	-4.91
24	24.67	26.91	-9.08	27.03	-9.56	24.18	1.97
RMAPE		6.56		11.44		2.52	

5. Conclusion

For the periodic, unsteady and fluctuant price time series of power market, this paper proposes an optimized algorithm which is based on NPGARCH-LSSVM, and the proposed method NPGARCH-LSSVM is used to forecast short-term electricity price of PJM, we can get the following conclusions:

(1) Nonparametric GARCH is used to correct the residual of prediction price. There is a heteroscedastic in the prediction price time series which is applied with LSSVM by ARCH test, but by using nonparametric GARCH to correct the residual can make it to be the stationary sequence.

(2) Nonparametric estimation can be used less assumption and has a better nonlinear fitting of conditional standard deviation, it can improve the prediction accuracy and the prediction of price spikes.

Acknowledgments

Work partially supported by Philosophy and Social Science Research Base of Hebei Province.

Reference

1. Pany Prasanta Kumar, Ghoshal S.P. Dynamic electricity price forecasting using local linear wavelet neural network[J].Neural Computing and Applications,2015.
2. Yu Fan, Shen Jiong, Liu Xichui. Day-ahead marginal price forecasting based on auto-regressive conditional heteroskedasticity-back propagation network model[J].Power System Technology,2008,32(8): 61–81.
3. Deng Jiajia, Huang Yuansheng, Song Gaofeng. Day-ahead electricity price forecasting using time series model based on nonparametric generalized auto regressive conditional heteroskedasticity[J].Power System Technology, 2012,36(4):190–196.
4. Kumar Aggarwal, Mohan Saini, Kumar. Electricity price forecasting in Ontario electricity market using wavelet transform in artificial neural network based model[J]. International Journal of Control, Automation, and Systems,2008,6(5):639–650.
5. Xie Peiyuan, You Dahai1, Zeng Ciling, etc. Clearing price forecast for day-ahead market based on genetic-algorithm-optimized BP network[J].Electric Power Automation Equipment,2004,24(3):53–56.
6. Tan Zhongfu, Zhang Jinliang. Day-ahead Electricity Price Forecasting Based on Multi-factor Wavelet Analysis and Multivariate Time Series Models[J]. Proceedings of the Chinese Society of Electrical Engineering, 2010,30(1): 103–110.
7. Zhang Jinliang, Tan Zhongfu, Yang Shuxia. Day-ahead electricity price forecasting by a new hybrid method[J]. Computers and Industrial Engineering,2012,63(3):695–705.
8. Zeng Ming, Zhao Yongliang, Zhang Jing. Electricity price forecasting methods combining ARIMA and GARCH with confidence intervals[J].East China Electric Power,2008,36(12):1–6.
9. Cifter Atilla. Forecasting electricity price volatility with the Markov-switching GARCH model: Evidence from the Nordic electric power market[J].Electric Power Systems Research,2013,102:61–67.
10. Andreou, Elena, Werker, Bas J.M. Residual-based rank specification tests for AR-GARCH type models[J]. Journal of Econometrics, 2015, 185(2): 305–331.
11. Zhang Rong-Mao, Sin Chor-Yiu, Ling Shiqing On. Functional limits of short-and long-memory linear processes with GARCH(1,1) noises[J]. Stochastic Processes And their Applications, 2015, 125(2):482–512.

College Students' Work Innovation Model Research under the Background of Internet

Shu-Yang Zhu

Jiangsu University of Technology; School of Management, Jiangsu University; Changzhou, Jiangsu, China

E-mail: zhushuyang@jsut.edu.cn

119429591@qq.com

Currently, college new media, represented by the internet and mobile media, have become major approaches for young people's learning and life .There are three main models as college new media in Universities in Jiangsu. This paper emphasizes, introduces and explores the construction approaches emphatically of the models, they are respectively College Student Work Innovation Model based on WeChat, College Students' Work Innovation Model based on PU Platform and College Students' work innovation model under the background of 'Internet+'. The student work in the background of Internet+ links school information release, student activity organization and second class development closely together, and meanwhile it can open up a welcome channel for freshmen to provide them with omnidirectional one-stop comprehensive guidance and services to explore a new mode of freshman education.

Keywords: Internet+, college students' work, innovation model.

1.Introduction

Currently, college new media, represented by the internet and mobile media, have become major approaches for young people's learning and life, provided important space for their gathering and contact, and served as integral platforms for their communication and interaction. College new media have also achieved breakthroughs to take a spot in the traditional media framework and play an increasingly prominent role in leading young people.

2. Student work innovation model based onWechat

WeChat is mobile chatting software launched by Tencent in early 2011 for sending voice messages, videos, pictures and words through internet, supporting voice communications among several people. According to Research on Economic and Social Influence of WeChat (2015), WeChat has developed through a primary stage, an open stage and an ecological stage. By the end of 2015, monthly active users of WeChat had increased to 0.697 billion at a year-on-year growth rate of 39% [1]. As new media, it took four years, 14 months and merely 10 months respectively for internet, microblog and WeChat to accumulate 5,000 users after their emergence [2]. Some university surveys have suggested that universities account for the largest proportion of users registering WeChat, mostly use WeChat and spend two hours or so online on average [3]. The Research Report on Chinese Users of the Public Platform WeChat in 2013 revealed that, although WeChat public platform was fairly popular, its practical marketing effects and user adhesiveness were below expectations. WeChat public platform isn't informative and practical enough, so secondary development is quite necessary [4]. In August 2012, WeChat launched its public platform, after which Huazhong University of Science & Technology opened the first WeChat public platform among Chinese universities and other universities of China successively opened their public platforms thereafter. Jiangsu University of Technology registered its WeChat platform in April 2013 and named it Assistant of Jiangsu University of Technology as its service account. After the launch of WeChat public platform, attempts have been constantly made in development. The services available from the platform have increased user adhesiveness to the greatest extent and guaranteed continuous growth of users. At present, there have been more than 30,000 users on that platform, and approximately 17,000 students pay attention to it. Among universities, it ranked Top 1 within Jiangsu Province and Top 2 all over China's universities at best.

The main objective of the system is to serve the undergraduates. Their personal information and student ID are connected, to satisfy their requirements of both life and study. Moreover, it is designed for serving graduates. After they graduated, their accounts are still connected for instant information of the university. Such continuous service satisfies the graduates' information requirements. Moreover, the upgrade and expansion functions of the system should satisfy the new requirements of students. On the other hand, the application of the system should promote to the other departments of the school, such as information center and security department.[5]

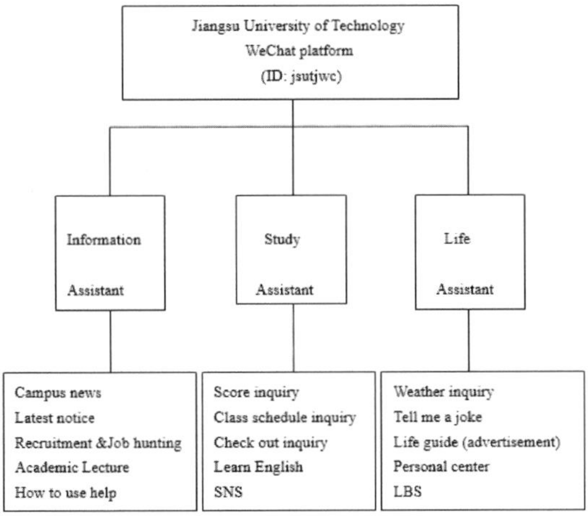

Figure 1. The Architecture of Jiangsu University of Technology WeChat Platform

3. Student work innovation model based on PU Platform

3.1. *Concept and characteristics of PU Platform*

Pocket University, a service platform for the practice and growth of university students, (known as the "PU Platform"), applies the mobile internet technology to concentrate the high-school education resources within Jiangsu Province so as to provide services for university students' second class and daily life. PU Platform also covers big data analysis, school information dissemination and school-wide interaction. The Platform is characteristic of openness, extensiveness, timeliness, virtuality, interactivity and equality[6]. The promotion of 'PU' Platform is the only innovation of universities of Jiangsu Province.

3.2. *Special planning has taken place to build the brand and improve cohesiveness*

Special activities are initiated that combine students' practical demands to attract young people, improve the cohesiveness of users and build brand segments.

First, CYL（Communist Youth League） day activities are held. The PU Platform is utilized for the following purposes: to publish regular activities participated by all youth league branches across the University, to select the most dynamic league branches and the most charismatic league secretaries, to develop online space for basic-level league branches, to innovate league branch management model and growth approaches, and to expand league branches' coverage and influence. Second, star associations are selected. The PU platform is utilized for the selection of star associations. The University has timely revised the "Standards for Selecting Star Associations for Jiangsu University of Technology". Under the standards, each association's situation of initiating activities is considered as a key evidence for becoming a star association. Such an approach is conducive to standardizing association management and building brand associations. Third, cultural and art festivals are hosted. The PU Platform is utilized to initiate cultural and art festivals, with the priority of developing high-participation and highly adored feature activities. As a result, the Platform has achieved high registration.

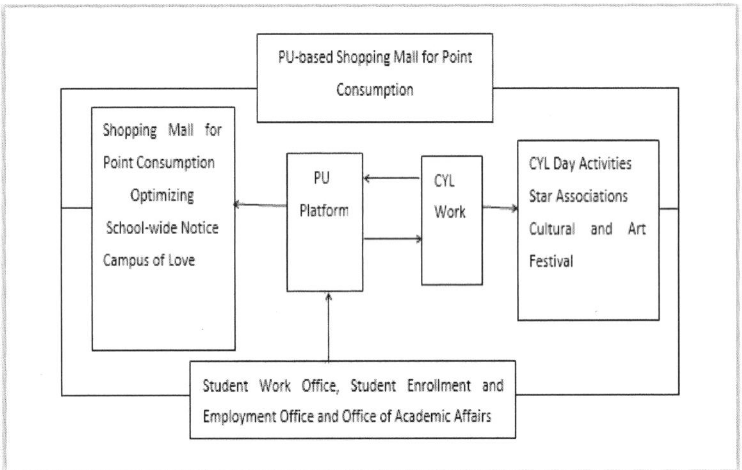

Fig.2: PU-based CYL Work Model

The PU Platform is developed into a multifaceted service community with three segments, namely, shopping mall for point consumption, university-wide notice issuance and campus of love, to meet students' practical demands and improve the cohesiveness of PU Platform. First, a shopping mall for point consumption is developed that combines online and offline activities. The mall integrates learning, sports, leisure and recreation and adopts a grade-based point exchange; regular offline activities feature "trading points for goods". Second, the issuance of university-wide notices is optimized. Such notices are divided into six segments: employment, lost and found, lectures, studying for master degrees, safety talks and gossip. Relevant information is regularly updated to provide students with all-round service. Third, the pilot program of "Campus of Love" has taken place. Efforts have been made to establish the "Fragrant Green Fruit" Fund and cooperate with the university benevolence supermarket. The PU Platform is also utilized for book donation.

3.3. *Joint efforts are made to enhance interaction and serve the University.*

The PU Platform is widely used by students, which has caught the attention of other school administrative departments who also joined the Platform by creating their own accounts. They introduce much-adored activities, including humanities lectures by the student work office, dormitory cultural and art festival, and a series of lectures of successful personalities organized by the student enrollment and employment office. Discussions are made with the office of academic affairs to design quality-development points and jointly promote the development of PU Platform. The Platform has been used for announcing activities, approval, publicity, comment and completion. As a result, students and teachers communicate more often on a widely used social network after Weibo and We Chat.

3.4. *Efforts are made on team-building, regular contact and training*

A pyramid-pattern work team is built, together with a regular contact mechanism and different categories of technical training so that the staff can make step-by-step progress and improve their professionalism.

First, a pyramid-pattern work team is built, with the top taken by PU working leadership team, the body composed of PU liaison personnel from college and university leaders for new media work, and the bottom formed by the managers of student organizations, associations and classes. This team covers student leaders across all colleges and CYL student leaders at all levels. Second, a regular contact mechanism is established, improving contact and communication via QQ groups, Fetion groups and We Chat groups, and providing technical guidance and macro control over the dissemination and management of the university-wide PU Platform. Third, a variety of technical trainings are provided, including trainings on how to conduct background operation and design electronic posters for student tribe managers and trainings on how to activate their PU account and use the Platform for average students, and covering the student union, student associations and basic-level youth league branches with a step-by-step and layer-based approach.

4.College students' work innovation model under the background of Internet

The guiding institutions of new media operation team. Guiding institutions bring together Publicity Department, Students' Affair Office, Youth League Committee, Recruitment and Employment Office, Office of Teaching Affairs and other departments, and the two-way communication between these institutions and students is mainly achieved through new media team.

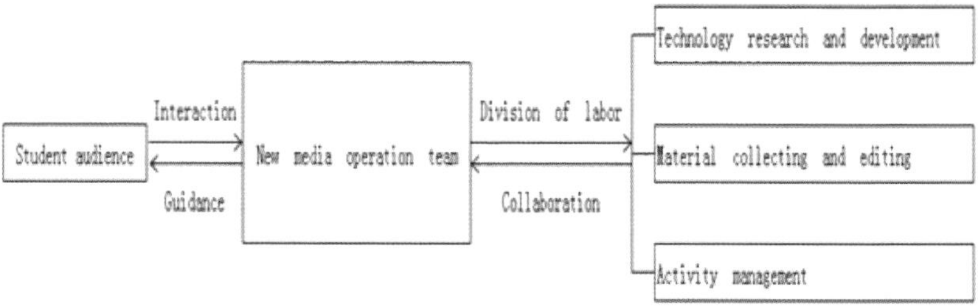

Fig.3: College students' work innovation model under the background of Internet

The main responsibilities of new media operation team. Focus on planning and operation of new media platform, it consists of WeChat, Microblog and PU Platform three operation teams, they have both division of labor and cooperation.

Construction of new media operation team. Technology development team focuses on conducting development and perfection on new media platform by using related network technologies. Team construction is in the form of stepped training, it is three layer echeloned team training and construction by taking professional teachers as guidance, student backbone as main body and new media enthusiasts as members; taking skills training as starting point, carry out many trainings related to new media, and enhance team accomplishment and work ability. Members of the new media operation team mainly consist of technology research and development, material collecting and editing as well as activity management three groups. School level new media operation team trains and guides the next level of teams, and the next level of teams provide materials for school level new media operation team.

5. Conclusions and Suggestions

The student work in the background of Internet+ links school information release, student activity organization and second class development closely together, and meanwhile it can open up a welcome channel for freshmen to provide them with omnidirectional one-stop comprehensive guidance and services to explore a new mode of freshman education. This has completely changed the "Pareto's law" existing in the activities of the communist youth league and student union, so that the students have greatly enhanced their enthusiasm in attending the activities of the communist youth league and student union, showing the elegant demeanor of the communist youth league and student union in the new era, reflecting the appeal and cohesiveness of the league organization. New media platforms are flourishing with each passing day under the background of Internet+, so student work directors need to do the following well:

Jump at the chance, improve the system, and standardize administration. Make systematic planning at the trial stage to make a related system for standardized management.

Build a team, building relationship, and carry out training. Build a pyramid-shaped work team, establish a common relationship mechanism, carry out diversified technical training, and conduct classified training to improve professional qualities step by step. Plan activities, create brands, and enhance stickiness. Launch thematic activities and create a brand forum in accordance with students' actual demand to enhance application stickiness to gather and attract young people.

References

1. Industry and Planning Institute, China Academy of Telecommunication Research, Research on Economic and Social Influence of WeChat (2015) [M], P: 6-9.

2. Tang Xujun,Report on Development of New Media in China (2013) [M]. Beijing: Social Science Academic Press, 2013, Journal of HeFei University (Social Science), 2014 (31):101.

3. Jin Xiaofang. Research on Impacts of WeChat upon University Students' Thoughts and Behaviors [J]. Journal of HeFei University (Social Science), 2014 (31): 101.

4. Limedia Research,"Investigation Report of China Wechat Official Account Admin Platform Users ", 2013,PP 1-2.

5. Zhu Shuyang, Shi, Study on system design of Wechat developer in a university based on C#,Asia-Pacific E-business and E-Government Conference (2014),P:1386-1390.

6. Wang Wenjie, Yuan Wen, Yang Lei and Wang Xiuyan, Innovation Research on Working Method for College Communist Youth League under the New Media Environment, Xuexiao Dangjian Yu Sixiang Jaoyu (School Communist Party Construction and Idological Education), 2013.2.

An empirical research of the input-output model on demand elasticity based on the energy consumption *

Yan Zhang†, Xin Liu

Hebei University of Science and Technology
Shijiazhuang Yuhua District, 26th Street Yuxiang, Heibei province, China
E-mail: 794389191@qq.com, www.hebust.edu

It was emphasized that building a resource-conserving friendly society during the 13th Five-Year Plan period. It is a major measure for building a moderately prosperous society that setting up the resource views of saving, economical, intensive and recycling. It is related to the developing sustainably and healthily of national economy. This paper construct the input-output model on demand elasticity of sub-sector energy consumption increasing energy consumption factors basing on the input-output table of 42 departments. The empirical analysis part of the main economic departments in Hebei province by the model calculation of induction degree coefficient, influence coefficient, energy consumption of each sector is analyzed on the basis of ripple effects. The results show that the energy spread effect exist diversity in different departments, so we should set out different energy saving measures to all sectors of the economy.

Keywords: Input-Output Model; Energy Consumption; Spreading Effectiveness; Influence Coefficient; Response Coefficient.

1. Introduction

This section only introduces the researches of the scholars in China with the input-output method is widely used in energy, environmental and economic problems in recent years. Ai-Jun Li [3]has set up input-output model on 13 departments based on input-output table on 64 departments of Hubei province in 2002,the results show that the energy saving strategies should focus on the coal utilization aspect of raw material industry, energy and manufacturing industry . Yi-Liang Ma and Qi-Yun Liu [4]have analyze the contribution in different stages on Structure factors, technical factors in energy consumption ,and then establish the input-output optimization model for economic growth and energy saving

* This work is supported by D2015001030, SB151016, and the Bohai sea in Hebei province granary demonstrative project of science and technology, agriculture and animal husbandry combined with circular economy development model and key technology research project.

double objectives. Yan Xia and Cui-hong Yang [5]have optimize the total energy saving goal according to the different industry based on the input-output optimization method. Ming-Qiu Liao[6]. Have develop input-output model based on energy conservation and emissions reduction including energy, environment and economy, it provides specific input and output table type to research energy conservation and emissions reduction problems .Ying-Zhi Xu and Quan-Zhen Zhang[7] have analyzed the energy consumption ripple effects in every section by calculating the influence coefficient and sense degree coefficient based on the input-output table in 2007 of China .This article mainly analyzes the ripple effects of energy consumption for various departments based on input-output in 2012 table of 42 departments in Hebei province, and then develop targeted energy saving policy.

2. The input-output model

This article build the input-output model on demand elasticity of energy consumption based on the basis of the energy input-output model, as equations follow.

$$E = e_1 x \tag{1}$$

E stands for energy consumption of various economic sectors ,e1 stands for energy consumption of Unit value added in the department, x stands for Unit value added ,as in Eq.(1).We can deduce the Eq.(2), according to input-output model of Wassily Leontief .

$$x = (I - A)^{-1} y \tag{2}$$

$(I-A)^{-1}$ stands for leontief inverse matrix ,y stands for Column vector of each department final demand in Eq.(2). Energy consumption departments can be expressed as Eq.(3),combined with Eq.(1) and Eq.(2).

$$E = e_1 (I - A)^{-1} y \tag{3}$$

We can assume that each department the added value of unit energy consumption and Leon if inverse matrix in the comparative period is constant,a1 stands for final demand rate of various department. So the energy consumption variation of various department ΔE as Eq.(4):

$$\Delta E = e_1 (I - A)^{-1} y a_1 \tag{4}$$

We define S for the proportion of final demand in total output ,So final demand column vector can be represented as Eq.(5):

$$y = \hat{x} s \tag{5}$$

$\hat{x}$ stands for diagonal matrix of the columns in various departments in Eq.(5), Merge Eq.(4) and Eq.(5) is as follows:

$$\Delta E = e_1 (I - A)^{-1} \hat{x} s a_1 \tag{6}$$

We can arrive at Eq.(7) stands for final demand rate of various department, left by E-1at the same time on both sides of the Eq.(6).

$$E^{-1} \Delta E = E^{-1} e_1 (I - A)^{-1} \hat{x} s a_1 \tag{7}$$

Each department unit value added calculation is as follows:

$$e_1 = E f \hat{x}^{-1} \tag{8}$$

f stands for Each department energy accounted for the proportion of the total energy consumption, then bring Eq.(8) to Eq.(7), on the left side of the equation expressed in ε, it is concluded that:

$$\varepsilon = f \hat{x}^{-1} (I - A)^{-1} \hat{x} s a_1 \tag{9}$$

And because:

$$\hat{x}^{-1} (I - A)^{-1} \hat{x} = (I - D)^{-1} \tag{10}$$

D stands for Partition coefficient matrix, it Will be concluded Eq.(11) after bringing Eq.(10) to Eq.(9).

$$\varepsilon = f(I-D)^{-1}\hat{x}\,a_1 \quad (11)$$

We should assume that final demand rate a1is 0, and the f diagonalization matrix form in order to calculate conveniently.

$$\varepsilon = \hat{f}(I-D)^{-1}\hat{x} \quad (12)$$

So ε_{ij} stands for the final demand change of elasticity of energy consumption to I department on j department, to constitute energy consumption of 42 * 42 demand elasticity matrix .The influence coefficient stands for matrix column, represented by ε_j.The response coefficient stands for matrix row, represented byε_i and the equations as follows:

$$\varepsilon_j = \sum_{i=1}^{42} \varepsilon_{ij} \quad (13)$$

$$\varepsilon_i = \sum_{j=1}^{42} \varepsilon_{ij} \quad (14)$$

Influence coefficient and response coefficient reflect the energy consumption of a department and other departments, so we can fully understand the various departments in the economic system ripple effects through the two aspects of energy consumption. There are three kinds of influence coefficient values: $\varepsilon_j > 1$; $\varepsilon_j = 1$; $\varepsilon_j < 1$. It effect the other departments over social influence on average lever if influence coefficient greater than1.The influence coefficient is bigger, the greater the pull function of the department to other departments, the greater the pillar role in the national economy stronger. The response coefficient values also have three types, it effect the other departments over social response on average lever if response coefficient greater than1.The response coefficient is bigger, the greater the said the department by induction, the greater the degree of demand which the more basic role in the national economy. In a word, the size of the influence coefficient and response coefficient depended to various industry departments for ripple effects influence the whole national economy, its value is bigger shows that the more obvious ripple effects of the industry.

3.The analysis of computing result

This article are based on data from 2012 input-output table of hebei province and Hebei province economy yearbook from year to year, and then according to the above down the energy consumption of the input-output model to calculate the response coefficient and influence coefficient of 42 departments. In order to further analysis, Take the middle value of influence coefficient and response coefficient as boundary points, ε_l stands for the middle value of influence coefficient(ε_l =1.05), ε_s stands for the middle value of response coefficient(ε_s =0.69), so the 42 departments can be divided into four categories, A, B, C and D. It will be shown in rectangular coordinate plane distribution location, as shown in Fig 1, to make the classification more intuitive.

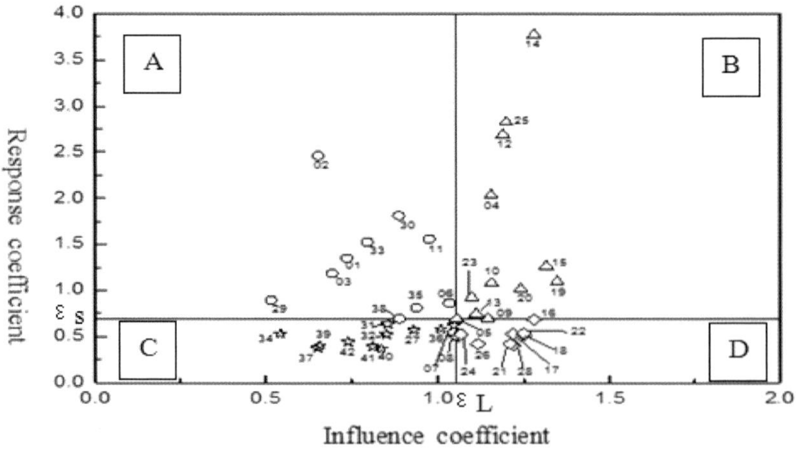

Fig. 1.The industrial sector classification figure

It mainly includes mining and washing of coal, extraction of petroleum and natural gas, foods and tobacco, processing of petroleum, coking, processing of nuclear fuel, traffic, transport, storage and post of class A department. Referred to these department as "bottleneck", its development speed will directly influence the development of the economy as a whole system, so the scale of production for the department of energy saving policy can be binding influence on other departments of production, in the process of industrial structure adjustment to fully guarantee the department in proportion in the economic system.

It mainly includes mining of metal ores, mining and processing of nonmetal ores and others, manufacture of nonmetallic mineral products, Manufacture and processing of metals, manufacture of metal products of class B department. The department response coefficient and influence coefficient were both larger, was taken as the "key sectors", their final demand growth will drive the large increase of energy consumption, in other departments at the same time, other departments final demand growth also drive the growth of energy consumption of this department. So for this kind of department to perform active energy-saving policy, in the process of energy saving to give attention, put it in the point of energy saving, not only to appropriately control the scale of production, and to supervise and urge the department in a timely manner to improve its production technology, reducing the energy consumption caused by the added value of the unit.

It mainly includes manufacture of textile, manufacture of textile, apparel, production and distribution of water, hotels and catering service, information transmission, software and information technology services of class C department. The response coefficient and influence coefficient of this class division are relatively small, its final demand growth for the energy consumption is not obvious, the influence of other departments when energy saving policy accordingly don't need to pay attention to these departments.

It mainly includes manufacture of general purpose machinery, manufacture of special purpose machinery, manufacture of transportation equipment, manufacture of measuring instrument, manufacture of others of class D department. The department has obvious influence coefficient, but response coefficient are smaller, that the department of final demand growth will drive the whole economic system large increase of energy consumption. We should strictly control the final demand of this department in the energy conservation policy form the numbers. From the perspective of the actual of Hebei province, the energy conservation policy for this type of departments should be on improving production technology, improving energy utilization to effectively reduce the energy consumption of the department.

4. The conclusion

In this paper, we calculate influence coefficient and response coefficient of the economic system in Hebei province by building energy consumption demand elasticity of input-output model, and analyze the ripple effects of each department. The results show that the ripple effects of different departments, on the whole of the second industry spread effect is bigger, the first and third industry spread effect is relatively small. We should strengthen the construction of energy-saving measures, for the second industry sector energy consumption is larger. So it is necessary to take into account different divisions for scientific, in view of the differences between the different department of energy conservation policy, achieving the goal of making smaller energy saving target. For the third sector, the proportion of energy consumption is relatively small, tending to ignore the department in the energy conservation policy, the ripple effects of individual services are significant but according to the result of model calculation analysis, so we should formulate the corresponding energy saving policy, avoiding the occurrence of energy-saving "blind spots" for this department.

References

1. Xi-Kang Chen and Cui-Hong Yang, The Technique of Input-Output [M], Science Press (2011).
2. Yan Wang and Hui Song, To Study on the Calculation Methods of Diffusion Coefficient and Inducing Coefficient [J], Value Engineering (2007).
3. Ai-Jun Li, The energy intensity decomposition based on input-output model of Hubei province [J] , Statistics and Decision (2008).
4. Yi-Liang Ma and Qi-Yun Liu, Difficulty in Achieving the Dual Objectives of Rapid Economic Growth and Energy Saving [J], Journal of ShanXi Finance and Economics University (2008).
5. Yan Xia and Cui-Hong Yang, Analysis of Industry-level Energy-Saving Potential and Goal: An Input-output Optimization Model [J], Management Review (2010).
6. Ming-Qiu Liao, Research on the Input-Output Model Based on “Energy Saving” [J],China Industrial Economics (2011).
7. Ying-Zhi Xu and Zhen-Quan Zhang, Energy Consumption and Adjustment of Industrial Structure: An Analysis Based on Input-Output Method [J], Journal of Nanjing Normal University (Social Science 2012).
8. Qi-Yun Liu, Research on input-output coefficient method for structure analysis [J], Statistical Research (2002).
9. Qiao-Ying Wang, New Exploration to the Influence Coefficient and Reaction Degree Coefficient Calculation Method [J], Theoretical Exploration.
10. Hao Xiao and Qiao Zhu, Evaluation and Improvement of the Diffusion Coefficient and Inducing Coefficient[J], Managment Review (2015).

Research on Network Public Opinion Evolution Model Based on Mobile Cellular Automaton*

Chao-Bin Wang
Department of Network Center, China West Normal University
Nanchong, SiChuan, China
E-mail: cbwang@cwnu.edu.cn

Qian-Ren Mao†
College of Computer, China West Normal University
Nanchong, SiChuan, China
†*E-mail: maoqianren@yeah.net*

In order to study how the mobility of entities affect the public opinion evolution in network public sentiment dissemination, we design a model of network public opinion evolution based upon mobile cellular automaton. Based Margolus affecting neighborhood, we propose cellular mobile rules, evolution rules, the Attitude Consistency Group and the Phase Exchange Interval, aiming to explore the influence of the scope and speed of entities' movement to the public opinion propagation speed. The simulation results show that, in the type of Margolus affecting neighborhood, entities' mobility promotes more communication among entities, making public opinion spread faster and more widely, while groups of views and opinions present a phenomenon of agglomeration. Meanwhile,In the process of public sentiment dissemination, there comes a region with coexistence of assentient phase and opponent phase. We get the changing range of phases through many experiments in the random initial state with different initial support rate.

Keywords: Network Public Opinion; Cellular Automaton; Mobility Rules; Structure of Neighborhood; Matlab Simulation.

1. Guidelines

Cellular automaton (CA) is a discrete and dynamical system in time, space and state, which is used by lots of scholars in researching some complex, nonlinear system such as network public opinion system. There are some classical models, such as, Janusz a Holyst [1], et al. Propose social influence model based on cellular automata. Paper [2] proposes Sznajd Model, which simulates interaction among individuals through setting the local characteristics of the cellular, and finally presents three steady states in public opinion evolution system. Paper [3] proposes Bounded Confidence rules to constrain cellular exchange, and establishes the Deffuant Model which is a model of opinion exchange with continuous opinion interval.

* This work is supported by *NSFC (Natural Science Foundation of China), No. 61379036*

In order to thoroughly simulate the actual dissemination and evolution process of the network public opinion, many scholars develop and improve these classical models. Papers [4][5] elaborate the effect of the entities' mobility and personality on network public opinion evolution based on cellular automaton model. Papers [6–8], combine the cellular automaton model of network public opinion dissemination with the model of MR (Majority rule), and apply this combination to the Von Neumann neighborhood (with five affecting neighbors). These papers analyze the influence of environmental factors to the attitude tendency of entities and to the dissemination of public opinion. While, paper [8], extending the research method of paper [7], proposes environmental factors influencing the susceptibility (the degree of public opinion as a whole spread toward " disagreement" and "agreement") to move closer to the majority. Most of the researches of classical public opinion evolution on cellular automata conversing entities' state are based on the type of Von. Neumann neighborhood [9] and Moore neighborhood[10], which the structure and the influence range of affecting neighbors are not enough to simulate the complexity of the real network environment. Without ignoring the complexity of the cellular mobility. We define a public opinion model with a new mobile rule and entities' attitude state transition rules, through analyzing main factors influencing the tendency of public entities' attitude, to explore the formation and evolution of network public opinion. Experimental results show that this new model effectively simulates the rapidly spreading and complicatedly space arranged public opinion evolution proposes.

2. Mobile cellular automaton based on the Margolus neighborhood

2.1. *Mobile cellular space and neighborhood*

The Margolus neighborhood structure takes adjacent 2*2 cellular as a cellular block to simulate cellular movement, which increases the complexity of the mobility, and some cellular can move to any position of the whole cellular space. Every grid block in the Margolus neighborhood structure has 4 cells, and the location of each cell in the grid block is respectively denoted as UL (upper left), UR (lower right), LL (left lower) and LR (lower right). The grid block varies with time changing, among which the upper left 4 cells, UL, UR, LL and LR, form the mutual neighbor relationship at even numbered time step (the evolution of the second time, the fourth time, etc.), and right below 4 cells ,UL,UR,LL and LR, form the mutual neighbor relationship at the odd time step (the evolution of the first time, the third time, etc.). As Fig.1 shows, labeled cellular LR will change into UL in next blocks due to block changing in the next time step. Four cells rotate and change repeatedly in the square grid, making information of each cellular stepping over 2*2 grid block boundaries for transmission. The range of cellular's movement becomes wide, because most of cellular may walk around everywhere in whole space.

2.2. *Cellular mobile rules*

In every 2*2 grid block, two diagonal cells exchange their locations at each time step. Moreover, in each block, mobile cellular must follow the following rules:

Rule 1: All of cellular follow the same mobile rules, so that cellular's information can move across the 2*2 grid block.

Rule 2: These rules can only reflect the movement trend of a certain direction. In order to reflect each cellular at each time step in the random direction, we need to consider two special cases, one is that cell collides the cell, another is that cell collides the wall. The two diagonal cells, must rotate counterclockwise by 90 degrees after a collision. Two mobile rules as shown in Fig.2:

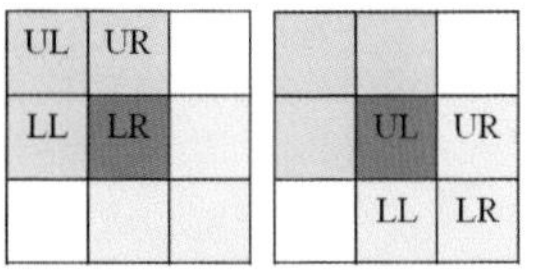

Fig.1. Margolus cellular structure

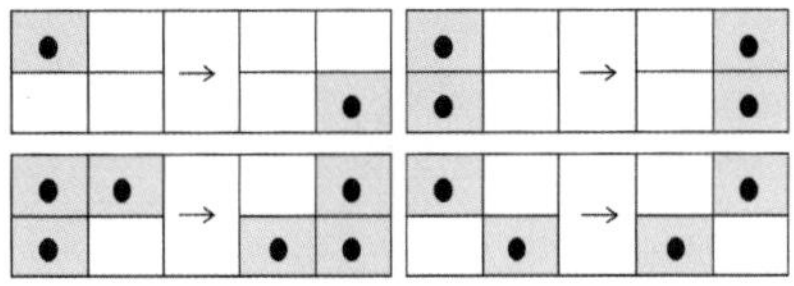
Fig.2. Cellular mobile rules

3. The model of public opinion propagation based on mobile cellular automaton

3.1. *Related Definitions*

The definition of this new mobile cellular automaton model consists of, mobile cellular (the entity of public opinion), mobile cellular space (the whole grid), neighborhood, cellular mobile rules, attitude state of mobile cellular and the evolving rules of cellular automata (state transformation function).

Definition 1:(the entity of public opinion). Every cellular is an entity, which represents an internet user participating and interacting in network affairs.

Definition 2: (attitude groups). Attitude group refers to a collected entities with a consistent conceptual and emotional tendency in the network affairs.

Definition 3:(entity attitude threshold f).The entities' attitude deviates to assentient or opponent .We set a value to indicate this deviation, namely, the attitude change threshold. The values are represented by a set of $\{f_2, f_3, f_4\}$. In the calculation results of the transfer function, there are three kinds of critical values, which are *2,3,4*, so the threshold value's subscript are defined by *2,3,4.*

Definition 4: (the rate of public opinion support).The initial support rate in network public opinion and network public opinion support rate after T times, are respectively expressed in algebra as $K_1(0) = N_1(0)/N$, $K_1(t) = N_1(t)/N$. $N_1(0)$ represents the number of how many entities' value is 1 at initial moment.

3.2. *The model of public opinion propagation*

3.2.1. *Mobile Cellular*

Considering an entity of public opinion may take a different approach to participate in the discussion network affairs, we define the mobile characteristic as an abstract concept. Entities can join the discussion in different ways as much as they want, such as the portal, online media and website of the government. This is one of the differences our model have comparing with the classical models.

3.2.2. *Cellular's Mobile Space*

The whole collection of square-section grid is cellular's mobile space that being determined by cellular's number. Each grid represents a person, the entity of public opinion.

3.2.3. *Affecting Neighborhood*

The type of Margolus effecting neighborhood structure is used to define the cellular's mobility. Similarly, the change regulation of entities' attitude state can also be defined by Margolus neighborhood cellular structure. We put the cellular with attitude value into every Margolus neighborhood of cellular grid in an odd number moving time step and even number moving time step before cellular moving. At odd time step the cellular configuration is that, UL= $e_{i-1,j+1}$, UR= $e_{i,j+1}$, LL= $e_{i-1,j}$, LF= e_{ij}. At even time step the cellular configuration is that, UL= e_{ij}, UR= $e_{i,j+1}$, LL= $e_{i,j-1}$, LF= $e_{i+1,j-1}$. Then, cellular transformation rules only consider, at each time step, the six neighbor cellulars with distance of r=1 around the central entity in Margolus neighborhood structure. The smallest unit of the central entity's state evolution is these 7 cells, 3 neighbors in the odd step time, other 3 neighbors in next even step time and the central entity. If a central cellular is $e_{i,j}$, cellular neighbors respectively are

$e_{i-1,j+1}$, $e_{i,j+1}$, $e_{i-1,j}$, $e_{i+1,j}$, $e_{i,j-1}$, $e_{i+1,j-1}$.

3.2.4. *Cellular state*

Cellular state is defined to signify the degree of entities' attitude tending to agree or disagree towards network affairs. To simplify this attitude, only two possible choices are assigned, one is assentient attitude, signified with 1, another is opponent attitude, signified with 0. Entity's attitude state is determined as it takes the assigned value.

3.2.5. *Attitude state evolution rule*

In the view of mathematics, attitude evolution rule, is a function, codetermined by the cellular current attitude state and other effecting neighbors' attitude state[11][12]. The Margolus central entity's local state can be expressed as a formula ,which is as in Eq. (1) ,

$$e_{i,j}^{t+1} = \phi(e_{i-1,j-1}^{t}, e_{i,j-1}^{t}, e_{i-1,j}^{t}, e_{i,j}^{t}, e_{i+1,j}^{t}, e_{i,j+1}^{t}, e_{i+1,j+1}^{t}) \quad (1)$$

The value of ϕ ,decided by Margolus effecting neighborhood,being irrelevant to i and t, refers to that the cellular state in the next time is determined by the current state and neighbors' attitude state. Setting cellular's attitude state is $e_{i,j}^{t+1}$, the attitude state evolution rule is expressed as a mathematical formula, which is as in Eq. (2).

$$e_{i,j}^{t+1} = \begin{cases} 1-e_{i,j}^{t}, & \phi \geq 4 \\ e_{i,j}^{t}, & \phi \leq 3 \end{cases} \quad (2)$$

Among which,

$$\phi = \left|e_{i-1,j-1}^{t} - e_{i,j}^{t}\right| + \left|e_{i-1,j}^{t} - e_{i,j}^{t}\right| + \left|e_{i,j-1}^{t} - e_{i,j}^{t}\right| + \left|e_{i+1,j}^{t} - e_{i,j}^{t}\right| + \left|e_{i,j+1}^{t} - e_{i,j}^{t}\right| + \left|e_{i+1,j+1}^{t} - e_{i,j}^{t}\right|$$

3.2.6. *Attitude consistency group*

According to the evolution rule of cellular, in the process of the system evolution, entity with unwavering attitude will not change throughout evolution. In order to study the effect of neighbor's attitude to the central entity in Margolus type neighborhood, at the initial moment of evolution, we define the Unwavering Attitude Entity as $\phi \leq (f_2=2)$, indicating, 4 and more neighbor cells' attitude are identical with its own attitude. Unwavering attitude entity' attitude will not change all the time. If $\phi \geq (f_4=4)$,with 4 and more neighbor cells' attitude differing from the central one, the central entity is defined as conformism entity with opponent attitude at next time step. If $\phi=(f_3=3)$, around the central entity, the number of these neighbors with assentient attitude have the same number of these with opponent attitude, the one called ordinary entity and its attitude could be assentient or opponent at next time step.

4. Simulation and analysis of results

In order to explore relationship between Margolus neighborhood structure and the evolution of public opinion, we take Von Neumann neighborhood model as a comparison. In order to explore the advantage of the proposed mobile rules in the public opinion evolution, we respectively simulated the evolution of two cases, cellular without moving rules and cellular with moving rules under the structure of Margolus neighborhood. Fig. 3 is a simulation of opinion evolution in type of classic Von Neumann effecting neighborhood with no cellular mobile rules. We can see from Fig.3, in the initial time, two attitude state values are randomly assigned 50% of the probability to all entities. In this case, influence degree between cellular and neighbors is very weak. In clustering area, entities have consistent opinion, but it is difficult to cluster a large-scale if cellular don't move, and the gathering area is very fragmented after the simulation. Around the "cluster group" , has a lot of discrete entities can not dwell together.

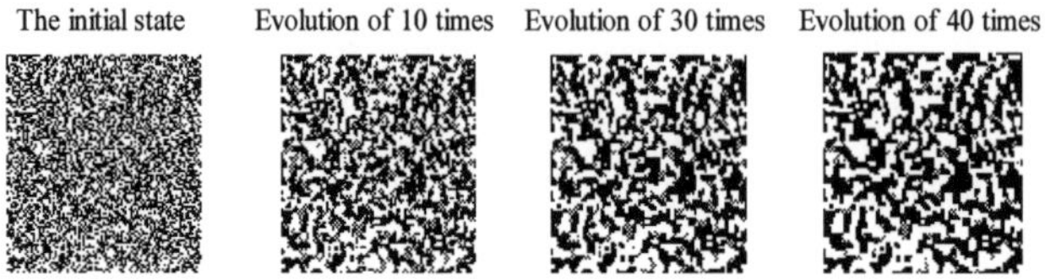

Fig.3. Public opinion evolution of Von Neumann neighborhood without cellular mobile rules with w=100,K1(0)=0.5.

Comparing with Fig.4 and Fig.5, the public opinion evolution with Margolus neighborhood cellular structure, we find out, in the process of evolution, with the number of iterations increasing, two kinds of attitude will reach their steady state. Public opinion evolution with cellular movement has a faster convergence to the steady state. As we can see from Fig.5, the cluster area becomes greater than those in Fig.3 and Fig.4. Entities interaction shows the neighborhood structure influence in the formation and change of network public opinion. The more widely the entities move, and the more frequently they communicate, the faster the "cluster group" will form and the wider their range will be.

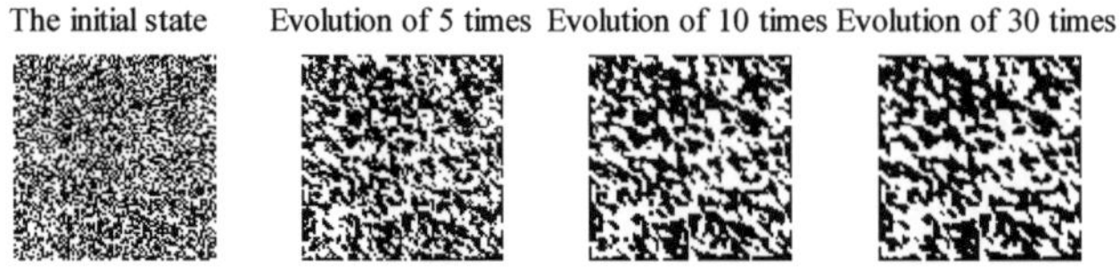

Fig.4. Simulation experiment on the Margolus type of public opinion evolution without cellular movement under w=100,K1(0)=0.5

The public opinion model with Von Neumann effecting neighborhood basically reaches the steady state with 40 steps of evolution, while Margolus neighborhood public opinion model reaches the steady state with only 10 steps. This kind of "clustering effect", as we can see form Fig.4 and Fig.5, under Margolus effecting neighborhood with cellular movement, will be maintained at a relatively stable boundary until the edge is gradually eroded and the boundary becomes smooth, and it becomes larger and larger just as Fig.5 shows.

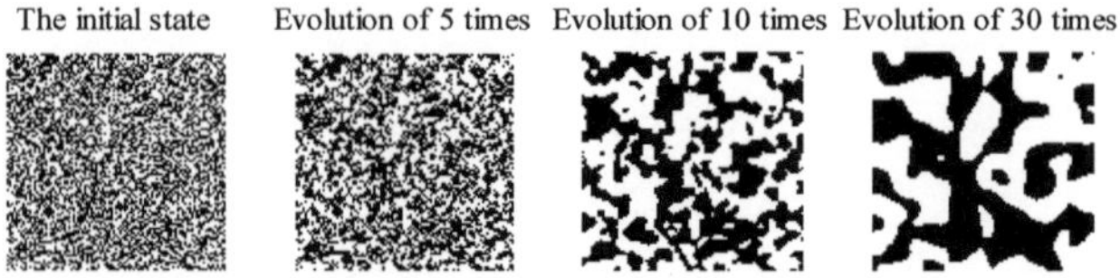

Fig.5. Simulation experiment on the Margolus type of public opinion evolution with cellular movement under w=100,K1(0)=0.5

Above is the evolution of public opinion, in which the initial approval rate of K1 (0) is been controlled at 0.5. If K1 (0) has been placed in this interval [0~1], we also get assentient and opponent phase[13] in the evolution of public opinion based on Margolus neighborhood structure. If most of the entities hold favourable views, the system is named the assentient phase, if the majority of entities hold opposing attitude, the system is named the opponent phase. A large number of experiments show that, the final approval rate changes with the change of initial approval rate as is shown in Fig.6.

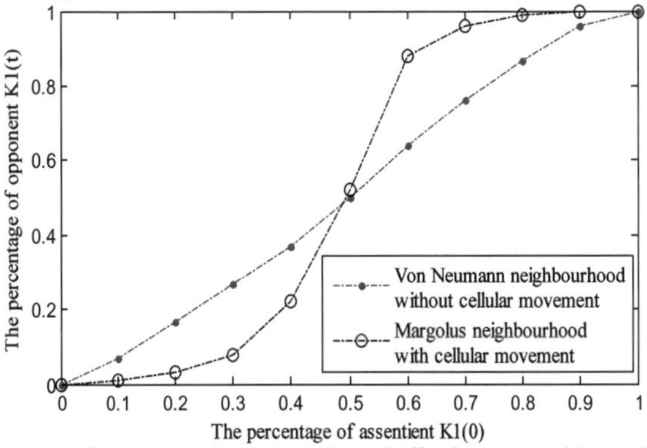

Fig.6. The evolution of public opinion in two kind of effecting areas with w=100,$K_1(0) \in [0,1]$

Comparing these two curves, conclusions can be obtained as follows. When the initial approval rate is less than 0.5, the last state approval rate is less than 0.5 as well. If the initial approval rate is greater than 0.5,the final state approval rate is greater than 0.5 too. At the same time, we can see, if $K_1(0) \le 0.5$, at the curve of the evolution of public opinion in Von Neumann type of effecting neighborhood structure, the approval rate of the final state is higher than that of the final state in the Margolus structure model. The reason is that, a simple neighborhood structure lack of mobility cause entity's lack of the opportunity to exchange information with other entities, therefore, most of the entities maintain their original attitude. Similarly, in the contrast diagram, when $K_1(0) \ge 0.5$, in the Margolus type of effecting neighborhood structure, the final approval rate of the public opinion system is generally higher than that of the Von Neumann structure model. The reason is that all single entity's movement enhance the propagation velocity and range of pubic opinion. It is easier for the system to reach a consensus with the increase of initial approval rate. In summary, public opinion propagation system based on Margolus structure model has a faster rate of convergence to assentient phase or opponent phase.

To analyze the evolution of mobile cellular separately, it is clear that when $K_1(0)>0.5$, at the initial time, the greater density of assentient entities accompanying unwavering attitude is, the higher $K_1(t)$ will be. Because the assentient entity ($\phi \le (f_2=2)$)accompanying unwavering attitude is not easy to change its attitude, instead, it makes the conformism entities($\phi \ge (f_4=4)$) and ordinary entities($\phi=(f_4=3)$) change theirs, which makes the final approval rate have a rapid convergence to the assentient phase. When $K_1(0)<0.5$, the greater density of opponent entities with unwavering attitude is, the lower $K_1(t)$ will be, which makes the system have a rapid convergence to the opponent phase.

There exists a Phase Exchange Interval in the evolution of the system. While in the phase of out of Phase Exchange Interval, the system will get to a state of complete agreement or completely disagreement. In the public opinion system with K1(0) being part of the Phase Exchange Interval, the outside media can guide entity's attitude, which will have an effect on the evolution of the whole public opinion system. The guidance will adjust the public opinion to a state of complete agreement or disagreement. Moreover, The "homogenization phenomenon[14][15]" of network public opinion only appears in the assentient phase and the opposition phase. Most of the final attitude of the entities are consistent with the attitude of the entities which has a larger proportion at the initial time, showing a highly-identified phenomenon.

Table 1. Statistical results of $K_1(t)$ with different initial approval rate of K1 (0)

T	K1(0)								
	0.4208	0.4562	0.4832	0.5032	0.5229	0.5338	0.5562	0.5647	0.5832
1	0.3820	0.5006	0.4700	0.4945	0.4894	0.5698	0.6302	0.6931	0.7833
2	0.3420	0.4732	0.4932	0.4885	0.4818	0.6043	0.6438	0.7268	0.8311
3	0.2812	0.4663	0.4928	0.4853	0.4754	0.6132	0.6732	0.7516	0.8459
4	0.2806	0.4640	0.4926	0.4829	0.4720	0.6345	0.6877	0.7531	0.8461
5	0.2798	0.4636	0.4923	0.4822	0.4699	0.6421	0.7009	0.7642	0.8506

For a deeper research into the influence of initial approval rate to the Phase Exchange Interval, the initial approval rate has been controlled to be around 0.5. the final approval rate and initial approval rate are shown in the Table 1.

7 times of statistics experimental results show that, with the initial approval rate of K_1 (0) being about 0.5, the possibility of a disguised interval is great. When $K_1(0)$ is at 0.4562, in the 5 traversal results, K_1 (t) are about 0.4562 (0.5006 to 0.4636), and $K_1(0)$ is at 0.5229, 5 times the results of $K_1(t)$ is also about 0.5229 (0.4894 to 0.4699). But we can find, when the $K_1(0)$ =0.4208, $K_1(t)$ is in the range of 0.3820 to 0.2798, when $K_1(0)$ =0.5562, $K_1(t)$ is in the range of (0.6302→0.7009). When the $K_1(0)$ =0.5832, $K_1(t)$ is among (0.7833→0.8506). Among these traversal results,$K_1(t)$ is significantly greater than or less than $K_1(0)$. It can be estimated, when the order of magnitude is 10^{-1}, the deviation range of Phase Exchange Interval is about $5*10^{-2}$. That is to say, the range of Phase Exchange Interval is about 0.45 to 0.55. The evolution data can be seen from the initial approval rate of K1 (0) from 0.5229 to 0.4562. At the initial time, the number of assentient is as many as the opponent. With the movement of the main body and the evolution of public opinion, entities' attitude state will change. In pace with public opinion evolution to about 10 times, the final state will not change and approach to an invariant state, that is to say, the entities will eventually be clear and definite about their attitude. In the Phase Exchange Interval, the change of final approval rate will not be a sharp increase or a sharp decrease.

5. Summary

The propagation and evolution of network public opinion is a complex process. In order to reflect this complexity, in this paper, we set up mobile rules and public opinion evolution rules based on the cellular automaton with Margolus type of effecting neighborhood structure. Mainly in the fifth traversal times or so, the public opinion evolution based on the Margolus neighborhood structure will appear a steady state, which is faster than other cellular automata model. Phase Exchange Interval, proposed by this paper, is a area where public opinion guidance playing an important part. The next research of the network public opinion, we can set a public opinion guidance parameters, deeply study, in the case of the entity's mobility, the guidance effect of website on the propagation of the network public opinion.

References

1. Galam S, Minority Opinion Spreading in Random Geometry, J. Eur. Phys. J. B., Vol.25(2002), pp.403–406.
2. Sznajd-Weron K and Sznajd J, Opinion Evolution in Closed Community, J. International Journal of Modern Physics, Vol.11 (2000), pp.1157–1165.
3. Deffuant G, Neau D, Amblard F, et al., Mixing Beliefs Among Interacting Agents, J. Advances in Complex Systems, Vol. 3(2001), pp.87–98.
4. Hai-lin Xiao and Min-yi Deng, et al., The influence of personnel movement on the spread of mobile communication in the public opinion model of automaton, J. Journal of systems engineering, Vol.03(2005), pp.225–231.
5. Qing-Feng Wu and Ling-jiang Kong(eds.), Public opinion propagation model of cellular automata in personnel personality influence, J. Journal of Guangxi Normal University (Natural Science Edition), (2004), pp.5–9.
6. Galam S. Sociophysics, A Review of Galam Models, J. International Journal of Modern Physics C, Vol. 19 (2003), pp.409–440 .
7. Xian-Kui Zeng, Majority rule and cooperation rule of cellular automaton, Master's thesis, Guangxi Normal University (2007).
8. Wei Fang and Liu-Jin He and Liang Song, Collaborative cellular model of internet public opinion communication, J. Computer Application, Vol.02(2012), pp.399–402.
9. Yassine Daadaa, et al., Network Decontamination with Temporal Immunity by Cellular Automata, in Proc, (ICSC, Ascoli Piceno, Italy, 2011.

10. Xiao-Chao Dang and Chun-Jiao Zhang, et al., Based on Fuzzy Cellular Automata network public opinion propagation model, J. Computer Engineering,Vol.04(2014), pp.209–213.
11. Ding Chunlong and Wei Chaofan, et al ., Study on propagation model of network public opinion based on fuzzy cellular automata Measurement, Information and Control (ICMIC),Vol.02(2013), pp. 1009 – 1013.
12. K. Kacperski and J. A. Holyst, Phase transitions and hysteresis in a cellular automata-based model of opinion formation, J. Journal of Statistical Physics.Vol.84 (1996), pp.169–189.
13. Li-Sheng Zhang and Min-Yi Deng, et al., Automaton Cellular Simulations for Target Waves in Excitable Media, J. Communications in Theoretical Physics, Vol.01(2010), pp.171–174.
14. Ryan J. Hammer and Thomas W. Moore, et al. The Role of Community Structure in Opinion Cluster Formation, J. Complex Sciences. Vol.126 (2013), pp.127–139.
15. Wei Fang and Liu-Jin He, et al., Using the network public opinion propagation model of cellular automata, J. Journal of Computer Applications, Vol.30 (2010), pp. 751–755.

Rule and Maximum Entropy Model Based Method of Temporal Expression Recognition

Jie-Jun Li[†], Qing-Qiu Cai

Faculty of Computer, Guangdong University of Technology,

Guangzhou, 510006, China

[†] *E-mail: 357038355@qq.com*

Hao Tang

Synergy Innovation Institute of GDUT,

Heyuan, 517000, China

E-mail: 122260851@qq.com

Temporal relation exits between concepts of time and event. It provides a natural clue for information organization. This paper studies the method of temporal expression recognition on the basis of information extraction and temporal denotation. We analyze the features of temporal expressions and conclude that the temporal expressions are constructed by temporal expression baseline and temporal expression director, which respectively indicate time reference and time point or time range and are the theory reference of recognition rules. Rule based method is used to extract explicit time information because of the certainty feature of it. Machine learning based method is used to recognize the implicit time information because there are no definite rules in it. The experiment shows the blended method is of relatively high accuracy and has higher recall rate than an individual method.

Keywords: Maximum Entropy; Entity Extraction; Rule Matching; Temporal Expression; Feature Model.

[†] Work partially supported by grant 2-4570.5 of the Swiss National Science Foundation.

1. Introduction

Temporal expressions commonly exit in all kinds of text. Events in articles are linked to each other by temporal information. For instance, time plays an important role in news text. When indexing news text, we can easily know the post date of the news by extract the time stamp of it, but the duration of the news story is not that easy to extract. So temporal expressions in the news story are imperative when analyzing such information.

Nowadays, temporal information is relatively mature. Paper [1] descript how a temporal expression is expressed in computer language and discussed divide temporal expression into time duration and time point. Entity extraction is applied in various industries, brings about huge improvement to production in papers [2–6,9,10]. However, in the field of temporal expression recognition, the task becomes more difficult because there are both explicit and implicit forms in temporal expressions. Paper [7] achieve the temporal expression extraction based on Conditional Random Fields model. They explore the features of temporal expression and use them in training the model. Paper [8] developed a rule based method to recognize temporal expression and provides conditional random fields based approach to extract temporal relationship. It indicates that rule and machine learning approaches are wildly used in temporal expression extraction. However, there are no researches found to combine the two kinds of methods. But as the feature of temporal expression shows in the follow chapters, the different between explicit and implicit temporal expression shouldn't be ignored. So in this paper, we use the regular expression rule to recognize explicit temporal expression and maximum entropy model to extract implicit temporal expression.

Section 1 describes the problem of temporal expression recognition and give the features of both explicit and implicit temporal expression and the explicit temporal expression recognition rule set. Section 2 introduces maximum entropy model and gives the features model we used in the complement recognition task. Section 3 shows the experiment and analyzes the result in detail. Section. The overall conclusion and improvement direction is shown in Section 4.

2. Rule Set

We only gives the core elements constituting short temporal expression, but solve the long temporal expression recognition problem, and such kinds of temporal expression is quite commonly exist in the corpus. Therefore the rules we gave are in order to constitute long temporal expression. Table 1 shows the element set we defined.

S represents a word segmentation set, t_b, t_d respectively identify the word to be temporal expression baseline and temporal expression director. S_t^p , S_t^n represent the front and rear part of a shortest temporal expression.

Table 1. Definition of temporal expression element set

Set	Description
S_p	common temporal expression prefix
S^n	number representing morpheme
$S_{t_b^n}^u$	temporal expression unit act as time front word baseline
$S_{t_b^p}^u$	temporal expression unit act as rear temporal expression baseline
$S_{t_d^p}^d$	direction representing morpheme act as front temporal expression director
$S_{t_d^n}^d$	direction representing morpheme act as rear temporal expression director
$S_{t_d}^p$	pronoun morpheme act as front temporal expression director
S_s	temporal expression suffix
S_c	auxiliary words and conjunctions
S_f	holiday word

We use a combination of temporal expression baseline and temporal expression director establish a temporal expression to construct rule set resource set for classification, rather than regarding a temporal expression as the minimum granularity. The benefits of doing so is when users adding temporal expression baseline and temporal expression director to the resources set, the proper temporal expression baseline and temporal expression director will match up each other automatically, it greatly eases the repetition operations for users and reduces omissions when construct the resources rule set.

Table 2. Definition of Resource Rule Set

Set	Description
Rule1	$(w_i \in S^n, w_{i+1} \in S_{t^n}^u)$
Rule2	$(w_i \in S_{t^p}^u, w_{i+1} \in S^n)$
Rule3	$(w_i \in S_{t^p}^d, w_{i+1} \in S^n, \mathrm{w}_{i+2} \in S_{t^n}^u)$
Rule4	$(w_i \in S_{t^p}^u, w_{i+1} \in S_{t^n}^d)$
Rule5	$(w_i \in S_{t_d}^p, w_{i+1} \in S^n)$

We define Vector (w_p, w_n) as the minimum temporal expression, S_t as temporal expression, and $(w_p, w_n) \in S_t$. Assuming the words segmentation vector in the test corpus is S_w , and $S_w = (w_1, w_2, \cdots w_i, \cdots w_n)$.

According to the rule constitutes a short explicit temporal expression is a front temporal expression baseline and a rear temporal expression director or a front temporal expression director and a rear temporal expression baseline, we classify the temporal expression morpheme and give the rule shown in Table 2 for matching temporal expression.

After recognizing explicit temporal expressions, we can reuse the explicit temporal expressions to acquire common explicit temporal expression matching rule. As long temporal expression is commonly exists in the corpus, therefore explicit long temporal expression should be under consideration. Also, long temporal expression has the feature of certainty, so it's better to use rule based method to recognize it.

Long temporal expression matching rule is described below with regular expression:

$$S_p\{0,2\}S_t\{1,\}(S_c\{0,1\}S_t\{1,\})\{0,\}S_s\{0,1\} \quad (1)$$

3. Complement recognition with maximum entropy model

In order to enhance the efficiency of entity recognition, we have reached all kinds of feature, including temporal expression internal feature, context feature and feature of the influence by other temporal expression recognition. We analyze the usage of these features in the classification model, and confirmed the function of each features. On this basis, combine these features properly and used by the classification model effectively makes the efficiency of temporal expression recognition enhanced.

The feature set we used is shown in Table 3.

Table 3. Definition of Feature Models

Order	Feature model	Description
1	w_{i-2}, w_{i-1}	left two words
2	t_{i-2}, t_{i-1}	types of left two words
3	w_{i-1}	left word
5	t_{i-1}	type of the left word
6	p_{i-1}	left punctuation
7	w_{i+1}, w_{i+2}	right two words
8	t_{i+1}, t_{i+2}	types of right two words

9	w_{i+1}	right word
10	t_{i+1}	type of the rightword
11	p_{i+1}	right punctuation
1	w_{i-2}, w_{i-1}	left two words

4. Experimental Result and Analysis

In the experiment, the corpus we used contains 1 million words and is automatically produced by web clawer. We select 14222 sentences in the downloaded content to train and test the classifier.

Decided the combination of the features, we divided the training set and testing set into 2 to 1, and retest the efficiency of the feature model.

The error in the experiment we found in Table 4 could categorized into two points:

When punctuations continuously occurred more than once in one temporal expression, we misrecognize it as the border of two temporal expressions. But this kind of expression is rather rare.

Table 4. Experimental Results

	Precision (%)	*Recall* (%)	*F − score* (%)
articles categorized as news(4233 sentences)	86.28	83.04	84.62
articles categorized as fiction(3562 sentences)	72.75	73.11	73.03
articles categorized as blog(6427 sentences)	79.08	75.28	77.13
total	79.37	77.14	78.26

Long explicit temporal expression contained unexpected simple, will probably be recognized as two or three temporal expression. This is caused by the resource rule set lack of universality.

5. Conclusion

This paper introduced the research of temporal expression extraction using rule and maximum entropy model based approach. We gave the distinct features in temporal expression to help construct the resource rule set which is for explicit temporal expression recognition. The experiment shows that with introducing rule based approach not only increase the recall rate but also greatly reduce the misrecognition when using machine learning based approach to recognize implicit temporal expression as complement. But the experiment also shows that the ambiguity exists in temporal expression often leads to misrecognition, which also needs to be improved.

References

1. N. Leach and K. H. Cheng, "Recognizing and understanding time-terms in communica-tion," Systems, Man, and Cybernetics (SMC), 2011 IEEE International Conference on, Anchorage, AK, 2011, pp. 3023-3028.
2. M. h. Siu, T. Vessenes, I. Bulyko and O. Kimball, "Improved named entity extraction from conversational speech with language model adaptation," Spoken Language Technology Workshop (SLT), 2010 IEEE, Berkeley, CA, 2010, pp. 418-423.
3. P. Sutheebanjard and W. Premchaiswadi, "Thai personal named entity extraction without using word segmentation or POS tagging," Natural Language Processing, 2009. SNLP '09. Eighth International Symposium on, Bangkok, 2009, pp. 221-226.
4. K. S. Yang, Chun-Cheng Chen, Yuen-Hsien Tseng and Zih-Ping Ho, "Name entity extraction based on POS tagging for criminal information analysis and relation visualization," Information Science and Service Science and Data Mining (ISSDM), 2012 6th International Conference on New Trends in, Taipei, 2012, pp. 785-789.
5. S. Volkova, D. Caragea, W. H. Hsu, J. Drouhard and L. Fowles, "Boosting Biomedical Entity Extraction by Using Syntactic Patterns for Semantic Relation Discovery," Web Intelligence and Intelligent Agent Technology (WI-IAT), 2010 IEEE/WIC/ACM International Conference on, Toronto, ON, 2010, pp. 272-278.
6. D. Deng, G. Li and J. Feng, "An Efficient Trie-based Method for Approximate Entity Extraction with Edit-Distance Constraints," Data Engineering (ICDE), 2012 IEEE 28th International Conference on, Washington, DC, 2012, pp. 762-773.
7. Hua Zhao and Xiaowen Ji, "English temporal expression recognition based on Conditional Random Fields," Natural Computation (ICNC), 2013 Ninth International Conference on, Shenyang, 2013, pp. 1088-1092.
8. Zhou Xiaojia, Li Haomin, Lu XuDong and Duan Huilong, "Temporal Expression Recognition and Temporal Relationship Extraction from Chinese Narrative Medical Records," Bioinformatics and Biomedical Engineering, (iCBBE) 2011 5th International Conference on, Wuhan, 2011, pp. 1-4.
9. Mingli Wu, Wenjie Li, Qing Chen and Qin Lu, "Normalizing Chinese temporal expressions with multi-label classification," Natural Language Processing and Knowledge Engineering, 2005. IEEE NLP-KE '05. Proceedings of 2005 IEEE International Conference on, 2005, pp. 318-323.
10. Fu G. Chinese Named Entity Recognition Using a Morpheme-Based Chunking Tagger[C]// International Conference on Asian Language Processing, Ialp 2009, Singapore, December. 2009:289-292.

Feature Mining and Sentiment Orientation Analysis on Product Review

Jie-Jun Li†, Han Yang

Faculty of Computer, Guangdong University of Technology,

Guangzhou, 510006, China

† *E-mail: 357038355@qq.com*

Hao Tang

Synergy Innovation Institute of GDUT,

Heyuan, 517000, China

E-mail: 122260851@qq.com

In order to help producers improve the quality of products and service provider higher quality services, a research on the feature and sentiment orientation mining of product review is necessary. In this paper, we introduce tasks that relate to the problem of product review sentiment classification and point out the main task of which is finding the sentiment orientation of a review is positive or negative. We propose a client perception based product feature mining algorithm using the sentiment analysis technology and realize the semantic analysis, dynamic extraction and general information mining on the product features and sentiment orientation. The feature and sentiment orientation of products are sorted by the weight of user interest, which helps customers access the conclusive review orientation more effectively. The experiment is based on real product review corpus downloaded from the internet, and it proved the effectiveness of the approach.

Keywords: Product Review; Review Mining; Sentiment Orientation Analysis; Word Segmentation; Product Features.

1. Introduction

In the recent years, with the rapid development and popularization of the internet, network gradually becomes the important platform of posting review information. The number of reviews on the internet is increasing at an

† Work partially supported by grant 2-4570.5 of the Swiss National Science Foundation.

alarming speed. Product reviews represent customers' attitude toward the product and show extinct appraise evaluation. For the producers and customers, the effective way to access the overall sentiment orientation of such a large number of reviews is a new problem. Review mining and opinion mining is one of the solution to it.

Review mining has attracted the attention of many researchers. Review mining is a new field of unstructured information mining. It mainly involves sentiment analysis, feature mining and subjective content recognition. In the field of English review, researchers have received initial research results, but it's in its infancy in the field of Chinese review. Because of the cultural differences, their ways of expression are different, which makes the research result in English field can't directly apply to Chinese field. Therefore we explore the extraction technology of product feature, service quality and sentiment orientation of customers of Chinese review, aiming at providing more sophisticate review mining tool for customers and produces in Chinese field.

2. Research on Review Sentiment Analysis and Mining Problem

The definition of data mining is as follow:

- The source of the data must be real, in massive scale and noise contained.
- Find the knowledge users are interested in, and the knowledge found is acceptable, understandable and usable.

The advent of data mining technology provides a reliable means of transferring a large number of data into knowledge, useful information or news clues automatically and intelligently. And the purpose of product review mining is providing useful information for producer to improve their products and help the customers purchase the ideal product.

The main task of sentiment analysis in Chinese reviews contains extracting the properties customers are interested in, rank the products according to the frequency of occurrence, find the review sentence of that property and evaluate the sentiment orientation is positive or negative.

Four subtasks are included in review mining:

- Product feature extraction.
- Review opinion extraction.
- Evaluate the sentiment orientation.
- Summarize the review mining result and rank in order of opinions of users.

2.1. *Product Feature Mining*

In English review, due to the time when researchers realized the great impact of reviews of customers is early, researchers have achieved some preliminary results. Liu Hu et al [1] first proposed an association rule based approach to extract product features in English review and use an unsupervised approach to mining reviews of mobile phone and digital camera. The average recall score of which achieved 80% and the average precision achieved 72%. The subsequent research evaluate the sentiment orientation of the users [1,2]. There are other approaches are applied to the same task, for instance, [3] use a semi-automate cyclic approach to extract the product feature and users' opinion, but it also is labor intensive. Paper [4] extract the product feature base on Bayesian model, which improved the precision score but downgraded the recall score.

In Chinese review mining, product feature mining is mainly accomplished with supervised and semi-supervised method which requires human intervention. Paper [5] proposed build the product feature of auto-mobile base on ontology; Lujian et al [6] mined product feature and sentiment orientation with the building of relevant knowledge. Shi et al [4] also performed a research on Chinese product review but require manual building a concept model based on the properties of the product.

2.2. *Sentiment Orientation Analysis*

The purpose of sentiment analysis is evaluating the speaker's sentiment orientation, the general class of which is positive, negative and neutral. Currently, technologies of sentiment analysis are mainly includes machine learning based approach and semantic based approach.

Paper [7,8] use machine learning based approach to mine sentiment orientation, using naive Bayesian model, maximum entropy model and SVM to classify review and achieved relatively high precision sore. While [9–11] classify reviews using an unsupervised method based on semantic orientation.

3. Sentiment Analysis based Product Feature Mining

This paper will use sentiment orientation analysis technology to accomplish the task of product feature mining in Chinese reviews. The approach we designed is based on association rule and combining the sentiment analysis approach. The algorithm we proposed accomplished the sentiment analysis, opinion extraction and mining of product feature and service quality.

The steps of the algorithm shown in Fig.1 are as follow:

(1) Data preprocessing, parameters initialization.

(2) Extract product feature phrases from product reviews.

(3) Extract customers' opinions from product reviews.
(4) Evaluate the polarity and the intensity of customers' opinions.
(5) Rank the product features in order of the frequency of occurrence.
(6) Dynamic analyze the sample data with the addition of timeline.

The detailed research process and approach is as follow:

(1) Experiment data preparing and data preprocessing. The experiment data is preprocessed from the raw data crawled in the review section of zol.com, amazon.com and taobal.com.
(2) The process of preprocessing includes review extraction, word segmentation and part-of-speech recognition. We use a Chinese analysis tool accomplish word segmentation and part-of-speech tagging, including Chinese noun phrase. The tool we use is ICTCLAS created in Chinese Academy of Sciences.
(3) Train the classifier base on the classification category and evaluate the sentiment orientation taking sentence as unit. The natural language analysis tool we use is LingPipe. Apply the classifier trained in the experiment corpus to classify the sentiment orientation of the review and store the classified data into opinion database.
(4) Build the transaction file of association rule using the part-of-speech tagged corpus.
(5) Find the frequent item-set as candidate product feature set S_0.
(6) Trim the candidate product feature set S_0 base on nearest-neighbor rule.
(7) Amend the candidate feature set S_1 base on independent support rule, forming the candidate feature set S_2.
(8) Construct the frequent Chinese noun phase and non-product feature set. Filter S_2 into feature set S_3. Remove the candidate item of the noun phrase with single character, forming the last product feature set S.
(9) Search the sentence contained the product feature in opinion database, extract the sentiment orientation of the sentence.
(10) Rank the result in order of the frequency of the occurrence of the product feature, and summarize the number of the positive sentences and the negative ones.

Assume S_p is the frequent item-sets, and S_p contain n noun phrases, a sentence s contains S_p and the order of the words occurred in S_p is $w_1, w_2, \cdots, w_n$. If any adjacent noun phrases is w_i 和 w_{i+1}, that is to say S_p is adjacent.

Traverse the two dimension and three dimension item-set S_p. If S_p occurred in m sentences in review database, and it is adjacent in at least two

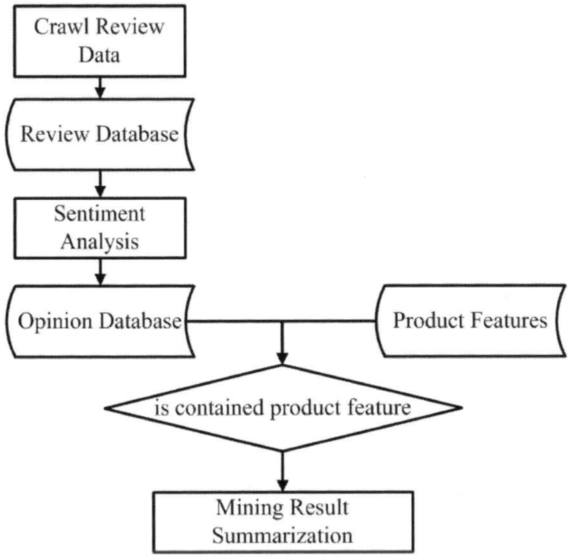

Fig. 1. Chinese Product Review Mining Algorithm Process

sentences, that is to say S_p is an adjacent noun phase and then add it to the candidate set S_1.

4. Experiment Verification and Evaluation

In the experiment, we crawled 600 review documents from zol.com and tag the reference properties set manually. The research target we choose is mobile phone because of its extinct feature of the property and the application of mobile phone is of wide range. Recognize and label the product properties mentioned in reviews. Build a minimum feature set base on maximum coverage rule, in order to cover the properties in all review documents. We Choose 300 sentences as training set and the rest of those as testing set. The result of product features mining is shown in Table 1.

Table 1. Properties Mined from Reviews

Rank	Property	Number of Positive Sentences	Number of Negative Sentences	Ratio of Positive and Negative Sentences(%)
1	Iphone	68	23	74.7
2	Moto	19	54	26.0
3	Samsung	32	29	52.5
4	Camera	39	20	66.1
5	Smart Phone	42	31	57.5

5. Conclusions

This paper combines the sentiment analysis technology and product feature mining in Chinese reviews. We proposed an effective approach to mine a large number of reviews on the internet base on the product feature information which includes the number of positive review sentences, the number of negative review sentences, the ratio of them and the ranking of the feature. The approach we proposed provides an scientific and practical tool for producers and customers.

Currently we had made some achievements in product review mining, but it also has some deficiency. The review corpus is not standard enough because as a new research field, there are no public and standard review data for the mining test.

References

1. Pang B, Lee L, Vaithyanathan S. Thumbs up?: sentiment classification using machine learning techniques[C]//Proceedings of the ACL-02 conference on Empirical methods in natural language processing-Volume 10. Association for Computational Linguistics, 2002: 79-86.
2. Tang H, Tan S, Cheng X Q. Research on sentiment classification of Chinese reviews based on supervised machine learning techniques[J]. Journal of Chinese Information Processing, 2007, 21(6): 88-94.
3. Tang H, Tan S, Cheng X. A survey on sentiment detection of reviews[J]. Expert Systems with Applications, 2009, 36(7): 10760-10773.
4. Zhu Y L, Min J, Zhou Y, et al. Semantic orientation computing based on HowNet[J]. Journal of Chinese Information Processing, 2006, 20(1): 14-20.
5. Turney P D, Littman M L. Measuring praise and criticism: Inference of semantic orientation from association[J]. ACM Transactions on Information Systems (TOIS), 2003, 21(4): 315-346.
6. Ye Q, Shi W, Li Y. Sentiment classification for movie reviews in Chinese by improved semantic oriented approach[C]//System Sciences, 2006. HICSS'06. Proceedings of the 39th Annual Hawaii International Conference on. IEEE, 2006, 3: 53b-53b.
7. K. Mouthami, K. N. Devi and V. M. Bhaskaran, "Sentiment analysis and classification based on textual reviews," Information Communication and Embedded Systems (ICICES), 2013 International Conference on, Chennai, 2013, pp. 271-276.
8. Sheng-Chun Ding, Ting Jiang and Neng Wen, "Research on sentiment orientation of product reviews in Chinese based on cascaded CRFs models," Machine Learning and Cybernetics (ICMLC), 2012 International Conference on, Xian, 2012, pp. 1993-1999.
9. Mingli Wu, Wenjie Li, Qing Chen and Qin Lu, "Normalizing Chinese temporal expressions with multi-label classification," Natural Language Processing and Knowledge Engineering, 2005. IEEE NLP-KE '05. Proceedings of 2005 IEEE International Conference on, 2005, pp. 318-323.
10. Z. Wang and W. Jiang, "Online reviews sentiment analysis applying mutual information," Fuzzy Systems and Knowledge Discovery (FSKD), 2012 9th International Conference on, Sichuan, 2012, pp. 1315-1319.

Design of Raptor Code for Space-to-Ground High-speed Communication*

Bao-Li Hong†, Zhang-Wei

Technology and Engineering Center for Space Utilization, Chinese Academy of Sciences,
Beijing, Haidian Zone, China
†E-mail: baolihong13@csu.ac.cn, zhangwei@csu.ac.cn
www.csu.cas.cn

Li-Hong Bao

Communication, Strategic Planning Laboratory, No.9 South Dengzhuang Road
Beijing, Haidian Zone, China
E-mail: baolihong13@csu.ac.cn

Relay communication by space network is the main method of future space-to-ground communication, however, space network has low stability and huge transmission distance. A kind of Raptor code for space network was designed by using simplified LDGM as outer code and truncated LT as inner code, besides, on the basis of the truncated distribution function raised by Shokrollahi, a degree distribution function for small file was designed to adapt to on-board processing capability of relay satellites. Simulation results show that this Raptor code has low complexity, and its decoding overhead is less than 15%, it can effectively adapt to spatial channels and reduce transmission delay. This Raptor code can meet the requirements of high-speed communication in space-to-ground relay communication.

Keywords: Relay communication; LEC; Raptor code; distribution function; LDGM

1. Introduction

The complexity of the spacecraft platform and payload is constantly improving with the continuous advancement of space science and technology. For one thing, the data generated by spacecraft and its payload are becoming more and more big and diversity, and these data require real-time transmission. For another, the development of relay satellites and on-board processing technology promotes the build of space network, which greatly expands the communication coverage. Moreover, with the development of space network and space data system, space users will increase and produce large amounts of real-time user data, which will challenge the space data service.

The consultative committee for space data system (CCSDS) protocol provides standard support for space data system. Compared to the traditional point to point space-to-ground communication, spatial network will bring a lot

*This work is supported by advanced research project in field of manned spaceflight (050102) .

of problems. Firstly, the nodes of space network are spacecraft, so the network is unstable. The same with ground network, space network ensure the reliability of transmission with the feedback mechanism. If the data are lost or there are some uncorrectable errors in the transmission via binary erasure channel (BEC), the receiver needs to send a feedback signal to the sender to request retransmission. If there is a poor channel quality, the receiver will send a lot of feedback data to the sender, then the sender will repeatedly send many data, especially in case of broadcast or multicast, any receiver does not receive data correctly, the sender needs to resend, which reduces transmission efficiency, and increases the burden of network. Secondly, the huge transmission distance between satellite and ground increases the transmission delay, which makes ARQ has low efficiency.

CCSDS raised a draft recommendation named long erasure correcting (LEC) to solve the above problems in 2011, LEC equals packet switch channel using error detection mechanism to the erasure channel. In LEC, the sender firstly encodes the data packets, and then sends the encoded data and original data; the receiver would recovery the lost or uncorrectable original data with the help of encoded data.

There are three kinds of codes in accordance with LEC thought: Reed Solomon codes (RS), Low Density Parity Check codes (LDPC) and digital fountain codes. RS codes are not suitable for high-speed data because of their small code length range. The efficiency of maximum likelihood decoding algorithm of LDPC can approach Shannon limit, but the complexity of the decode algorithm is $O(N^3)$ [1]. Digital fountain codes are rateless forward erasure codes for BEC, they have low encode and decode complexity and can approach Shannon limit with arbitrary probability.

Digital fountain codes have two notable advantages; firstly, they can reliably transport data without feedback channel, which can greatly reduce transmission delay. Secondly, they are rateless, any change of channel conditions only affects the number of codes that the sender sends, without affecting its decoding performance. However, the decoding efficiency of traditional erasure codes (like RS, LDPC) can change with channel conditions because their code rate is stable.

Digital fountain codes are mostly used in terrestrial wireless multicast system[2], reference[3] proposed a relay cooperative communication method for deep space communication with digital fountain codes, and analyzed its decoding performance and energy cost. Based on [3], this paper will propose a kind of Raptor code for space-to-ground relay communication, and analyze its performance.

2. Space-to-ground communication model

Fig.1 shows space-to-ground communicating model. This model is based on the data transmission link and methods CCSDS system proposed.

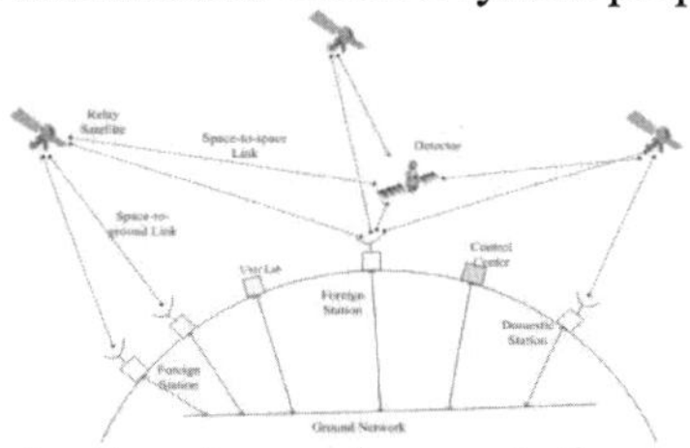

Fig.1. Space-to-ground communication model

In this model, there are two ways to transport data. One way is to transport data to ground receiving station through space-to-ground link, and then the received data will be transported to domestic station through ground network. This way is inefficient because data start to transport only when the spacecraft fly over the ground station. The other way is to use relay satellites, first transport data to relay satellites through space-to-space link, and then transport data to ground receiving station through space-to-ground link. If Relay Satellites could cover the whole earth, we can seamlessly transport data through this way.

In relay communication, transporting the data information from spacecraft to ground station is generally divided into two stages. Firstly, the source node encodes original data with digital fountain codes, then transport the encoded information to relay nodes. When the encoded information received by a relay node is enough to be decoded, the relay node starts to decode it, if the received information is successfully decoded, the relay node would send a termination signal to source node. When every relay node has decoded successfully, source node stops to transport the encoded information, the first stage end. Secondly, each relay node encodes the decoded data with digital fountain code again, and then sends the encoded data to the target node.

In the wireless communication, with increase of carrier frequency, the atmospheric attenuation of signal increases, exception the special band about 60GHz. So in the first stage of the relay communication, we use Ka band with higher frequency to reduce link attenuation, in the second stage, we use X band with lower frequency to reduce atmospheric attenuation. Besides, in the first stage, channel SNR will in the range 12~18dB, the modulation can be either BPSK or QPSK, channel error rate can be controlled in the range about 10^{-4}~10^{-6}, but in the second stage, the channel SNR would be smaller because of atmospheric attenuation[4][5].

3. The design of digital fountain code

3.1. *Code Select*

LT codes invented by Luby are the first practical realization of digital fountain codes[7]. The encoding progress of LT codes can be divided into 3 steps. First randomly choose the degree d of the encoding symbol by sampling a predetermined degree distribution, then choose uniformly at random d distinct original symbols as neighbors of the encoding symbol, and set the value of the encoding symbol to be the XOR of the values of the d neighbors. Lastly, the value of each encoding symbol, together with the information about its degree and neighbors, are then put into a packet and sent to the receiver.

The main decoding algorithm of LT codes is BP algorithm, which can be divided into hard decision algorithm and soft decision algorithm. The hard decision algorithm is shown in Fig. 2, unlike hard decision, likelihood information is transmitted between original symbols and encoding symbols in the soft decision algorithm. The decoding process is: first find all encoding symbols whose degree is one, decode these symbols, then update the encoding symbols associated with these decoded symbols, and start the next decoding until all original symbols were decoded or we can't find a encoding symbol whose degree is one.

There are some highly connected symbols in LT encoding symbols, these symbols ensure a good cover of input symbols, but consume a lot of XOR operations in encoding and decoding progress, and therefore reduce the rate of low connectivity symbols, increase the probability of stopping sets in decoding process at the same time, thus reducing the decoding success rate. Raptor codes can solve these problems via truncating the distribution function of LT codes and adding an outer code to ensure the cover of original symbols. Experiments show that Raptor codes can greatly reduce the complexity of encoding and decoding by a small increase in redundancy. Fig. 3 shows the encoding structure of Raptor codes[8].

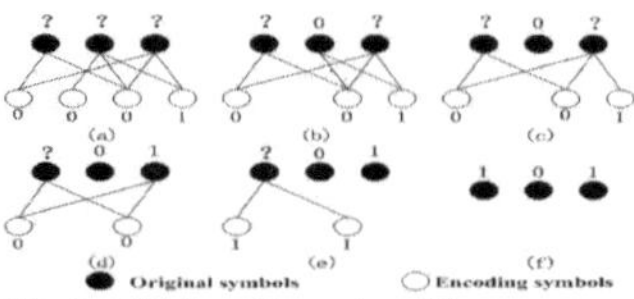

Fig.2. BP decoding algorithm

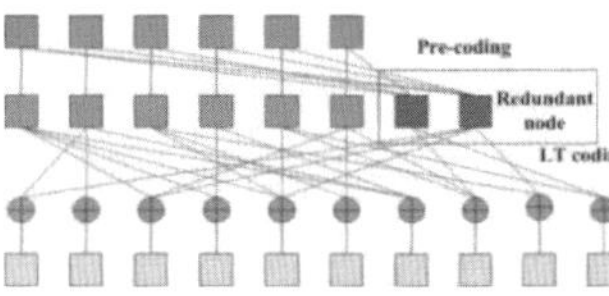

Fig.3. The encoding structure of Raptor code

The key factor which determines the performance of LT codes is the degree distribution used in the encoding procedure. A good degree distribution should meet the following two goals: firstly, the average number of encoding symbols required to ensure success recovery of original file should be as fewer as possible; secondly, the average degree of the encoding symbols should be as small as possible. The commonly used distribution function is robust soliton distribution (RSD), which exhibits good behavior in practice.

Theoretical analysis shows that: under RSD distribution, if we use the number of XOR operations to measure, the complexity of LT codes are $O(k \ln k)$, where k is the number of original symbols. However, the complexity of Raptor codes are about $O(k \ln 1/\varepsilon)$, ε is the redundancy. So we choose the Raptor codes in this paper because their low complexity would better adapt to the on-board processing capability of relay satellites.

3.2. *Design of Raptor Code*

3.2.1. *Outer Code Design*

The outer code used in this paper is simplified irregular low density generator matrix (LDGM) codes to reduce the complexity of encoding and decoding algorithm. The design ideas of this code are: considering the complexity and decoding efficiency, choose 2 as column weight, and construct LDGM parity check matrix by Mackay structuring method, then choose the line weight of this matrix to be the degree of codes, and generate LDGM codes like LT codes. If outer code rate is $R = (1+\varepsilon/2)/(1+\varepsilon)$, the size of this LDGM parity check matrix is $(1-R)k \times (2-R)k$,where k is the number of original information symbols. Let every information symbol have two check symbols as its neighbors, that is, each column of the check matrix has two 1, so the average degree of the LDGM is $2/(1-R)+1$.

The same as LT code, the decoding algorithm of LDGM is BP algorithm. The difference is, the decoder only decodes those codes that the inner code failed to decode, if the percentage of those codes is less than 2%, the number of XOR operations of outer code decoding is less than $0.02k(2/(1-R)+1)$. Let $\varepsilon = 0.1$, then R=0.955, so the numbers of XOR operations of outer encoding and decoding process are about $2.045k$ and $0.91k$.

3.2.2. *Inner Code Design*

The inner code of Raptor code is weakened LT code, that is, remove the highly connected packages in LT codes, when reflected in distribution function, the distribution function should be truncated. Reference[8] mentioned a heuristic deduction raised by M.Luby about degree distribution, after ensuring the cover of original packets, the number of newly appeared codes whose degree is 1 should be not less than the input packets to be recovered each time after the decoding:

$$k(1-x-e^{-\omega'(x)(1+\varepsilon)}) \ge c\sqrt{(1-x)k}\ . \qquad (1)$$

For $x \in (0,1-\delta)$, the inequality can be manipulated to yield:

$$\omega'(x) \ge \frac{-\ln(1-x-c\sqrt{\frac{1-x}{k}})}{1+\varepsilon}\ . \qquad (2)$$

Where c is a positive number, δ needs to be larger than $c/\sqrt{k}$. Based on (2), Shokrollahi designed several improvements of truncated distribution function for different values of k .When $k < 65536$, a set of empirical data of his truncated distribution function are given as follows:

d=[1,2,3,4,5,8,9,19,65,66], $\omega 1$ =[0.007969,0.493570,0.166220,0.072646,0.0082558,0.056058,0.037229,0.055590,0.025023,0.003135].

For small k, the simulation shows that the decoding success rate of this function is not ideal. In the relay communication using digital fountain codes, relay satellites need to decode and recode, besides, in the BP decoding algorithm of digital fountain codes, receiver would start decoding until it has received $k+\varepsilon$ codes. Considering the large code length, it will challenge the on-board processing capability of relay satellites, so under the circumstances of $k < 10^4$, this paper will make a further improvement to the truncated distribution function raised by Shokrollahi. Let:

$$y = \frac{-\ln(1-x-c\sqrt{\frac{1-x}{k}})}{1+\varepsilon}\ . \qquad (3)$$

Take different values of k, make 9 order Tayler expansion of y at $x = 0$, take the average of Taylor expansion coefficients for different k, we can obtain:

y= $\omega'(x)$ =0.0099+0.8377x+0.4200x^2+0.3128x^3+0.2385x^4+0.1689x^5+0.1469x^6+0.1211x^7+0.1059x^8+0.0943x^9+o(x^9). (4)

The integration of $\omega'(x)$ can be given by:

$\omega(x)$ =0.0099x+0.4189x^2+0.1400x^3+0.0782x^4+0.0477x^5+0.0281x^6+0.0210^7 7+0.0151x^8+0.0118x^9+0.0094x^{10}+ o(x^{10}). (5)

It can be deduced by (2) that the distribution function should be greater than $\omega(x)$, so we can appropriately increase the probability of small degree codes, add several coefficients of $\omega(x)$ together to get intermediate degree codes probability, and give a certain percentage of big degree codes, at the same time, all coefficients of designed degree distribution function should be added up to 1. After repeated simulation and improvements, a set of improved empirical data of degree distribution function are obtained:

d=[1,2,3,4,5,8,9,19,20,65,66], $\omega 2$ =[0.0110,0.5012,0.1554,0.1063,0,0738,0.0424,0.0300,0.0418,0.0051,0.0258,0.0045].

In summary, the Raptor code in this paper would concatenate simplified irregular LDGM with truncated LT code.

4. Simulation results

4.1. *Distribution function simulation*

Fig. 4 shows the comparison between two truncated distribution functions, where $\omega 1$ is proposed by Shokrollahi for $k < 65536$, $\omega 2$ is our improved one for $k < 10^4$, k is the number of initial information packets. For $x \in (0, 1-\delta)$, comparing the derivative of $\omega 1$ and $\omega 2$ with (3), the simulation results are shown in fig.4a, with the increase of derivative value, the codes that each step decodes increase. We can see that the derivative value of $\omega 2$ is slightly larger than that of $\omega 1$ and both larger than (4). Analyzing the difference of decoding success rate between $\omega 1$ and $\omega 2$ on condition that k equals to 1000 and 5000 respectively (shown in Fig.4b), we can draw the conclusion that decoding success rate of $\omega 2$ is more quickly approach 1 than that of $\omega 1$, and with the increase of k, the decoding overhead decrease. If $k = 5000$, we can decode successfully with the probability close to 1 by only 5% overhead.

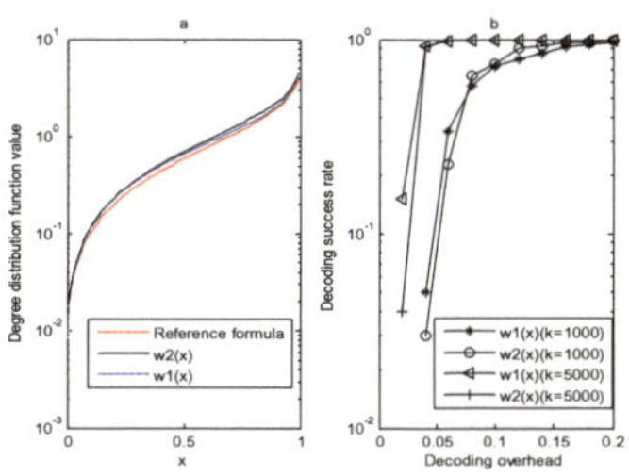

Fig.4. Performance comparison of $\omega 1$ and $\omega 2$

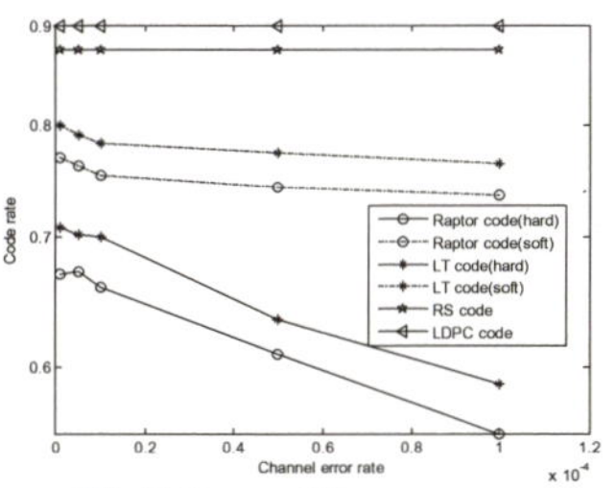

Fig.5. Rates of four erasure codes

4.2. *Comparison of Several Erasure Codes*

Under the same software and hardware environment, encode an image whose size is 256×256 with RS, LDPC, LT, and Raptor code separately, and transport these codes with error channel. Define code rate is the value that effective bits number divided by total bits number sent by sender, Table 1 shows the decoding time of four erasure codes, under different channel error rate, the code rates of these erasure codes are shown in Fig. 5.

Table 1. Decoding Time of Four Erasure Codes

Code name	RS	LDPC	LT *Hard*	*LT Soft*	Raptor *Hard*	*Raptor Soft*
Decoding time(s)	529.22	200.03	3.28	78.46	2.25	55.32

Combining Fig.5 with table I, we can see that although the rates of LT and Raptor code are lower than that of RS and LDPC, the decoding complexity is greatly reduced only by about 10%(soft decision) and 30%(hard decision) increase of decoding overhead. Compare Raptor code with LT code, we achieve a 30% improvement in decoding performance with a dinky decoding overhead.

4.3. *Raptor Code Simulation*

Use the model shown in Fig.1 to simulate the Raptor code proposed by this paper, assume that channel error rates of the first stage is 10^{-6}~10^{-4}, and the channel SNR of the second stage is 8~12dB. We use hard decision and soft decision of BP decoding algorithm in the first and second transmission stage respectively, the head of the Raptor code is 20 bits. In the first stage, some codes would be deleted because of transmission error, the k is the number of initial information packages, and the length of an information package is 2Kbits. In the second stage, we use soft decision of BP algorithm to resist channel error, k is also the number of initial information packages, but the length of an information package is 1bit, the simulation results are shown in Fig. 6 and Fig. 7.

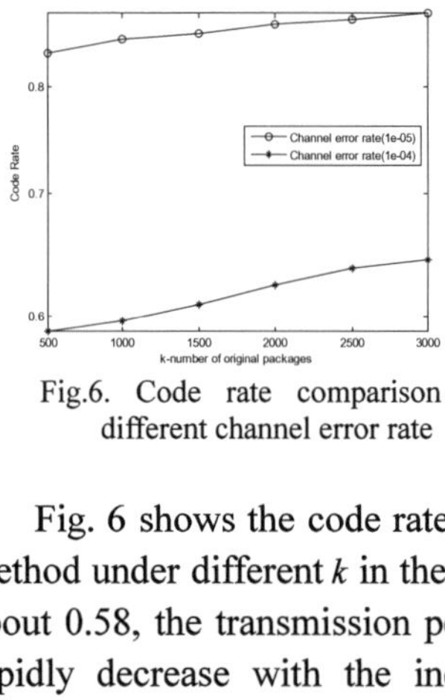

Fig.6. Code rate comparison of different channel error rate

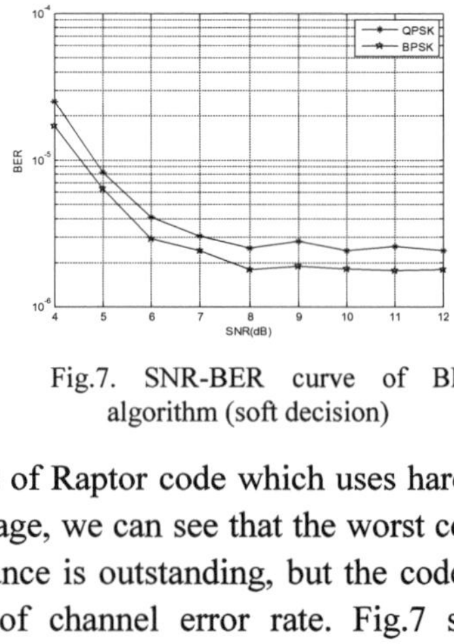

Fig.7. SNR-BER curve of BP algorithm (soft decision)

Fig. 6 shows the code rate curves of Raptor code which uses hard decision method under different k in the first stage, we can see that the worst code rate is about 0.58, the transmission performance is outstanding, but the code rate will rapidly decrease with the increase of channel error rate. Fig.7 shows the SNR-BER (bits error rate) curves of Raptor code (k =2048) which uses soft decision, under different modulation method, the BER after decoding is about 10^{-6} when SNR>8dB, which can meet the requirements of space-to-ground communication system in the aspect of BER.

5. Conclusion

Raptor codes has been widely concerned and applied because its advantages of low complexity and no feedback channel. This paper designed a kind of Raptor code for relay communication in space network by concatenating simplified irregular LDGM with truncated LT code, the Raptor code improved the truncated degree distribution raised by Shokrollahiti and get a set of empirical data of distribution function for small file to adapt to the on-board processing ability of relay satellites. Simulation results show that the Raptor code can not only greatly reduce transmission delay, but also spontaneously adapt to channel quality in relay communication, furthermore, its decoding time is about 1/90 of LDPC's under the same conditions, and its decoding overhead is less than 0.2. the Raptor code designed in this paper has good performance in both transmission efficiency and quality in space-to-ground communication.

References

1. Michael Luby, Tiago Gasiba, Thomas Stockhammer and Mark Watson, Reliable Multimedia Download Delivery in Cellular Broadcast Networks [J]. IEEE Transactions on Broadcasting, 2007, 53(1):235–247.
2. Tiago Gasiba, Thomas Stockhammer, and Wen Xu, Raptor codes for reliable download delivery in wireless broadcast systems[A]. Ln: Consumer Communications and Networking Conference[C]. Las Vegas: IEEE, 2006, 192–197.
3. Lei Wei-jia, Xie Xian-zhong, and Li Guang-jun, the Scheme and Performance of Wireless Cooperative Relay System Using Digital Fountain Codes [J]. Acta Electronica Sinica, 2010, 38(1):228–233.
4. Li Ying-tao, Hao Chuan, Zhang Zhong-zhao, and Zhang Nai-tong, Satellite mobile communication channel model and error rate characteristics analysis [J]. Communications, 2006, 17(6):22–30.
5. Shang Shang, Zhang Qin-yu, and Li Hui, Research On Relay Downlink Characteristics And Simulation Analysis In Deep Space Communication [J]. Journal of Telemetry, Tracking, and Command, 2007, 28(4):57–63.
6. Li Xu, and Zhang Qin-yu, Latency Budget of Prompted Negative Acknowledgement CFDP Protocol in Deep Space Communication [J]. Aerospace Shanghai, 2008, 29(4):18–22.
7. M. Luby, "LT codes" in Proceedings [A]. Ln: The 43rd Annual IEEE Symposium on Foundations of Computer Science [C]. Canada: IEEE, 2002. 271–280.
8. Shokrollahi A, Raptor codes [J]. IEEE Transactions on Information Theory, 2006, 52(6) : 2551–2567.
9. CCSDS. Erasure correcting codes for use in near earth and deep space communications [EB/OL]. [2011 07 27]. http://ccsds.org.
10. Zhou Lin-kai, and Diao Ming, Influence of The Chopped Degree Distribution on The Decoding Performance in LT Codes[J]. Applied Science and Technology, 2013, 40(2):243–250.

Fund Risk Early Warning System Design of Power Grid Enterprise Group

Lin-lin Wang

State Grid Energy Research Institute

Beijing, China

E-mail: wanglinlin@sgeri.sgcc.com.cn

Construction of comprehensive and systematic fund security early warning system has great significance for preventing fund risks and improving financial soundness aiming at large enterprise groups. In the paper, interaction characteristics of power grid enterprise group business flow and fund flow are firstly combed. On the basis, power grid enterprise group fund risk information database is established. Fund risks of power grid enterprises are divided into quantitative and non-quantitative fund risks. Corresponding early warning indicator system and administrative rule systems are constructed in order to dynamically and comprehensively preventing fund risks of power grid enterprise group.

Keywords: Power grid enterprise group; Risk early warning; Indicator system; Administrative rule

1. Introduction

Fund is the core resource for enterprises. Fund security is the priority in fund management. Enterprise group fund is characterized by large flow, wide distribution and complex hierarchy. Most groups set up fund centralized management systems with financial company as carrier. Internal capital market with similar financial characteristics is formed. When fund operation efficiency is improved, corresponding fund risk can not be ignored. Traditional financial early-warning theory is mostly targeted to single enterprise main body (Carmichael, 1972) in the aspect of study objects. Early warning contents are concentrated in disclosing gradual financial distress caused by management risk and improper use of financial leverage (Wu Zhanchi, 2013). Early warning methods are mostly based on early warning system of traditional financial indicators (Altman, 1968). Fund risks caused by internal control failure and improper operation are rarely involved. How to establish a set of corresponding, complete, scientific and operational enterprise group fund security early warning monitoring indicator system, strengthen early warning, control, supervision and timely treatment on enterprise monetary fund security risk, guarantee fund security and liquidity, improve financial management level and risk control capacity have important practical significance aiming at enterprise group-a special organization form.

2. Fund flow characteristics of power grid enterprise group

Operational characteristic of power grid enterprises is main factor to affect group fund security. It is also the main reason of interaction among business flow, information flow and fund flow. Business flow of power grid enterprise can be divided into three parts, the first part refers to electricity purchase and sales, the second part refers to power grid construction, alteration and maintenance, and the third part refers to research, training, logistics, service and other work. Large-scale fund flows are formed in ongoing power sales. Operating cash income of power grid enterprises refers that mutual transformation between current and fund flow is continuously implemented mainly through transmission networks at different voltage grades. Large-scale cash demand is produced by power grid infrastructure construction investment and electricity purchase cost. Investment and financing cash flow is manifested as transformation among cash, credit and debt, equity and power grid tangible assets.

2.1. *Strong exogeneity of operating cash flow*

Power grid enterprises are different from ordinary enterprises. Market capacity of power grid enterprises can not be changed because of changes in enterprise competition relationship or products themselves due to natural monopoly operation and electricity product homogeneity. Electricity sales amount mainly depends on factors such as economic conditions, seasonal climate, etc., which is affected by electricity price level to certain extent. However, current electricity price is formulated by the government. Power grid enterprises have own suggestion rights, but they do not grasp the initiative rights. Therefore, power grid enterprises can't increase the electricity sales amount through own marketing activity, and can not take the initiative to formulate the prices according to own needs. Purchase-sales electricity amount and purchase-sales electricity price have very strong exogeneity.

2.2. *Large short-term fluctuation range of cash flow*

Payment progress is earlier than income progress in power grid enterprises due to influence of the following factors: regulation department's requirements on electric charge payment progress, electricity sales collection at the end of month, etc., therefore funds are fluctuated downwards greatly within the month, and the month-end deposit is warped. The cash flow change within one month is shown in a concave parabola, the cash flow is higher at the beginning and end of month, and the cash flow reaches the lowest point. The decreasing time is more concentrated. Short-term fluctuation range of power grid enterprises is increased by the change rule, thereby bringing larger challenge to fund liquidity management.

2.3. *Large investment and financing cash flow*

Power grid enterprise suffers from larger fund pressure with constantly improving power grid construction standard and unceasingly expanding investment scale. Power grid enterprises have weaker profit ability due to influence of national policy. Self-accumulation fund is increased more slowly. Internal financing with main sources of profits and depreciation can not meet growing financing demand of the company. External financing is still important fund source of company investment increase, thereby resulting in more long-term investment, larger investment scale, high financing demand and larger fund flow capacity of power grid enterprises.

3. Identification of power grid enterprise group fund risk

Risk identification is important basis to build fund security risk early warning model. It is important basis for enterprises to decide whether countermeasures can be adopted and the most appropriate strategies are selected or not. In the paper, the following risk identification methods are adopted as reference in order to improve scientific property and rigorousness of fund risk evaluation work, such as 'Central Enterprise Comprehensive Risk Management Guidelines', 'Enterprise Risk Management Guidelines', COSO-ERM model and 'GBT 27921-2011 Risk Management and Risk

Assessment Technology', etc. Fund security is regarded as guidance. Business process is regarded as main line, the whole process of fund operation is comprehensively considered. Risks of funds in assembly, transfer, distribution and use within enterprise group are identified. Meanwhile, fund link of enterprise group with external world is focused, and enterprise group fund risk information database is formed as shown in Figure 1.

Fund risks of power grid enterprise group are decomposed into three major levels layer by layer from top to bottom, which respectively correspond to level I risk domain, level II risk domain and level III risk domain. Level I risk domain contains two categories of liquidity risk and security risk, wherein fund flow is regarded as main line of liquidity risk. Fund liquidity risk and financing risk of companies are sorted. Business is regarded as main line in security risk. Three level II risk domains of interest rate risk, credit risk and operational risk are sorted. Level III risk domain is composed of one or several risk scenarios, which is divided into quantitative risk scenarios and non-quantitative risk scenarios according to whether risk scenario has quantitative basis or not. Quantitative risk scenarios are decomposed of risk early warning indicators, and non-quantitative risk scenarios (including finance management risk, liquidity management risk, client credit risk, bill management risk and management process risk) are decomposed into administrative rules in the design of risk early warning indicator system.

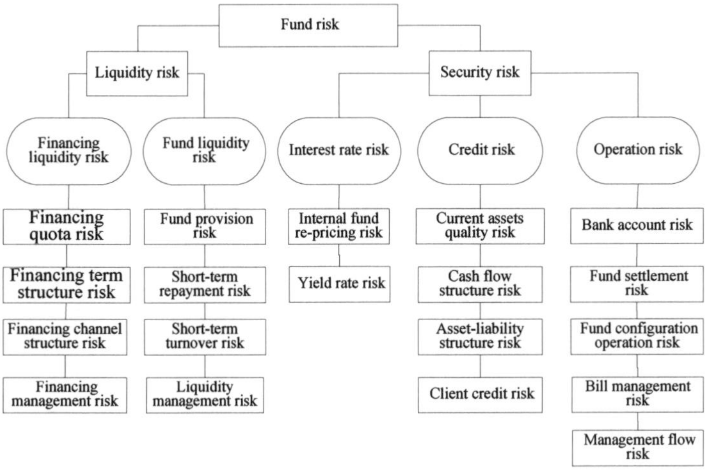

Fig. 1. Enterprise group fund security risk information database

4. Design of power grid enterprise group fund risk early warning indicator system

Design of power grid enterprise group fund risk early warning indicator system is based on the principles of 'science sensitiveness, key focus and comprehensive management'. Risk driving method is regarded as main method. Quantitative early warning and management control are implemented synchronously. Indicator selection, threshold setting and application effect are organically combined, which are continuously perfected in continuous dynamic application, thereby realizing fund risk early warning, handling and control. Risk early warning is gradually transferred from post-inspection to process management and prevention. Quantitative risk early warning indicators and non-quantitative risk early warning administrative rules are respectively designed aiming at the above-mentioned sorted fund security risk information database according to the importance of risk scenarios, whether the scenario is quantitative or not, and scenario features.

4.1. *Design of quantitative risk early warning indicator*

Corresponding risk early warning indicators are respectively designed for 32 quantitative risk scenarios in power grid enterprise group risk information database as shown in Table 1. They also belong to monitoring early warning indicators. In practical application, indicator selection should be focused according to management requirements of group headquarters and subordinate member units.

4.2. *Threshold setting*

Risk early warning refers to a risk identification behavior for judging the damage degree through observing risk representation indicators on the basis of established risk category. When some representation risk indicator value is higher than certain limit, early warning is implemented on the indicator representation risk. In the paper, parameter principle method, comparative method and expert consultation method are respectively adopted for determining indicator thresholds according to different indicator categories and early warning significance.

4.2.1. *Threshold determination through parameter principle method*

Some indicators belong to relatively common early warning monitoring indicators for electric power industry or state-owned large enterprise in early warning risk indicator. General benchmark level and general early-warning standards recognized in the industry are concluded and summarized in long-term practice. We fully adopt industry standards for threshold zoning as reference aiming at the indicators, such as 'Enterprise Performance Evaluation Standard Value in 2015' formulated by The State Council SASAC Financial Supervision and Examination Evaluation Bureau, annual report of large state-owned power enterprise listed companies, electric power industry research and analysis report of financial institutions, etc.

4.2.2. *Threshold determination through comparative method*

Since power grid enterprise groups have special industry status, a part of indicators can not be used for accurately determining true risk levels through directly adopting industry standards as reference easily. Data calculation can be implemented on the indicators, and corresponding early warning standards can be designed according to actual situation.

Financing strategy radicalness is adopted as an example. Financing strategy radicalness refers to the difference value of long-term liabilities to current liabilities ratio and long-term assets- current assets ratio. If the indicator is closer to zero, it is obvious that deadline proportion degree of assets and liabilities is better. If the indicator is greater than zero, it is obvious that the liability structure is mainly based on long-term liabilities, assets structure is mainly based on current assets. The company has stable fund flow. However, the fund can not be fully operated, thereby possibly leading to financial resource waste. If the indicator is lower than 0, it is obvious that the liability structure is mainly based on short-term liabilities, and the assets structure is mainly based on fixed assets, and the enterprise has larger short-term debt pressure possibility. It is obvious that short-term financing is excessively depended for supporting longer-term investment in order to pursue financing cost reduction. We calculate the radicalness of financing strategies in all units on the basis of monthly financial data of 27 member units of some group enterprise in 2015. The results show that annual average value of the indicator in only two units can reach the level higher than -4. Overall performance of 27 units for 12 months shows that the condition with indicator higher than -4 accounts for 5%, the condition higher than -7 accounts for 25%, and condition lower than -12.5 accounts for 25%.

We define the following criteria according to the above situation:
It is safe if financing strategy radicalness is -7 or higher than -7;
It is generally risky if financing strategy radicalness is -12.5 (including -12.7) to -7(excluding -7);
It is seriously risky if the financing strategy radicalness is lower than -12.5 (including-12.5).

4.2.3. *Threshold determination through expert consultation method*

We adopt the mode of interviewing and discussing with many related business leaders of the company and industry experts, suggestions of all related personnel are summarized and confirmed for many times, and all thresholds of other indicators are finally determined aiming at indicators whose effective threshold can not be obtained through industry bench-marking or fluctuation analysis.
Threshold settings of enterprise group fund risk early warning indicators are shown in the following table.

Table 1. Enterprise group fund risk early warning indicator system

Level 3 risk domain	Risk scenario	Early warning indicator	Serious risk early warning threshold	General risk early warning threshold	Security threshold
Financing quota risk	Remaining financing quota is insufficient	Credit application progress	≥50%	[30%,50)	<30%
Financing term structure risk	Financing term is unbalanced.	Financing strategy radicalness	≦-12.5	[-7,-12.5)	>-7
	Maturity payment is too concentrated.	maturity payment concentration ratio	≥60%	[30%, 60%)	<30%
Financing channel structure risk	Financing channel is unbalanced.	Financing channel proportion	≦3	[3,5)	>5
Fund provision risk	Bank account balance is rational.	Bank account minimum balance	≦10 million	[10 million, 100 million)	>100 million
	Short-term financing occurs frequently.	Short-term financing frequency	≥5	[3,5)	<3
	Fund provision is insufficient.	Provision rate	≥60%	[30%, 60%)	<30%
	Fund gap is expanded.	Monthly budget gap month-on-month growth	≥50%	[30%, 50%)	<30%
	High-frequency and high-quota payment	Daily high-amount payment frequency of the month	≥10	[5,10)	<5
Short-term repayment	Repayment ability	Cash current liabilities ratio	≥1.50%	[1.50%, 6.00%)	<6.00%

Level 3 risk domain	Risk scenario	Early warning indicator	Serious risk early warning threshold	General risk early warning threshold	Security threshold
ability risk		Cash debt coverage ratio	≦0.70%	[0.7%, 2.50%)	>2.50%
		Cash interest cover ratio	≦-15	[-15,1)	>1
	Financial flexibility	Current proportion	≦20%	[20%, 45%)	>45%
Short-term turnover ability risk	The degree of replacing cash income with bill income	Bill income proportion	≥50%	[20%, 50%)	<20%
	Fund turnover efficiency is lower.	Accounts receivable turnover rate	≥3	[1,3)	<1
Internal fund re-pricing risk	Re-pricing gap	Interest rate re-pricing gap	≥100 million	[10 million, 100 million)	<10 million
Yield rate curve risk	Financing is rapidly increased.	Weighted average financing interest rate	≥10%	[5%, 10%)	<5%
		Financing interest rate increase extent	≥10%	[5%, 10%)	<5%
Current assets quality risk	Current assets quality is worse.	Bad debt loss rate	≥0.50%	[0.20%, 0.50%)	<0.20%
Cash flow structure risk	Cash flow structure is not rational.	Operating cash inflow structure ratio	≦50%	[50%, 65%)	65%
		Operating cash outflow structure ratio	≥78%	[70%, 78%)	<70%
Asset-liability structure risk	Asset-liability structure is not rational.	Asset-liability ratio	≥80%	[70%, 80%)	<100%
		Current liabilities proportion	≥75%	[50%, 75%)	<50%
Bank account risk	Redundancy bank account is available.	Redundancy account list	≥10	[5,10)	<5
Fund settlement risk	Payee unit is abnormal.	Payee unit abnormity quantity	≥5	[3,5)	<3
	Payment amount is abnormal.	Payment amount deviation ratio	≥0.50%	[0.20%, 0.50%)	<0.20%
	Electric charge amount payment abnormity	Electric charge amount abnormal payment	≥1 million	[100 thousand, 1 million)	<100 thousand
	Critical time-point collection	Critical time-point fund collection amount	≥5 million	[1 million, 5 million)	<1 million
	Payment operation error	Repeated payment quantity	≥5	[3,5)	<3
Fund configuration operation risk	Loan-to-deposit ratio is higher than limit.	Financial company loan-to-deposit ratio	≥50%	[50%, 75%)	<75%
	Mid-term and long-term loan	Financial company mid-term and long-	≥75%	[50%, 75%)	<50%

Level 3 risk domain	Risk scenario	Early warning indicator	Serious risk early warning threshold	General risk early warning threshold	Security threshold
	proportion is too high.	term loan proportion			
	Investment term structure is not rational.	Financial company long-term and short-term investment proportion	≥75%	[50%, 75%)	<50%

4.3. *Design of non-quantitative risk early warning administrative rule*

Process control and early warning promotes can be implemented through designing risk early warning administrative rule, system control, system screening and inspection modes can be adopted for prevention aiming at administrative rule as shown in table 2 aiming at non-quantitative risk scenario in fund security risk information database.

Table 2. Enterprise group fund risk early warning administrative rule

Level 3 risk domain	Risk early warning rules	Implementation mode
Financing management risk	Financing implementation structure is not consistent with financing plan, and there are no examination and approval record in the system.	System screening
	Financing plan adjustment applicant =financing plan adjustment approver; financing plan adjustment approver =financing plan adjustment auditor; Financing plan adjustment auditor =Financing plan adjustment applicant	System screening
	Financing standing account record month >financing activity record month; financing standing account record time-financing activity record time > 14 days	System screening
	Financing actual performance factor is not consistent with financing contract factor.	System screening
liquidity management risk	Bank deposit balance of member unit at the end of month > security provision quota	System screening
Client Credit risk	Large client list with account age longer than 30 days in the month	System screening
	Accounts receivable with account age longer than 30 days in the month> monthly operation fund *5%	System screening
Management flow risk	Bank account increment of the term - bank account approval increment of the term>0	System screening
	Payment application approver =payment application applicant; payment application approver =payment application auditor; payment application auditor =payment application applicant	System screening
	Payment amount corresponding to current procurement order ≠ invoice amount corresponding to current procurement order	System control
	Reprocessing successful time - payment failure time < the longest	System screening

	processing time limit (system log)	
	System actual closing time is earlier than system normal closing time (system log)	System control
	System restoration operation time - system failure occurrence time > the longest stop time limit of the system	System screening
	Acceptance draft endorsement discontinuity	Inspection
	Bill recipient = bill checker, there is no inspection record of the second person.	System screening
	Monthly balance of acceptance draft issued by clients with arrearage record at the end of month is not 0.	System screening
	Acceptance bill without bank trusteeship	Inspection
	Balance of commercial acceptance draft> 0	System screening
	Invalid cheque is not provided with invalid tags and properly kept in the safe.	Inspection
	Bill checking and account checking are not implemented timely at the end of month.	Inspection
	Bill settlement date-bill safekeeping date>regulated term	System screening
	Actual amount of bills ≠ carrying amount	System screening
	System record cheque number is discontinuous.	System screening
	Financial system employee permission list is matched with permissions mutually exclusive matrix, and the inconsistent items are included into the list.	System screening

5. Conclusion and prospect

Power grid enterprise group is characterized by multiple management levels, more business blocks, huge business type and complex fund transfer process. Scientific and accurate early warning should be implemented for fund risk. A set of scientific, comprehensive and sensitive fund risk early warning indicator systems must be established. Traditional financial early warning theory is mostly targeted to gradual financial difficulties of single enterprises. The risk sources are more diversified aiming at group-based enterprises with concentrated fund. Risk representation is more complex. After operation risks of branches are reduced, fund operation risks are further amplified at the group level. In the paper, all risk sources of enterprise groups are divided into quantitative and non-quantitative risk scenarios on the basis of traditional financial ratio modeling. Early warning indicators are established aiming at quantitative risk scenario. Indicator thresholds are determined through parameter principle method, comparative method an expert consultation method aiming at enterprise group features. Related ideas of internet control are combined to form risk early warning administrative rule aiming at non-quantitative risk scenario, especially decision-making risk and operation risk. They are realized through system control, screening and inspection mode. The above two aspects are combined for finally realizing fund risk early warning indicator system covering the whole process of fund flow. Fund risks experience dynamic and quantitative management as well as online monitoring. Of course, it is inevitable that fund management of enterprise groups also suffers from new problems and characteristics with in-depth progress of financial reform and change of market environment. Indicator systems should be constantly updated and perfected in practice, thereby being closer to reality management demand of enterprises.

References

1. Wu Zhanchi, Li Xiaolong. Research on enterprise group fund security early warning system. Accounting Research, 2013,02:63-68+95.
2. Li Wenning. Research on enterprise group financial crisis early warning system. Statistics and Decision, 2014,16:176-178.
3. Zhang Yongchun, Zhang Li, Ma Jinhui. Research on frame of capital project cross-border fund flow monitoring and early warning system. North China Finance, 2009,06:45-48.
4. Cao Yajun. Discussion on petroleum enterprise fund risk early warning and control problem. Journal of Yangtze University (social science edition), 2009,06:40-43.
5. Zhu Dan. Research on PDCA-based enterprise group fund security early warning system. Accountant, 2015,16:16-17.
6. Zhu Jianfeng. Enterprise financial security early warning and risk prevention. Friends of Accounting (the first ten-day), 2008, 11:38-40.
7. Altman E.I: Financial ratios, discriminant analysis and the prediction of corporate bankruptcy, Journal of Finance, 1968,(4).
8. Carmichael, D. R. 1972. The Auditor's Reporting Obligation: The Meaning and Implementation of the Fourth Standard of Reporting. New York, NY: AICPA.
9. Edward, E.Deakin: A discriminant analysis of prediction of business failure, Journal of Accounting Research, 1972.
10. Salchenberger L M, Clnar E M, Lash N A. Neural Networks:A new tool for predicting thrift failures, Decision Sciences,1992,(4).p

Layer aggregation and structure detection of interconnected networks*

Quantong Guo†, Xin Jiang, Yanjun Lei and Zhiming Zheng

School of Mathematics and Systems Science, Beihang University, Beijing, 100191, China

†*E-mail: quantong@buaa.edu.cn*

There has been growing interest in exploring the topology structure of networks, especially multi-layer networks. However, with the rapid development of social media and technology, it has shown that modeling such multi-dimensional systems leads to a huge complexity in dealing with them. A fundamental open question is then how many layers are really needed to represent the systems, which has also attracted many scientists from various fields. Here we adopt the Von Neumann entropy theory to measure the information contained in network and through defining a distance based on it, we are able to reduce the number of network layers, as well as retaining the main characteristics of the whole system. Besides, in order to validate the method, we have built a benchmark networks which consists of three different kinds of networks to realize the aggregation process. Surprisingly, the method can not only reduce the redundancy but also distinguish the structure diversity, since it first merged the same kind of networks. Our results provide a better understanding of entropy theory and could also lead to interesting future research about how to reduce the structural reducibility of complex systems.

Keywords: Network; Information; Entropy; Complex system; Structure.

1. Introduction

Multiplex or interdependent networks[1][2] are effective ways to describe complex systems where different layers are used to represent various interactions between the entities of network. Recent works[3–6] have revealed that the dynamic processes on multiplex networks are different from

* This work is supported by NSFC under No. 11201017 and 11290141, Cultivation Project of NSFC (No. 91130019).

that on single-layer network, which points out the necessity of considering interconnection networks as multiplex networks in reality. But a good description of multiplex networks may require lots of layers to show all the possible connections between nodes, which is inevitable containing redundancy information in the complex system[7]. This leads to a fundamental question that how much information that the system truly contains and if it is possible for us to reduce the redundancy information by aggregating some layers together.

Recently, M. De Domenico[7] studied the layer aggregation on interconnected networks and proposed an useful way to make it. As the structure of each layer of multiplex networks may belong to different categories[8][9], for example, Scale-free (SF) networks, Erdős–Rényi (ER) networks and (WS) networks and so on, and different structures represent different nature of the networks[10]. Therefore, when we need to reduce the redundancy information, we should first distinguish the structure of them and then reduce the redundancy information of the same structure[11]. Here, in the paper, we provide an effective and efficient method which can not only reduce the redundancy information of multiplex networks but also distinguish the structure diversity of different layers.

2. Model Description

In order to carry out the layer aggregation process, here, firstly we use quantum information theory to measure how much information that each layer of multiplex networks has. Quantum information[12] is a field which has been expanding quickly as the theoretical successes in many different fields, such as, fast algorithms, teleportation and more. If we consider that each layer represents one possible state of the system, the whole multiplex system can be considered as an ensemble of states.

As for the question that how to measure the information that multiplex networks have, with the definition of density matrix ρ of a graph, we use the Von Neumann entropy S(ρ) to calculate the information that each layer has[7]. The density matrix of a graph G is defined as follows:

$$P = \frac{1}{2E} L_G \tag{1}$$

Where E represents the total edges that the graph G has and L_G means the Laplacian matrix associated to the graph, with $L_G = D - A$, D is the diagonal matrix of node degrees (the number of edges adjacent to it) and A is the adjacent matrix of the graph. After we have the density matrix of a graph, the Von Neumann entropy S(ρ) of the graph is:

$$S(\rho) = -\sum_{i=1}^{N} \lambda_i(\rho) \log_2 \lambda_i(\rho) \quad (2)$$

$\lambda_i(\rho)$ represents the eigenvalue of density matrix ρ. As the Laplacian matrix is a positive semi-definite matrix, all the total eigenvalues satisfy that $0 = \lambda_i \le \lambda_2 ... \lambda_N$. It is convenient for us to define $0 \log_2 0 = 0$. So the Von Neumann entropy can be calculated within the limits of log function.

The next question is how to reduce the redundancy information of multiplex network. In according to the above definition, the information that each layer has can be calculated. In order to reduce the information through layer aggregation process, a distance should be defined to evaluate the similarity between different layers. Since each layer is one possible state of the complex system, it is suitable for us to use the Jensen Shannon divergence[13] to quantify the distance between different layers, as defined by. The Jensen Shannon divergence between different density operators δ and ρ is:

$$D_{JS}(\delta \| \rho) = h(\theta) - \frac{1}{2}(h(\delta) + h(\rho)),\ \theta = \tfrac{1}{2}(\delta + \rho) \quad (3)$$

Noting that the definition of D_{JS} is symmetric and the range of so-called Jensen Shannon distance $\sqrt{D_{JS}}$ is [0,1], so we use it to represent the distance between different layers. In accordance with our definition of Von Neumann entropy $P = \frac{1}{2E} L_G$, here the distance between layer i and layer j can be defined as:

$$D_{JS}(\rho_i, \rho_j) = h(\theta) - \frac{1}{2}(h(\rho_i) + h(\rho_j)) \quad (4)$$

Because ρ_i and ρ_j are symmetric and positive semi-definite matrix, the matrix of θ also has the same property. So it is applicable for the matrix to use the Von Neumann entropy function.

Then, after the steps above, we have the information of each layer and the distance between different layers, we are able to carry out the layer aggregation process. Since the Jensen Shannon distance is used to measure the similarity of different layers, the smaller the distance is, the higher the similarity is. Therefore, we can aggregate the layer with the smallest distance between them without loss much information. If the layers of multiplex networks are M, then we just need to iterate the process for M-1 steps to get the final single layer network. At each step, we get a newly formed layer and update the distance between it with the other remaining layers according to the lance-williams algorithms[14]. The lance-williams algorithms are an infinite family of agglomerative hierarchical clustering algorithms for updating cluster distances at each step. In this paper, we update the distance according to the rules below:

$$D(G_m \cup G_k)=\frac{n_m+n_i}{n_m+n_k+n_i}D(G_m,G_i)+\frac{n_k+n_i}{n_m+n_k+n_i}D(G_k,G_i)-\frac{n_i}{n_m+n_k+n_i}D(G_m,G_k) \qquad (5)$$

G_m, G_k are the layers needed to aggregate and G_i is anyone of the remaining layers. n_m, n_k, n_i represents the number of layers in layer m, k, i, respectively. Each time the new distance matrix is calculated, we find the smallest distance pair and aggregate these two layers until we get the singer layer network. The whole procedure proposed above sheds light on how to reduce the redundancy information of multiplex networks, however, as in many cases, in order to have a better understanding of the system we deal with, multiplex or multilayer networks is the proper formalization other than singer layer network. This leads us to consider the critical condition where we should stop our aggregation process and then the remaining layers are the optimal configuration of the system instead of keeping reducing the network layer to one layer. Since we have the Von Neumann of layer i as h_i, which is used to measure the information of the layer, the total effective information of multiplex network h_{total} should be much less than the sum of all the layers due to the redundancy information. Hence we use the weighted average of all the h_i to obtain the total effective information h_{total} as follows:

$$h_{total}=\sum_{i=1}^{m} p_i h_i \qquad (6)$$

m is the total number of remaining layers, p_i is the weight of layer i. Because the structure of a graph G(V,E) is a combination of edges and nodes, the number of edges is an important indicator of the abundance of information when the nodes are the same. According to the assumption, we define the weight of layer i as

$$p_i = \frac{\text{number of edges of layer i}}{\text{the total number of edges of remaining layers}} \tag{7}$$

The weight reveals the importance of each layer, specially, if the multiplex network is made up of identical layers, thus the weight of each layer is the same. Then we define an information loss function in the following to measure to what extend that the mth step of layer aggregation process is proper:

□

$$Q(m) = 1 - \frac{h_{total}}{h_{\oplus}} \tag{8}$$

Where $\oplus$ is the associated single layer aggregated graph, $h_{\oplus}$ represents the Von Neumann entropy of the graph. In order to have the optimal configuration of layers, we just need to maximum the Q function through the hierarchical clustering process defined above. Every step we can get the Q value, and the whole process needs at most M-1 steps. Thus, we choose the step that have the maximum value of Q as our optimal configuration[7][13].

3. Numerical simulations

In the following section, we use a benchmark model to test our analysis and the results. First, we build a 15 layers multiplex networks. These networks are made of three types of structures, which means ER networks, SF networks, WS networks. The generating process of these networks are as follows: (1) we generate a ER network, a SF network and a WS network. (2) Through different rewiring probabilities, including 0, 5%, 20%, 60%, 100%, respectively, we make each type of network consists of five layers.(3)Through the hierarchical clustering process, we calculate the different values of Q function and cut off the hierarchical at the maximum point.

In Fig.1, we show the distance matrix of the 15 layers using the Jensen Shannon distance. It is clear that there exists 3 "community" blocks, each of which consists of 5 layers. The distance between the layers of the same "community" is much smaller than the distance between different "community". Besides, with the increasing of rewiring probability, the distance also becomes larger and larger. Interestingly, the SF networks has an extreme large distance with the other two types of networks, which shows the fundamental differences of the SF network structure with others. These results give us a better understanding that our definition our distance is proper for layer aggregation. Therefore, next we use our clustering definition to carry out the layer aggregation process also on the same multiplex networks.

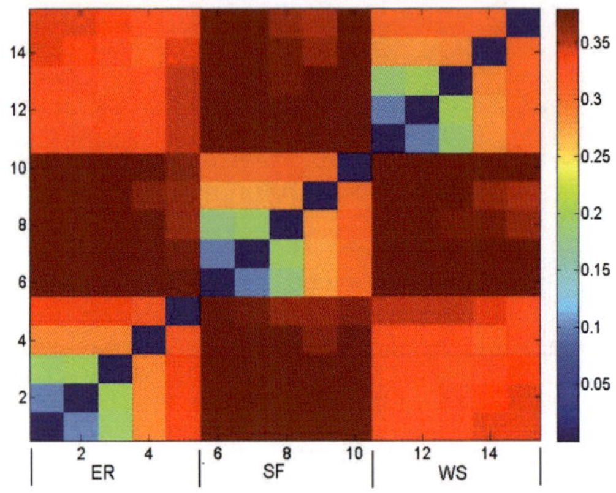

Fig. 1. The distance matrix of the 15 layers with 3 different types of network structures. The first 5 layers are ER networks, the second 5 layers are SF networks, and the last 5 layers are WS networks. Each type of networks is built up through different rewiring probabilities, 0, 5%, 20%, 60%, 100%, respectively. All the networks have 1000 nodes and the average degree is 8.

In Fig. 2, we perform the lance-williams algorithms on our multiplex network. The figure shows two significant characteristics: one of the characteristics is that the same types of networks are merged together at first; the other characteristic is that in the same type of networks, the layers with less difference of rewiring probability are merged first. In particular, the SF networks show an "interdependent" characteristic because in the whole process, only if the five layers have merged together can it merges with others. Besides, as the increasing of rewiring probability, the WS networks exhibit a strong similarity with ER networks. This is the reason why the 60%, 100% WS layers are aggregated with ER layers other than other WS layers. These results suggest that our layer aggregation process can not only effectively reduce the redundant information but also maintain the structure stability of networks.

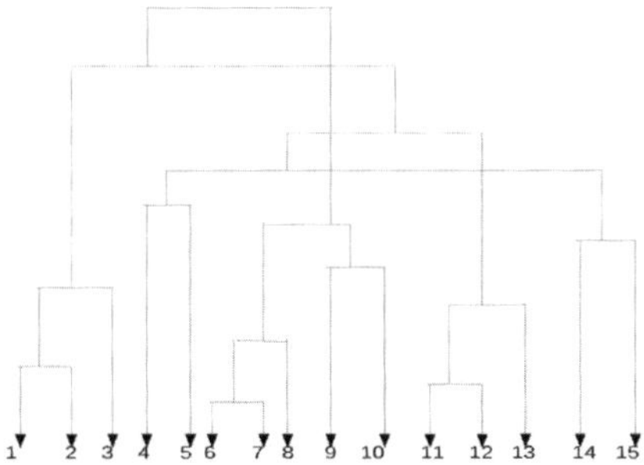

Fig. 2. The dendrogram of aggregation process on the multiplex network. The 1-5 represent the 5 ER layers, the 6-10 represent the 5 SF layers, and the last represent the 5 WS layers. Two layers are merged together at each steps and the process has a total number of 14 steps.

In Fig.3, we plot the Q function of the layer aggregation process. The maximum value is the 0 step, which means that the optimal configuration of the system is keeping all the layers distinct. This is easy to understand since our model is made up of 15 layers which are produced through rewiring the edges of three original network. Even the difference between the 0, 5%rewiring probability is small, they also have some special information that is not included in others. At the same time, we find that there is also a value at the fourth step which is close to the maximum value. This point means that the 6-7 pair, 11-12 pair, 1-2 pair, (6,7)-8 pair are merged together, which are the layers with the larger percentages of overlap information. The result suggests an useful balance between loss of information and complexity as we are able to decrease the complexity by reducing some layers without so much loss of information.

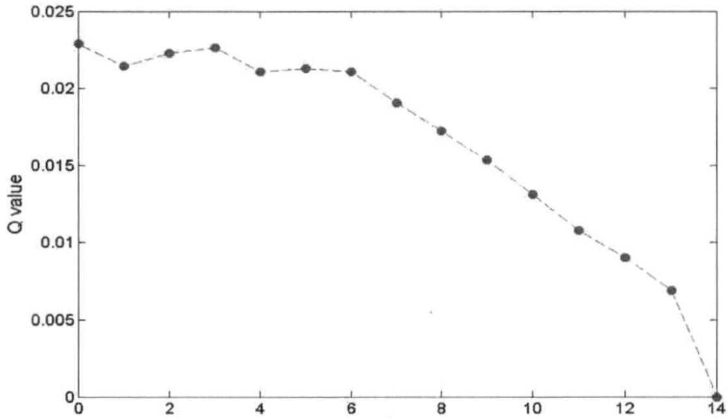

Fig. 3. The fitness function values versus the steps of layer aggregation. The maximum value is Q(0) which indicates all layers are kept distinct.

4. Conclusion

As a summary, in this paper, we have extended the Von Neumann entropy theory to measure the information contained in network. Then, with the help of Jensen-Shannon distance between graphs, we are able to realize layer aggregation process. The results show that this method can efficiently reduce the redundancy of interconnected networks by hierarchical clustering algorithm. Furthermore, through building a benchmark interconnected networks, we also find that this method could help us recognize different structure of networks. Although the method we proposed is a quite easy model, our findings give us a proper way to consider and deal with the problem. Furthermore, since how to measure the information contained in the complex system is still an open question for scientists, this would also lead us to explore many other methods to make the aggregation process more efficient in our future work.

References

1. P. J. Mucha, T. Richardson, K. Macon, M. A. Porter, and J.-P. Onnela, Science 328, 876 (2010).
2. M. Szell, R. Lambiotte, and S. Thurner, Proc. Natl. Acad. Sci. U. S. A. 107, 13636 (2010).
3. Z. Wang, L. Wang, M. Perc, Phys. Rev. E 89, 052813 (2014).
4. K.-M. Lee, J.Y. Kim, W.-K. Cho, K.-I. Goh, and I.-M. Kim, New J. Phys. 14, 033027 (2012).
5. Q. Guo, X. Jiang, Y. Lei, M. Li, Y. Ma, and Z. Zheng, Phys. Rev. E 91, 012822 (2015).
6. W. Li, S. Tang, W. Fang, Q. Guo, X. Zhang, and Z. Zheng, Physical Review E, 92(4), 042810 (2015).
7. De Domenico, M., Nicosia, V., Arenas, A., and Latora, V. Nature communications, 6. (2015).
8. M. E. J. Newman, Phys. Rev. E 66, 016128 (2002).
9. Y. Moreno, R. Pastor-Satorras and A. Vespignani, Eur. Phys. J. B 26, 521 (2002).
10. A. Motter, Y. Lai, Physical Review E 66 (6), 065102 (2002).
11. M. E. J. Newman, SIAM Rev., 45, 167 (2003).
12. Dirac, P. A. M. The Principles of Quantum Mechanics 27 (Oxford University Press, 1981).
13. Lamberti, P., Majtey, A., Borras, A., Casas, M. & Plastino, A. Metric character of the quantum Jensen-Shannon divergence. Phys. Rev. A 77, 052311 (2008).
14. Kaufman, L. & Rousseeuw, P. J. Finding Groups in Data: An Introduction to Cluster Analysis (John Wiley and Sons, New York, 1990).

Design of Network Security Distributed Pre-Warning System

Ya-Lin Ye†1, Wei-Feng Sun[1], Kun Han[1], Quan Yuan[1], Ke-Li Yang[1], Ning Shan[2]

1.Xi'an Communications Institute, Xi'an, 710106, China

2.Equipment Engineering College, Engineering University of CAPF, Xi'an, 710086, China

†E-mail: 137187781@qq.com

The paper presents a distributed architecture of pre-warning system to solve problems of the current network protection, such as low reliability of network events alarming, low efficiency and reliability of pre-warning and difficulties to monitor the whole network security status. The architecture includes three parts, detection domain, warning agent and local warning center. The system can percept and analyze all-round and multi-level security events by analyzing information from many monitoring equipment and safety probes deployed in the whole network. The system then forms the diagram of security situation and distributes pre-warning information immediately. The system applies all kinds of safety equipment and many technologies, such as event collection, synthesis analysis and data fusion technology to improve monitoring and pre-warning capability. The system can effectively detect anomalies, abnormal behavior and virus threats of the network, thus alleviates the harm of network intrusion and improves emergency response capability.

Keywords: Distributed Architecture; Network Security Protection; Network Security Pre-Warning System; Internet;. Regional Warning Center.

1. Introduction

With the rapid development of the Internet in recent years, network security has become more and more important. The existing network pre-warning capability has become unable to meet the current and future network applications requirements. Network security incidents get more and more complicated so far, but network security pre-warning technology is still in the stage of monitoring, such as the safety state of the network could not be grasped entirely and accurately[1], network pre-warning efficiency and reliability is too low[2], and etc., so the possible network security incidents can't be alarmed reliably and promptly, Therefore, it is very necessary to research network security pre-warning systems in order to protect network and improve the network security precaution ability. It is very important to improve the emergency response capability of network attacks, to alleviate the harm of the system, to improve the ability to anti-attack and to tolerant intrusion and disaster.

2. The overall design of the distributed network pre-warning system

According to the characteristics of Internet, such as large distribution range, large amount of data and complex data, therefore when we design network pre-warning system, it is necessary to take the real-time monitoring of network activity, monitoring and analyzing the behavior of network user and system[3], auditing the system configurations and vulnerabilities, assessing the sensitive system and the integrality of data[4], identifying attack behaviors, counting abnormal behavior, tracking and recognizing violation behavior, deploying decoy server and so on into account[5], and besides, we should audit records of network behavior, querying and analyzing various behaviors regularly, to discovery the potential danger and vulnerabilities, to obtain evidence, and to analyze the cause of the accident. We can distribute probes over different nodes of the network to acquire and analyze data real-timely, therefore, the system can detect the network intrusion and virus infection to evaluate the degree of network risk, to extract the characteristics of attack, or to find new attack mode. According to the local or global alert analysis results, the system can infer the current and the future network security state, and link to response system or to give local or global warning. Therefore, network pre-warning system is composed of detection field, warning agent(WA) and Regional warning center (RWC), as is shown in Fig. 1.

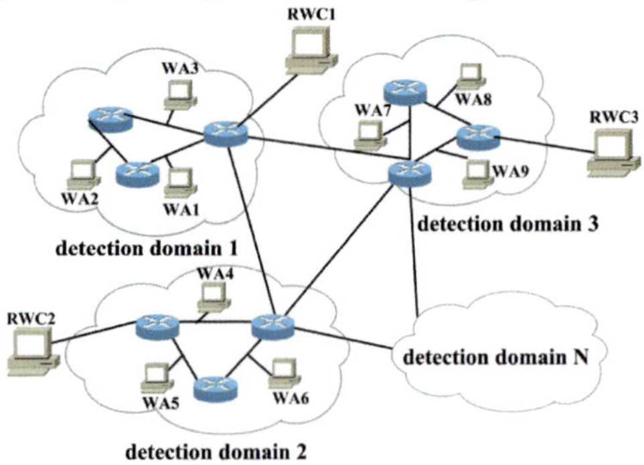

Fig. 1. Architecture of distributed network pre-warning system.

According to the different security tactics, Detection Domain divides a network into several segments, and the hosts, routers, switches, firewalls and kinds of application servers are bonded with their IP addresses. We add several warning agents to different network segment to detect potential threats and network attacks, and according to different tactics, the system realizes pre-warning function and sends alarm information to the warning center. Each detection domain repairs its vulnerabilities, and receives warning information from the warning agent and security policies from management module.

Warning Agent collects information from each network segment of detection domain, and real-timely detects and analyzes possible network attack behaviors, virus infections and network abnormal behaviors through anomaly detection and misuse detection based on data mining. It records invalid and aggressive data packets after treatment, and then it alarms and sends intrusion and anomaly information to the warning center, and receives the warning information from pre-warning center.

RWC collects alarm events from warning agents of each segment, then implements data fusion with local information from knowledge database, reveals the relationship between incidents and achieves alarms correlation to detect the distributed attacks, combined attacks, cooperative attack. By these methods, it can reduce the false positive rate and false negative rate of network intrusion alarm, and predict and identify future possible attacks. It can assess network risk and send pre-warning information to each detection domain through the warning agent. It can also respond to treat the warnings through response module.

3. Design of key components of the distributed network pre-warning system

The whole scheme of distributed network pre-warning system is shown as Fig. 2. The working flow of pre-warning system is as follows. Each warning agent acquires data packets through from safety equipment in each detection domain. If warning is got, alarm information is sent to the RWC; then, the threats are filtered and merged. And if threat is confirmed, the events will be sent directly to response module as intrusion by RWC. The abnormal information is integrated by correlation and fusion analysis module, thus, it can predict possible network attacks and assess network threats. If a new intrusion features found, it will be reported to the management module. The management module is also responsible for the task to adjust and update the filtering rules, add new intrusion patterns to intrusion characters database.

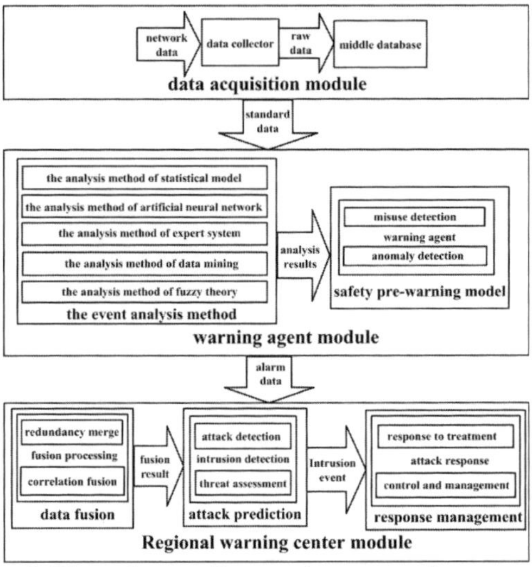

Fig. 2. The scheme of network pre-warning system.

3.1. *Design of Warning Agent*

Each detection domain comprises of several warning agent, a RWC and other network nodes. In each detection field, the warning agent in the different segments is responsible for collecting network data, analyzes and detects real-timely network attacks, virus behavior and abnormal network behavior using anomaly and misuse detection methods based on data mining. Design of warning agent is shown as Fig. 3.

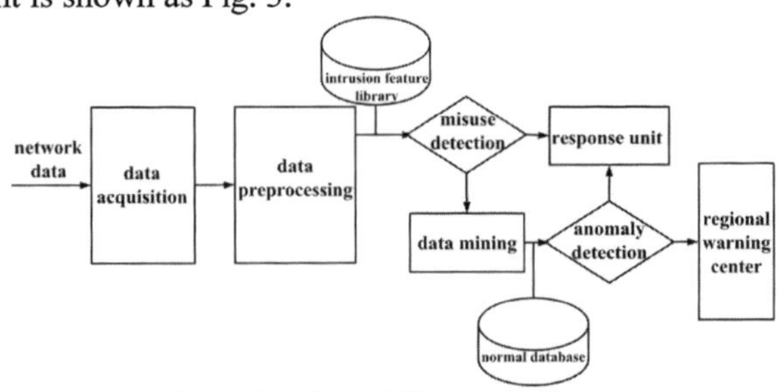

Fig. 3. Design of Warning Agent

Warning agent assumes that each data packet is potentially hostile. It captures analyzes packets. Packet capture is implemented by setting the network card in promiscuous mode. Packet analysis is performed by misuse detection module comparing attack characteristics grasped with attack feature library. Anomaly detection is then performed on unknown attacks and unformed attacks to discover network attack behavior and suspicious behavior. Meanwhile, warning agent performs real-time alarm, record the alarm and the network behavior. Warning agent can connect directly or indirectly with RWC. When a warning agent malfunctioned, its state can be detected by RWC connected and can be recover to normal working state.

3.2. *Design of Regional warning center*

RWC receives alarm events from each warning agent, and integrates the alarm information with local knowledge database to reveal the relationship between events, thus to correlate the alarms to discover distributed attacks, combined attacks and cooperative attack. Then, it can predict incoming intrusion behaviors, and distribute warnings to the monitoring points. Diagram of RWC is shown as in Fig. 4.

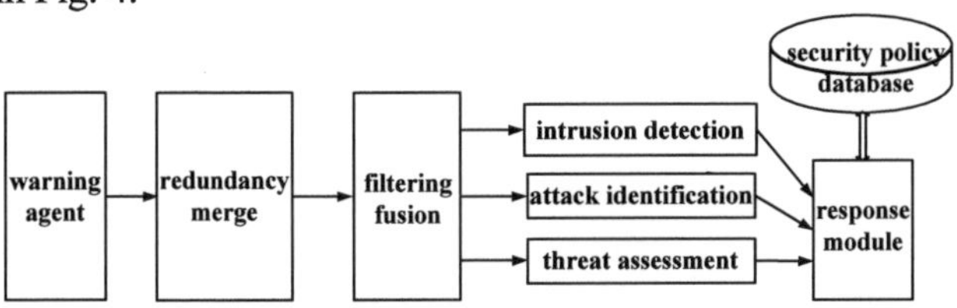

Fig.4. Diagram of RWC

Each RWC has a mapping table which records each detection domain and its IP address. Once the RWC generates pre-warnings, it sends these pre-warning to the corresponding protection domain according to the mapping table. RWCs can exchange pre-warnings over independent links between each other, thus distributed cooperation processing is realized. If one warning center failed, it can be detected and recovered by other RWC.

4. Key technologies of distributed network pre-warning

Network monitoring and pre-warning technology implements pre-warning and active defense function by monitoring the anomaly of network traffic, abnormal behaviors, and threats of virus, and integrated using of various types of safety equipment. Monitoring and pre-warning technology includes network events collection technology, events analysis technology and events fusion etc[6,7,8]. The key technologies of network events collection technology include Libpcap technology, WinPcap technology, NetFlow technology, DPI technology and etc[9,10]. The common event analysis methods include: detection and analysis method based on statistical model, detection and analysis method based on artificial neural network, detection and analysis method based on expert system, detection and analysis method based on data mining, and etc.

5. Conclusions

Research of distributed network pre-warning system is of great significance to improve the counterattack ability. The system applies all kinds of safety equipment and many technologies, such as event collection, synthesis analysis and data fusion technology to improve monitoring and pre-warning capability. The system can effectively detect anomalies, abnormal behavior and virus threats of the network, thus alleviates the harm of network intrusion and improves emergency response capability.

References

1. Chauhan S.Project Report:Analysis and Detection of Network Covert Channels[R].CMSC 691 IA.2005:08.
2. Wang Ruigang., Bakos G.Using sensor networks and data fusion for early detection of active worms[C].Proc of the SPIEAeroSense,2003.
3. He Xinwen, Wang Xuehua. Strategic Analysis and Choice for Forestry Enterprises Developing E-business[J]. Chinese Forestry Science and Technology, 2007(4):80—88.
4. The national strategy to secure cyberspace[EB/OL].2003.http://www.us-cert.gov/reading room/cyberspace strategy.pdf.
5. Kijewski P. ARAKIS-An early warning and attack identification system [C] //The 16th Annual First Conf. Dudapest, Hungary:[s.n.], 2004.
6. Liu J L, Yuan H Y. Secure Network Monitoring Technology in Mine Methane Monitoring[J]. Coal Technology, 2013.
7. Xing W. Analysis and Research of Network Monitoring Technology[J]. Science Mosaic, 2009.
8. Venkatesan K G S, Khanna V, Reddy S B A. Network Monitoring using Test Packet Generation[C]// International Journal of Engineering Research and Technology. ESRSA Publications, 2015:1-12.
9. Cidon I, Horn L, Sidi M, et al. Quality of service testing of communications networks: US, US 7835293 B2[P]. 2010. Luo T, Liu W. Design and Implementation of PPPoE Analysis System Based on Winpcap[C]// Proceedings of the 2009 Second International Conference on Future Information Technology and Management Engineering. IEEE Computer Society, 2009:385-388.
10. Tao L I, Fei L I, Zhang G. Study of Network Security Situation Assessment Based on NetFlow Technology[J]. Modern Electronics Technique, 2011.

Performance Evaluation of Different Grade INS Aiding GNSS Receiver Acquisition in Hot Start State

Zhou Qin†, Wen-Xiang Liu, Ying-Xue Su and Fei-Xue Wang

Satellite Navigation R&D Center, National Univ. of Defense Technology, Changsha, 410073, China

†E-mail: qinzhouhit@163.com

www.nudt.edu.cn

The uncertainty range of Doppler shift when GNSS receivers restart causes slow acquisition, in this paper we present an acquisition scheme with INS assistance to narrow down the Doppler search range. Doppler shift caused by user's movement along LOS (Line of Sight) can be calculated based on the velocity provided by INS. Details of normal acquisition under different receiver dynamic in hot start are expounded firstly. Then inertial sensors error models are described and different grade INS velocity outputs are given. Simulations and comparisons are made to show the performance improvement of acquisition time in hot start with and without INS assistance. It is concluded that the acquisition time decreases with the INS assistance and performance improves more with higher grade INS. In conclusion, it is unnecessary to use INS to assist acquisition for low dynamic receiver during the hot start interval. For middle dynamic receivers, the consumer grade INS can offer effective velocity value for about 195 seconds, the automotive grade can provide available velocity for about 7 minutes, the time for tactical grade is about 77 minutes, and the navigation grade INS is serviceable during the whole hot start period. For the high dynamic users, the consumer INS can be useful for 10 minutes, the automotive INS can be effective for 17 minutes, and the tactical and navigation INS are always effective. The acquisition time can reduce 98.95% when velocity error is 5 m/s under high dynamic, 89.14% even when the velocity error is up to 500m/s.

Keywords: Acquisition; Hot-Start; Receiver; GNSS (Global Navigation Satellite System); INS (Inertial Navigation System).

1. Introduction

With the extensive use of (GPS) Global Positioning System, more satellite navigation systems are developed including GLOSSNASS, GALILEO, BDS and other regional augmentation systems. GNSS (Global Navigation Satellite System) receivers are widely applied to different users and situations. However, weak GNSS signal and its vulnerability to interference restrict the use of simple GNSS receiver. Multiple methods have been developed such as HSGPS (High Sensitivity Global Positioning System), AGPS (Assisted Global Positioning System) and INS/GNSS integrated navigation in order to meet requirements of different users[1].

Among various ways to reinforce the availability of GNSS, the INS/GNSS integration navigation is becoming popular due to its anti-interference ability and high accuracy, so most GNSS receivers are equipped with INS nowadays. Using INS to aid GNSS signal acquisition and tracking can enhance the ability of receivers' performance in different application circumstances[2][3]. With different integration algorithms and different grade INS, the integrated navigation system can be widely used in different occasions including daily pedestrian navigation, precision agriculture, aircraft precision approach and landing, missile guidance and so on.

Differing from previous research about integrated navigation system, in this study we focus on different grade INS in aiding GNSS signal acquisition of hot start under different receiver dynamic in order to testify the availability of different grade INS, providing guidance for actual applications. The uncertain Doppler frequency range determined by almanac is still broad, that's why we utilize INS to provide velocity of user to narrow down the range. Performance of normal GNSS receiver and INS–aided GNSS receiver in signal acquisition is quantitatively analyzed by calculating the mean acquisition time in this study.

2. Principles of INS-aided GNSS signal acquisition

The GNSS signal acquisition process toward specific satellite is accomplished by the two-dimensional search of carrier frequency and code phase, the acquisition is fulfilled once the signal is obtained and confirmed. The two-dimensional search area is characterizing the uncertainties of carrier frequency and code phase. Frequency uncertainties are caused by unknown Doppler frequency shift and clock errors of users and satellites, while time uncertainties are dictated by the indicated time and user position.

A GNSS signal can be tracked when the locally generated code slip and Doppler frequency are less than T_{bin} and f_{bin}, where T_{bin} is the code slip and f_{bin} is the frequency bin. Therefore cells number of search area (N_c) is determined by[4]:

$$N_c = \left(\frac{2f_u}{f_{bin}} + 1 \right) \frac{2L_u}{T_{bin}} \tag{1}$$

Where f_u is the frequency uncertain range, L_c represents the code chip uncertain range.

The INS-aide signal acquisition is using INS to predict user Doppler ahead of time while the almanac provide the Doppler caused by the satellite dynamics and satellite oscillator error. Hence the Doppler shift deriving from relative movement of user and satellite can be compensated and the Doppler search range is narrowed. The accuracy of Doppler prediction from INS is mainly depended on the accuracy of information offered by INS and the availability of almanac. The almanac depends on the start state of receiver. The accuracy of INS is determined by the grade of INS, the higher grade INS will provide better assistance information such as more precise user velocity.

3. Analysis of general GNSS signal acquisition in a hot start

The hot start is usually defined as a start state which is less than 2 hours from last shutdown, so there are available ephemeris which provide latest satellite information, almanac which provides fewer items compared to ephemeris, receiver's accurate preliminary position and time. In this case, the error of receiver's clock and satellite's clock are already known, the receiver searches for only one code bin, the search area cells are mainly decided by the uncertainty of Doppler shift caused by the receiver[1]. The maximum Doppler shift caused by receiver and satellite dynamic is discussed which is vital to acquisition and tracking of GNSS signal.

We take MEO satellite of BDS as an example to analyze Doppler shift, other GNSS satellites can be studied in the same way with a few changes. For MEO satellites of BDS, the height above the earth is around 21528 km and the average radius of the earth is 6368 km, so the radius of the satellite orbit is 27896km, f_{B1} is 1561.098MHz. The maximum Doppler shift is about $\pm$4.5kHz for static user[4]. However, the user is not always stationary, movement of user should be taken into consideration especially when the speed is high. Analysis of Doppler shift is given according to different user dynamic which is showed in Table 1.

Table 1. Characteristics under Different User Dynamic.

	Low Dynamic	Middle Dynamic	High Dynamic
Velocity(m/s)	0-100	100-1000	>1000
Acceleration(g)	<1	1-4	>4

The Doppler change caused by user velocity can be calculated by:

$$f_d = \frac{v}{c} f_{B1} \tag{2}$$

where c is the speed of light and v is velocity vector along LOS (Line of Sight). It's easy to obtain that $f_{B1}/c \approx 5.2037$. Besides, the Doppler frequency change rate due to user acceleration can be calculated by:

$$\delta f_{dr} = \frac{dv_d}{dt} \frac{f_{B1}}{c} \tag{3}$$

The Doppler change rate is about 51.5Hz/s when the user moves towards a satellite with 1 g acceleration. So the Doppler change and change rate of Doppler frequency can be obtained according to discussions above:

Table 2. Doppler range and change rate under Different User Dynamic.

	Low Dynamic	Middle Dynamic	High Dynamic
Doppler Change f_{dr} (Hz)	0-520	520-5200	>5200
Doppler Change Rate δf_{dr} (Hz/s)	<51.5	51.5-206	>206

If the Doppler range f_{dr} is less than 500Hz which is typically set as the value of Doppler search bin, then it is no need to search the Doppler frequency. It can happen when the user speed is below 96.09m/s under low dynamic according to Table 2.

4. Performance assessment of different grade INS

Normally the inertial navigation system is comprised of gyros and accelerometers. Gyros measure the rotation rate or incremental rotation based on the principles of gyros. Accelerometers measure specific force.

The outputs of sensors can be described in a uniform model. It is decided by[5]:

$$s_m = (1 + s_f) s_t + b(t) \tag{4}$$

In this equation, s_m represents the measured output of sensors, s_t is the true value of what's measured. s_f here is a scale factor indicates corruption of sensors, $b(t)$ is the bias consisting of three parts:

$$b(t) = b_0 + b_1(t) + b_w(t) \tag{5}$$

Where b_0 is a constant null-shift referred to as the "null-shift" or "turn-on to turn-off" bias variation. The $b_1(t)$ is a time varying part which will be explained later. The term $b_w(t)$ is the output noise which can be modeled as a band-limited white noise.

Obviously, the main challenge in building the sensors' error model is determining term $b_1(t)$. It is proved that $b_1(t)$ can be modeled as [5]:

$$\dot{b}_1(t) = -\frac{1}{\tau} b_1(t) + \omega_{b1} \tag{6}$$

It is a first-order Gauss Markov process, τ represents the correlation time and ω_{b1} is the driving process noise.

It is vindicated that the vibrating structure rate gyros and most accelerometers can be modeled as noted above. In this paper we use this model to simulate the outputs of inertial sensors. Many different kinds of accelerometers and gyros can be used in INS, they are usually classified into four grades according to their stability performance: navigation, tactical, automotive and consumer. A table versus quality of different grades of gyros and accelerometers is constructed as follows[5–7].

Table 3. Characteristics of Different Grade INS.

Grade		Navigation	Tactical	Automotive	Consumer
Gyro	Bias $(\deg/hr)$	0.0075	5	180	360
	Scale Factor (ppm)	27.5	350	1000	2000
	Noise $(\mu g/\sqrt{Hz})$	0.0035	0.125	3	60
	Time (sec)	3600	100	300	300
Accel	Bias $(\deg/hr)$	7.5	350	1200	2400
	Scale Factor (ppm)	15	700	2000	4000
	Noise $(\mu g/\sqrt{Hz})$	7.5	300	1000	3000
	Time (sec)	3600	60	100	100

After the velocity resolving of static experiment, the errors of total velocity estimation of different grade inertial navigation systems in ENU frame are show as follows:

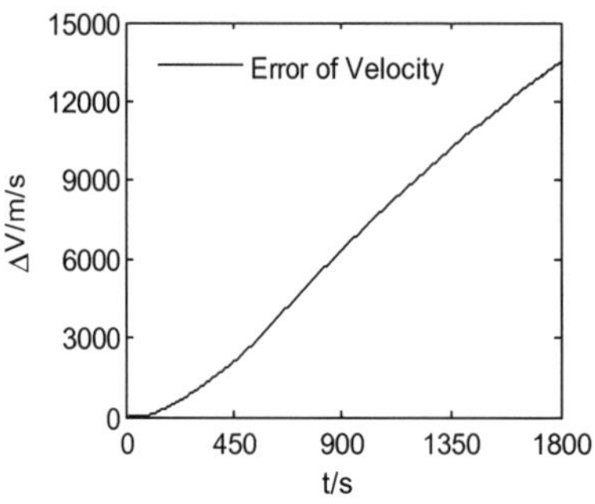

Fig. 1. Velocity Error of Consumer Grade INS.

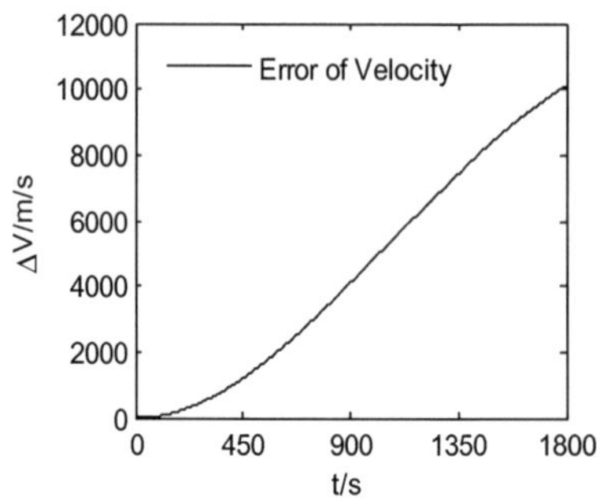

Fig. 2. Velocity Error of Automotive Grade INS

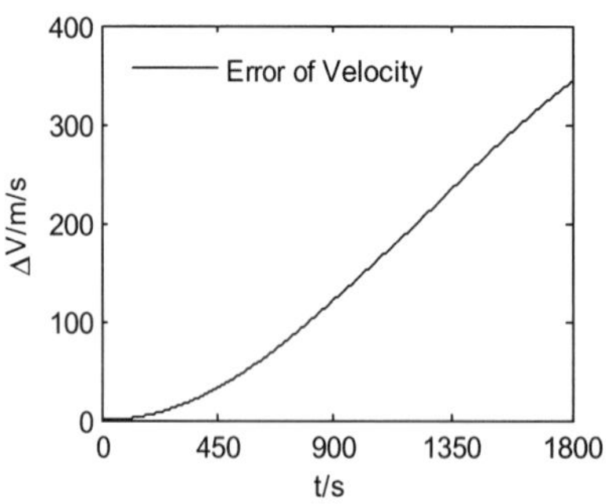

Fig. 3. Velocity Error of Tactical Grade INS

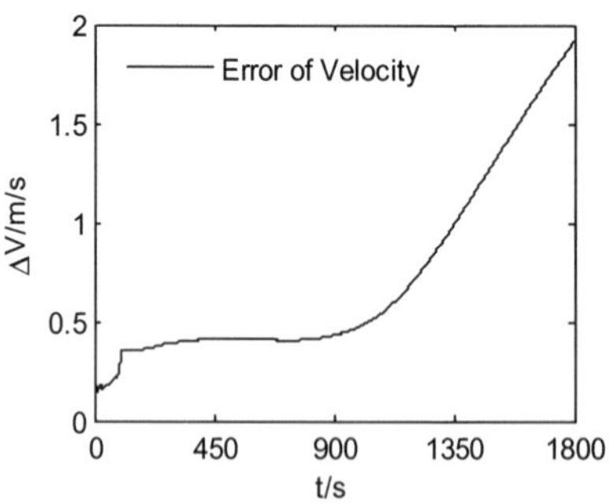

Fig. 4. Velocity Error of Navigation Grade INS

5. Performance of INS-aided GNSS signal acquisition in hot start

Acquisition time is crucial to assess the acquisition capability of receiver, in this paper we use serial search to assess the average signal acquisition time whose mathematical form is showed below[8]:

$$T_{acq} = \left(\frac{2-P_D}{2P_D}\right)\left(k_p P_{FA} + 1\right)\tau_D N_c \tag{7}$$

P_D represents the detection probability while P_{FA} is the false alarm probability. k_p is a penalty coefficient defined as $k_p = \tau_v / \tau_D$, where τ_v is the additional time compared to dwell time(τ_D) because of false alarm. The particular forms of P_D and P_{FA} are introduced in [9][10].

The Doppler frequency search range is different under different user dynamic when there is no INS assistance in hot start, and it narrows down to the value of INS velocity estimation error with INS assistance which varies according to different grade INS. The maximum user speed and corresponding Doppler frequency shift are given in Table 4:

Table 4. Maximum User Velocity and Corresponding Doppler frequency.

	Low Dynamic	Middle Dynamic	High Dynamic
Maximum Velocity v (m/s)	100	1000	5000*
Maximum Doppler Change f_{dr} (Hz/s)	520	5200	26000

*The maximum velocity under high dynamic is set 5000 for demonstration of the conclusion.

It indicates that if the INS velocity estimation error is less than the maximum velocity of different dynamic then the INS assistance is usable and it's not usable when velocity error is bigger than this value.

According to Fig. 1, Fig. 2, Fig. 3 and Fig. 4, we can get the corresponding time when velocity error exceeds 100m/s, 1000m/s and 5000m/s. The results are showed in Table 5:

Table 5. Time of Exceeding Maximum Speed for Different Grade INS.

	100m/s	1000m/s	5000m/s
Consumer	56s	195s	632s
Automotive	134s	417s	1024s
Tactical	838s	4611s	-*
Navigation	-*	-*	-*

*It means the INS is effective during the whole hot start period.

In order to quantitatively assess the improvement brought by the INS assistance, the mean acquisition time defined by Eq.8 is calculated based on the parameters given in Table 6:

Table 6. Search Parameters in Hot Start.

Receiver Type	$f_{dr}\,/\,Hz$			$f_{bin}\,/\,Hz$	P_{FA}
	Low	Middle	High		
Simple GNSS	520	5200	26000	500	0.1
INS-Aided GNSS	5.2* v	5.2* v	5.2* v	500	0.1

Now we take v as 5m/s, 50m/s and 500m/s as examples to illustrate the performance improvement compared to simple GNSS.

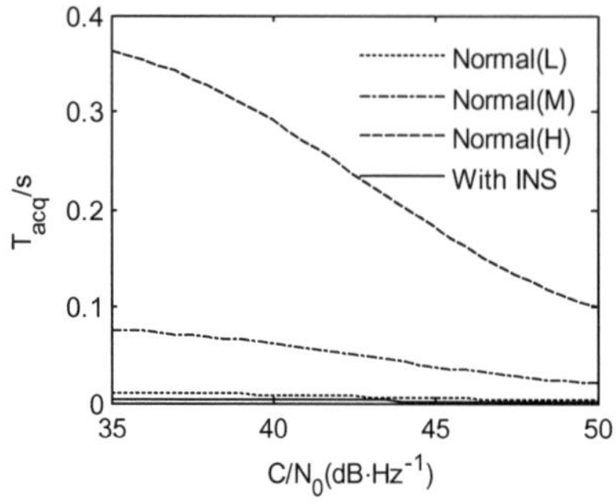

Fig. 5. Mean Acquisition Time (5m/s)

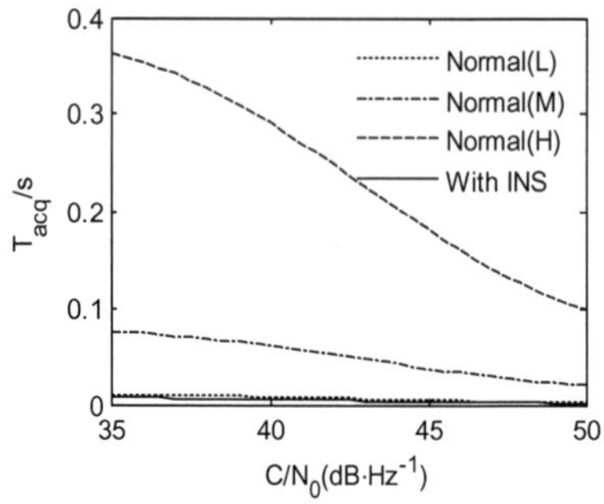

Fig. 6. Mean Acquisition Time (50m/s)

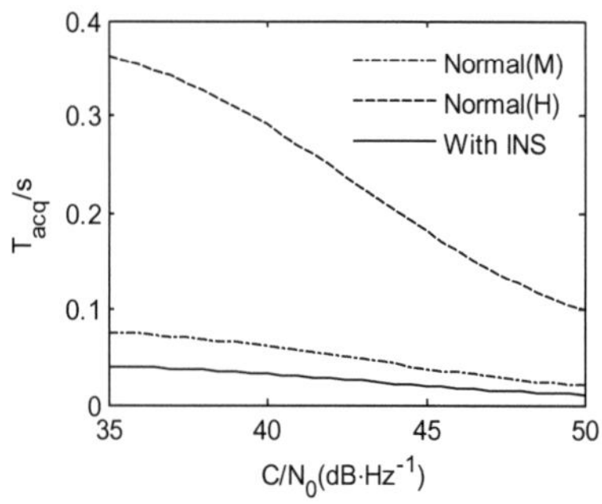

Fig. 7. Mean Acquisition Time (500m/s)

As we can see in the figures above, the mean acquisition time decreases with the assistance of INS compared to the simple GNSS receiver. With the increasing CNR, the mean acquisition time decreases as we already know. In order to show the specific improvement brought by the assistance of INS, we made Table 7 to show the percentage of improvement comparing normal GNSS receiver with INS-aided GNSS receiver. Clearly, the improvement is maximized when the velocity error is 5m/s which indicates an INS with high accuracy, the improvement percentage is decreasing with bigger velocity error or lower INS grade. It is even unable to assist GNSS signal acquisition when the velocity error is big enough.

Table 7. Improvement of Acquisition Time $\left(P_{FA} = 0.1, T = 1ms\right)$.

Speed Error/m/s	5			50			500		
Dynamic	Low	Middle	High	Low	Middle	High	Low	Middle	High
Mean/%	64.16	94.94	98.95	33.76	90.64	98.06	-*	47.71	89.14

* It means the INS can't assist GNSS signal acquisition.

6. Conclusion

This paper focus on the availability of different grade INS aiding GNSS signal acquisition under different receiver dynamic. The accuracy of Doppler prediction is mainly determined by the grades of INS sensors, so we analyze the velocity outputs of different grade INS and the availability of them during the signal acquisition. We draw the conclusion that there is almost no need to use INS to assist acquisition for low dynamic receivers. For middle dynamic receivers, consumer grade INS can assist acquisition for about 195 seconds, automotive grade INS can last for about 7 minutes, tactical grade INS can be useful for 77 minutes, and the navigation grade INS is effective during the whole hot start period. For high dynamic users, the consumer INS can be used for 10 minutes, the automotive INS can be available for 17 minutes, and the tactical and navigation INS are always effective. Besides, the performance improvement of GNSS signal acquisition is quantitatively analyzed by mean acquisition time, it shows that the acquisition time decreases about 64% when the velocity estimation error is 5m/s under low dynamic, the reduction is around 95% under middle dynamic and 99% under high dynamic. The time of acquisition cuts down 89% even when the velocity estimation error is up to 500m/s. Therefore INS assistance can truly improve the acquisition ability especially the reacquisition ability after a short interruption. With INS such as MEMS INS becoming more cost-effective and rapid developing of BDS, INS will be broadly implemented in BDS receiver. The analysis of INS-aided GNSS signal acquisition of cold start and warm start will be discussed in my other papers.

References

1. Tsui J B-Y, *Fundamentals of Global Positioning System Receivers: A Software Approach* (New York: John Wiley & Sons, 2005).
2. J. D. Gautier, GPS/INS generalized evaluation tool (GIGET) for the design and testing of integrated navigation systems, PhD thesis, Stanford University (2003).
3. G. J. Gao, INS-assisted high sensitivity GPS receivers for degraded signal navigation, PhD thesis, University of Calgary (2007).
4. X Gang. *Principles of GPS and Receiver Design* (Beijing: Electronic Industry Press, 2009).
5. D. Gebre-Egziabher, R. C. Hayward, and J. D. Powell, A low-cost GPS/inertial attitude heading reference system (AHRS) for general aviation

applications. In *Proceedings of the IEEE Position Location and Navigation Symposium*, PLANS 1998, pages 518-525. IEEE, 1998.

6. Eun-Hwan Shin. Accuracy Improvement of low cost INS/GPS for land applications, Master's Thesis, University of Calgary (2001).
7. S. Alban. Design and performance of a robust GPS/INS attitude system for automobile applications, PhD thesis, Stanford University (2004).
8. J. B. Lozow. Analysis of direct P(Y)-code acquisition, The Charles Stark Draper Laboratory, March 1997.
9. W. Zhuang, Modeling and analysis for the GPS pseudo-range observable, in *IEEE Transactions on Aerospace and Electronic Systems*, Vol. 31, No. 2 (April 1995).
10. Kaplan E D, Hegarty C J. *Understanding GPS: Principles and Applications*, 2nd edn. (London: Artech House, 2006).

Comprehensive Evaluation Index Model Construction and Application in the Transnational Power Transmission Project

Huang Qing*

State Grid Energy Research Institute,

Beijing, China

†*E-mail: huangqing@sgeri.sgcc.com.cn*

With the features of comprehensiveness, complexity and continuity, the transnational power transmission project is likely to be affected by the international politics, diploma, energy strategy, environmental protection and other factors, therefore, it is difficult to make the independent evaluation for the whole efficiency from a point of quantitative view, nor simply replace the comprehensive evaluation by the project economic profits, otherwise, reached evaluation conclusion is easy to be overgeneralization. This paper will establish the general efficiency evaluation index model about transnational power transmission project to measure the general efficiency via establishing the general efficiency evaluation index model, researching the progress of latest general evaluation model method from the viewpoint of general efficiency connotation of the international power trade project, after that, the general evaluation value of the project can be directly reflected, finally, the paper outlooks the model application value in the future.

Keywords: Transnational Power Transmission Project; Comprehensive evaluation; Energy strategy; Environmental protection.

1. Comprehensive evaluation theory and method

1.1. *Comprehensive evaluation theory*

Along with the increasingly large-scale, digitized, intelligent and integrated evaluation system in reality, it is difficult for the commonly-used comprehensive evaluation methods, such as, linear planning method, expert evaluation method, comprehensive index method, comprehensive index method, hierarchy analysis method and so on, to adapt to the multiple-layer and multiple-factor problems involved in the complicated system evaluation system. At present, Fuzzy Comprehensive Evaluation, Gray Comprehensive Evaluation and Combination

Evaluation have developed into the new effective methods for solving the actual evaluation[1–3].

Based on the fuzzy mathematics and aiming at the "unique solution" disadvantage in the traditional mathematics, Fuzzy Comprehensive Evaluation can make effective evaluation for "both the one and the other" of the objective things difference during the intermediary transition. Therein, the intermediary transition refers to the phenomenon when one degree is upgraded to another degree after passing through a continuous transient process (from quantitative variation to qualitative variation) without clear boundary. In such case, the fuzzy mathematics theory's advantages are reflected in solving the "uncertain recognition" problem with features of "definite connotation with unclear epitaxy" [4–6].

Based on the gray system theory, Gray Comprehensive Evaluation includes gray relevancy evaluation method, gray clustering analysis method and so on. With the feature of "less data modeling", the gray system theory takes the 'small sample", "insufficient information" and other uncertainty system with features of "known partial information and unknown partial information" as the research subject to extract the valuable information and realize the correct description and effective monitoring for the system operation behaviors and evolution rules via generating and developing the "partial" known information[7][8].

Combination Evaluation is a kind of technology which integrates two kings or more comprehensive evaluation results (or evaluation technology). Comparing with the single evaluation method, it is more rational and scientific because it can overcome the defects of different evaluation methods and absorb its advantages, as a result, which is able to make more comprehensive and rational evaluation for the evaluation subject. Besides, its main contents include the systematic research for the weighting combination, method combination and result combination in the evaluation process. In such case, the paper will build models and calculate comprehensive benefit by utilizing hierarchy analysis and fuzzy evaluation method[9].

1.2. *Comprehensive benefit connotation of transnational power transmission project*

Political efficiency

This paper mainly evaluates the project political efficiency from four perspectives:

I. Reflect the project political efficiency for the state safety;

II. Reflect the project efficiency for constructing the comprehensive strategy cooperation partner diploma relations;

III. Reflect the project efficiency for the energy diploma strategy improvement and expansion;

IV. Reflect the project political efficiency for enhancing the international influence and state prestige.

Energy efficiency

The project energy efficiency shall be evaluated from the following two aspects:

I. Energy safety efficiency, make evaluation from the energy safety diversity index perspective. The energy category and source diversity are considered as the important means for coping with energy supply risks. Meanwhile, the supplier diversity on the energy market has also developed into the effective means for coping with energy market risks. What's more, the diversity has been considered as the important basis for the energy supply departments of numerous countries to guarantee the energy supply and safe investment decisions.

II. Energy structure optimizing efficiency. Electricity import from other countries can replace the domestic thermal power generation for the accepting country to a certain degree, and generate the thermal power replacement efficiency.

Economic efficiency

Reflect the direct economic efficiency which is obtained in the economic efficiency generation, and economic drive efficiency. As an enterprise, the main purpose about operation activities lies in providing the commodity or service which can satisfy the social demands and creating more economic values to the society. Therefore, the favorable economic efficiency is considered as the basis and source for project investment and cooperation as well as sustainable development.

Take the economic efficiency into account from two aspects: I. Direct economic benefit, making the direct economic efficiency evaluation for the transnational power transmission project; II. Indirect economic benefit, making calculation and evaluation for the benefits when the transnational power transmission project drives the national electric equipment export and overseas project contracting business.

Social environment benefit

The environment costs for electricity generation are composed of resources consumption costs and pollution treatment costs. Therein, reflected by the resource tax, the resource consumption costs refer to the long-term adverse influence due to energy source consumption, including water resource consumption costs and coal resource consumption costs; the pollution treatment costs refer to the costs which are paid for treating the pollutants discharged in the power generation process, including the local treatment costs, regional treatment costs and global treatment costs.

Based on it, the project social environment efficiency shall include:

I. Resource consumption efficiency, making resource consumption cost evaluation for the transnational power transmission project; II. Pollution treatment efficiency, making pollution treatment cost evaluation for the transnational power transmission project[10].

2. Comprehensive evaluation index model about transnational power transmission project

2.1. *Design though for comprehensive evaluation index model about transnational power transmission project*

Construct the summary and review about research achievements based on the related comprehensive evaluation mode, analyze the connotation and compositions of the comprehensive evaluation about international electricity trade project, then, research and design the comprehensive evaluation index model for the international electricity trade project, followed by carrying out actually-measured research for certain power transmission project by applying this mode. The basic design thought is show in the following figure.

2.2. *Comprehensive evaluation index model construction in the transnational power transmission project*

(1) Model framework

Based on the foregoing analysis, it is necessary to establish a transnational power transmission comprehensive evaluation index model with four first-level dimension, ten second-level dimension and 16 special indexes. The basic framework of the comprehensive evaluation index model in the transnational power transmission project is shown in the following figure.

(2) Model index system and its weight

In order to avoid the weight setup subjectivity, the paper applies the hierarchy analysis method to serve as the weight calculation method, and invite the company experts to calculate the index weights on the basis of making pair-wise comparison and balance for every index. In addition, the original data will pass through the dimensionless treatment, as a result, transfer all data into centesimal system and enhance the comparability of different indexes and data.

Index system design. According to the theory analysis, the comprehensive evaluation index model framework in the transnational power transmission project as specified in the foregoing figure is refined.

Index weight calculation. In order to enhance the scientificity and objectivity of the weight calculation, analytic hierarchy process (AHP) is applied to calculate the weight. Besides, 20 experts related to international electricity trade, management and scientific research and 10 foreign experts are invited to make pair-wise comparison for the index importance, followed by utilizing yaahp6.0 software to calculate the index judgment and weight of every expert, after that, the final weight will be obtained by exerting arithmetic means for the judgment and weight results of the foregoing 30 experts. As for the weight distribution of the first-class dimensional “political efficiency”, “energy efficiency”, “economic efficiency” and “social and environmental efficiency” , the foregoing 30 experts will also be invited to make judgment. Finally, they are unanimous that the importance of the foregoing four efficiency are basically equivalent to each other, therefore, it is unnecessary to apply AHP method to determine that the weight is 0.25.

The comprehensive evaluation index system and its weight in the transnational power transmission project which is established by applying the foregoing method is shown as follows.

Table 1 Comprehensive evaluation index system in transnational power transmission project

	First-class evaluation dimension	**Second-class evaluation dimension**	**Special evaluation index**
omprehensive aluation index in the transnational power transmission project	Political benefit	National safety	1. National safety index
		Diploma strategy	2. Comprehensive strategic cooperation diploma relation index
		Energy strategy	3. Energy strategic improvement index
		International influence	4. International influence index
	Energy benefit	Energy safety	5. Energy safety diversity index
		Energy structure optimization	6. Thermal electricity installed capacity percentage replaced by imported electricity
	Economic benefit	Direct economic benefit	7. Differential between landing electricity price and benchmark electricity price
		Indirect Economic benefit	8. Equipment export drive efficiency
			9. Project contracting drive efficiency
	Social and environmental benefit	Resources consumption	10. Water resources consumption costs
			11. Coal resources consumption cost
		Pollution treatment	12. Wastewater treatment cost
			13. Carbon dioxide treatment cost
			14. Sulfur dioxide treatment cost
			15. Oxynitride treatment cost
			16. Solid particles PM treatment cost

3. Comprehensive benefit empirical evaluation in certain transnational power transmission project

3.1. *Practical measurement result*

Please measure and calculate the comprehensive benefit indexes of certain transnational power transmission project and domestic power generation respectively. The actually-measured results which are obtained by applying the foregoing evaluation model to measure the project comprehensive evaluation index are shown in the following figure:

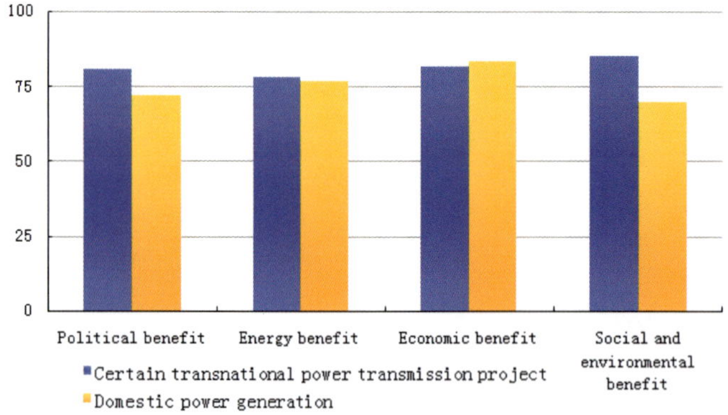

Fig1 Comprehensive evaluation result in certain power transmission project

As shown in the foregoing figure, comparing with the domestic power generation scheme, the certain transnational power transmission project is superior to the domestic power generation in the aspects of political efficiency, energy efficiency as well as social and environmental efficiency, however, which is only inferior to the domestic power generation in the economic efficiency. In the general index view, the comprehensive benefit index (including social and economic efficiency) of certain transnational power transmission project is higher than the index of domestic power generation scheme with superior social and economic efficiency, therefore, it is feasible.

3.2. *Second-class evaluation dimension scores*

In order to intuitionally reflect the comparison conditions between certain transnational power transmission project and the domestic power plant in terms of the comprehensive benefit indexes, radar chart is applied to present the second-class dimension changing conditions of the two schemes, as follows:

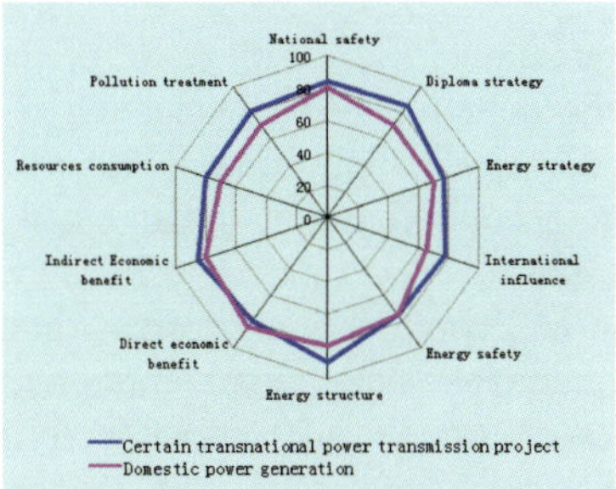

Fig 2. Second-class evaluation in transnational power transmission project

The actually-measured result indicates that the transnational power transmission project has the higher comprehensive benefit index (social and economic efficiency) than the domestic power transmission. In terms of the separate index (refer to the radar chart), other indexes of the transnational power transmission project are close to or slightly higher than the indexes of the domestic power generation except for the direct economic efficiency.

3.3. *Research analysis conclusion*

The comprehensive benefit of the international electricity trade project is mainly reflected in political efficiency, economic efficiency, energy efficiency and social efficiency, therein, the political efficiency is considered as the centralized reflection for economic, energy and social environment efficiency, and the highest reflection at country level for the other three efficiency because it looks at the strategic space between two countries cooperation and development at the present stage and the win-win cooperation potentials; as the basis for the political efficiency, the economic, energy and social environment efficiency can not only reflect the specific efficiency of an international cooperation project in every respect, but also lay foundations and expand space for the political efficiency acquisition in future.

According to the actually-measured results of economic and social comprehensive benefit index in the transnational power transmission project, it is indicated that the certain transnational power transmission project is equipped with more advantages in the comprehensive benefit (including social and economic benefit) than the domestic power generation.

4. Comprehensive benefit index model value expectation in the transnational power transmission project

The comprehensive evaluation index in the transnational power transmission project is established to provide the comprehensive and scientific analysis tools to carry out quantitative evaluation for the comprehensive performance of the international electricity trade project, and offer important application values to the comprehensive evaluation for economic and social efficiency in the international electricity trade project because it not only takes economic efficiency into account, but also thinks about the political efficiency, energy efficiency and special environment efficiency, what's more, it has reflected the unification and coordination between short-term target and long-term target, economic target and social target, enterprise target and government target, as well as microscopic target and macroscopic target, in such case, the actual strategic value and social significance of the international electricity trade project can be comprehensively evaluated. Besides, we can also accurately grasp the operation values and strategic contributions of various kinds of electricity trade projects which are encountered by the domestic electricity enterprises, find out problems in time and promote the continuous and healthy operation and development of the international electricity trade project via the comparing and analyzing the economic and social comprehensive evaluation indexes of different schemes. What's more, it also provides reference values to evaluate the domestic energy enterprises and balance the international energy trade project.

References

1. Jacques Savoy. Statistical inference in retrieval effectiveness evaluation[J]. Information Processing and Management, 2007, 33(4) :495-512.
2. Wall Mark J Torkel. Quality of research measured by citation method and by peer review) a comparison[J] . IEEE Trans on Engineering Management, 2013, 19(4) : 218- 222.
3. Saaty T L. Fundamentals of Decision Making and Priority Theory with the Analytic Hierarchy Process[M] . Princeton: RWS Publications, 2014. 121-127.
4. Schen K S. Avoiding rank reversal in AHP decision support models[J] . European Journal of Operational Research, 2009, 74:4607-4619.
5. Carrettoni F, Castano S, et al . RETISS: A real time security system for threat detection using fuzzy logic[A]. 2011. 247-269.
6. Chen S J, Hwang C L. Fuzzy Multiple Attribute Decision Making[M] . Berlin: Springer Press, 2015. 163-287.
7. Dimitras A I, Slowinski R, Susmaga R, et al. Business failure prediction using rough sets[J] . European Journal of Operational Research, 1999, 95: 24- 37.
8. Levine P, Pomerol J C. PRIAM, an interactive program for choosing among multiple attribute alternatives[J] . European Journal of Operations Research, 1986, 25(2) : 272 -280.
9. Dagli C, Schierholt K. Stock market prediction using different neural network classification architectures[A]. 1996. 72-78.
10. U.S Energy Information Administration. Russia analysis briefs: 2015.

The Construction and Use Of Implicit Relationship In Knowledge

Jing Chen, Yun-Qing Li, Hu Kuang

College of Computer Information and Engineering, Jiangxi Normal University, Nan Chang, Jiang Xi,330022, China

†*E-mail: chenxijingjing@163.com，934109337@qq.com*

www.jxnu.edu.cn

This paper research on construct an ontology to support the knowledge framework of "Network Operating System", add implicit knowledge point relationship and use this ontology to implement a learning assistant system. We use ICTCLAS Chinese word segmentation to help domain-specific term extraction. Seven-step method is used to build course domain ontology, then add properties and characteristics to ontology, so computer can get and analyze the implicit relationships in knowledge. Then, we construct a learning assistant system based on the ontology we build, deal with the message user submit and parse out knowledge points relevant to it. Gain leading and follow-up knowledge point lists through ontology search. Set mapping file between course domain ontology and course sources in linked data way, helping achieve the convert from knowledge points to course sources, and helpful information will be displayed in system interface.

Keywords: Ontology; Semantic Web; Word Segmentation; Implicit Relationship; Learning Assistant.

1. Introduction

With the development of Computer Sciences and Internet, our social is in huge revolution. It is no exception to education, a new teaching model named "Flipped Classroom" raise the concern of scholars and experts. Different from traditional classroom, "Flipped classroom" method require learner learn what teachers teach on traditional classroom before class, and do interaction study with teachers guidance on class.

While practice "Flipped Classroom" [1], scholars found how to get useful content from the Internet is very important for learners. Compared with the traditional classroom, the learning process of "Flipped Classroom" before class lack of teacher's guidance. Especially unfamiliar content, learners have to judge whether it contains knowledge points, or which part should have learned before this part. This consumes unnecessary time, energy and enthusiasm.

At present, the courses in some online course systems categorized by system definition, such as grade or subject, the content in course divided into chapter only. In fact, the knowledge points in each chapter are not standalone, there are implicit relationships between them. For example, before learning a knowledge point, learners should have learned some knowledge points before, we call this leading and follow-up relationships between knowledge points. Make use of this implicit relationship to structure an learning assistant system, is an effective way to reduce unnecessary work and improve learning efficiency.

Semantic web ontology is the technology to structure a course domain ontology that contain implicit relations between knowledge points, enable computer to understand and use.

2. Semantic Web and Ontology Theory

Ontology originated from philosophy, is “a system description of the objective existence” [2]. In computer science field, the concept “an ontology is a formal, explicit specification of a shared conceptualization” [3] is most widely quoted, illustrates the ontology feature of participatory, conceptualization, clarity, formalization. The Web presentation languages are SHOE, XOL, RDF(S), OIL, OWL and so on[4].

Most ontology construction method are based on application requirements, such as IDEF5 method, SKELETON method, TOVE method, METHONLOGY method, KACTUS engineering method, SENSUS method, Seven-Step method. Through comparison and analysis, the development process of Seven-Step method and METH-ONLOGY method are relatively complete, the ontology developed is more mature[5].

Semantic Web is the semantic extension by adding semantic that computer and people can both understand, and provide a generic framework for shared data between applications, enterprise and community[6]. Founding father of Internet, Tim Berners-Lee put forward Semantic Web hierarchy model according to semantic web design principles, as shown in Fig.1[7].

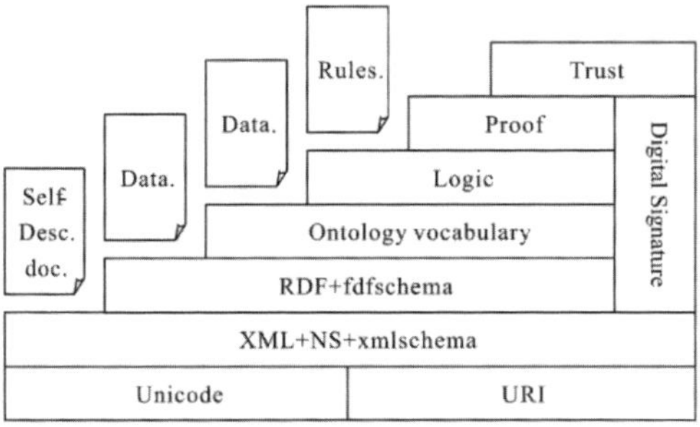

Fig 1. Semantic Web hierarchy model

As the core of the semantic web, ontology support with semantic web: domain ontology is the underlying data model to support computer interpretation of web data, web service description ontology use for describe semantic information of web service[8].

As a branch of semantic web technology, the research about semantic web of education domain is on the rise.

3. Structure Domain Ontology

We take "network operating system" course for example, use Protégé 4.3 and ICTCLAS as tools to build a network operating system course domain ontology partial automatic. The ontology built should have features of clarity, objectivity, completeness, consistency, maximum monotone extensibility and minimum commitment[9], ensure the ontology is reusable, no redundant, ease of expanding and merging. Build steps are as follow.

(1) Determine the category of ontology.

The concepts of course domain ontology comes from course knowledge points, analysis the similarities and differences between courses to determine the scope of knowledge points. Compared to other course content, the common knowledge area is operating system overview and basic functions of operating system, extended function of the network operating system and practice operation is different knowledge area.

The common knowledge area can be reused for similar course domain ontology, the different knowledge areas only can be reused for building other course domain ontology. In consideration of reusability and extensibility, the ontology built in this paper has four main branches are summary of operating system, basic functions of operating system, extended function of network operating system, practice of network operating system.

(2) Review existing reusable ontology.

The reusable universal ontology built by domain experts is not much at present, most of the existing ontology was produced for related system, generality is weak. Network operating system domain have no established universal ontology, the course domain ontology need to be built.

(3) List the terminology needed for ontology construction.

The terminology vocabulary is too much, get operating system terminology vocabulary by artificiality way not only require high knowledge base of worker, but also time-consuming and still much missing. Import course text to segment system and get segmentation result, deal with the segmentation result and get course vocabulary, domain experts screen out knowledge points words from course vocabulary as ontology term list.

We use ICTCLAS do the segmentation of course content text, the segmentation result of sentence "Command interface provide users the way to control computer directly or indirectly" is "Command interface/n_newword provide/v users/n the way/n to/ule control/vn computer/n directly/ad or/c indirectly/ad". Statistic with segmentation result, we found the words classified as noun exist most vocabulary of network operating system.

(4) Definition class and class hierarchy.

Class is the knowledge unit of ontology and stand for the collection of similar attribute. A higher hierarchy class is the superclass of its subclass and the subclass must have the attribute superclass have, but the subclass may have more attribute than its superclass. Define the hierarchy and the whole knowledge framework is built.

Classes derived from the ontology category determined in step one and the ontology term list result in step three, class level comes from the hierarchy of knowledge points. We take "operating system" as first level class, "summary of operating system", "basic functions of operating system", "extended function of network operating system", "practice of network operating system" as second level class, other knowledge points contained in those sorted classes, expand step by step.

(5) Define the property of class.

There are two import properties for ontology, object property and data property. We use object property to reflect leading and follow-up relationship between knowledge points. Define object properties and add to class to reflect relationship between classes of ontology. New "isPreKnowledgeOf" object property stand for leading relationship, new "isSubKnowledgeOf" object property stand for follow-up relationship, add to classes that have the kind of relationship.

Presume “user interface” is the leading knowledge point of “ interface of operating system”, add object property, we can see the broken line point to each other in protégé as shown in Fig .2.

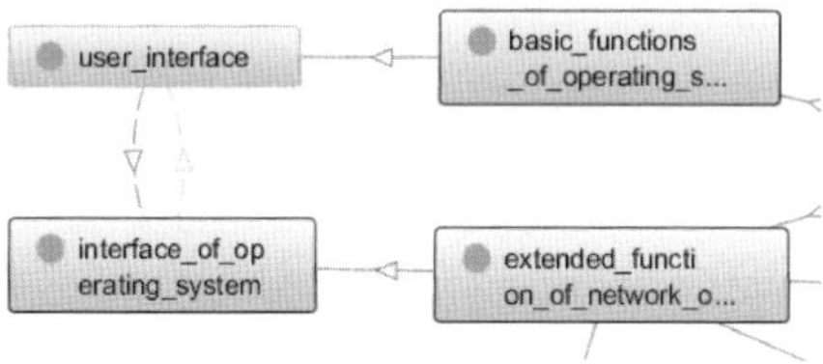

Fig 2. System structure

(6) Define facets of property.

The main characteristics of object property are functional, transitive, symmetric and reflexive.

The leading, follow-up relation between knowledge points has the feature that if class A has “isPreKnowledgeOf” property and point to class B, so class B must has “isSubKnowledgeOf” property and point to class A, so we should select “Inverse of” in description section of Protégé and add each other.

The leading, follow-up relation also have the feature if knowledge point A is leading/follow-up knowledge point of knowledge point B, meanwhile knowledge point B is leading/follow-up knowledge point of knowledge point C, then knowledge point A must is leading/follow-up knowledge point of knowledge point C, so we should select “Transitive” option in characteristics section of Protégé.

By adding this two kind of characteristic and description, we improved the logical legality of the two properties.

(7) add instances.

When knowledge point cannot be subdivided into other knowledge points, add it to ontology as instance, such as add “Windows”, “Unix”, “Linux” to “desktop operating system”.

We have construct a basic knowledge framework of “network operating system” course by define class and its hierarchy, by adding “isSubKnowledgeOf” object property and “isPreKnowledgeOf” object property, we add implicit relationship to the basic knowledge framework, the goal establish a knowledge framework with implicit relationship is achieved, save it as owl file, we will use it in learning assistant system.

4. System Design and Implementation

4.1. *Design Learning Assistant System*

The course domain ontology is established, the knowledge points with leading and follow-up relations exist as class or instance in ontology structured and hierarchy. Next, we design the system into four modules, they are user interface module, ontology search module, select handle module and course resource link module, the structure diagram as shown in Fig .3.

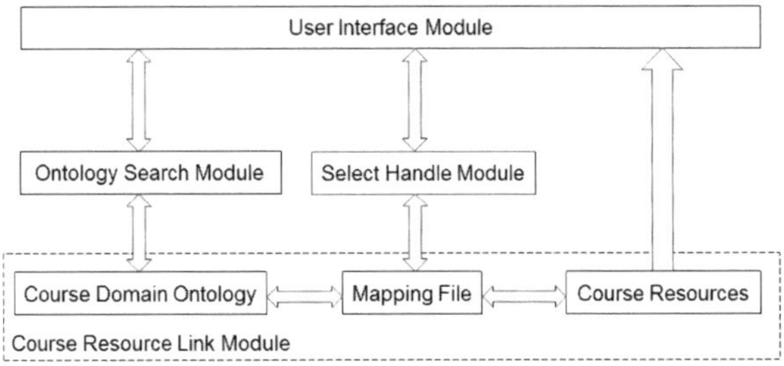

Fig 3. System structure

(1) User interface.

Show course content, relevant knowledge points list, leading and follow-up knowledge points list, chapter information to learner, operational components are provided.

(2) Select handle module.

Deal with the content learners chose. The content shown in system interface is enable to choose, while learners select some text, system receive selected text and return a knowledge point list relevant to chosen text, and the knowledge points in list will show in interface. Some other user action touch off system and return leading, follow-up knowledge points list, and the chapter of chosen knowledge point belongs to, also show in interface.

(3) Ontology search module.

Through ontology search in ontology search module, system get leading and follow-up knowledge points list, display it in user interface.

(4) Course resource link module.

The core of course resource link module is the module establish the relevant from knowledge points in ontology to course resource. A most directly way to achieve this is add data property to ontology, the value is the path link to course resource, but if there are several kinds of course resource for one knowledge point, the organization of data property may be mistiness, and it is hard to maintain the whole ontology. So we try to separate the course resource link from ontology.

A lightweight implementation form of semantic web to serve teaching and learning with linked data is to achieve storage and retrieval learning contents based on semantic linked data[10]. And one of the way to support linked data technology is to use management function of relational database system, build mapping files to map data table, row, column, value to class, property, resource, attribute value of RDF data in real time[11]. This inspire us to use a mapping file to organize the mapping from ontology to course content.

Analysis the relation between knowledge point and course content, we find it is a many to one mapping relation, because usually a unit of course content conclude at least one knowledge point, but a knowledge point usually detailed described in one unit of course content. So we can create a mapping file to storage table in database system as mapping file to achieve the link of knowledge points in ontology to course content.

Before create mapping file, we have to mark each course content into chapters. Then noted the knowledge points as keywords of each chapter, and each knowledge points must appear as keywords once but only once to ensure the mapping relationship is minimum redundancy but maximal covered. This requires the knowledge point can noted as keyword only when it is most detailed described in the chapter. For example there are knowledge points A and B, chapter N describe A and B, but A is the most detailed described, then find B is most detailed described in chapter M, the result is A is keyword of chapter N, B is keyword of chapter M but not chapter N.

In mapping file, the mapping is from knowledge point to its chapter, and each knowledge point also must appear once but only once. The form of mapping file is shown in Table 1.

Table 1. Form of mapping file

Knowledge point	Chapter	Resource path
Batch process system	Chapter 1	Path 1
Communication of net	Chapter 5	Path 2

4.2. *System implementation*

The software environment to implement system in this paper is Windows 7, JDK1.7, MySQL, development tools is Eclipse and Protégé. Code with Java, use ontology by methods in Jena library, structure a system with interactive interface.

The system interface is divided into course content area, chapter number display and change area, relevant knowledge points list area, leading and follow-up knowledge points list area, target knowledge chapter information area. Learner trigger system function by mouse click, drag and drop. System functions are as follows.

4.2.1. *Display course content*

The content of a chapter is shown in course content area, a slider is provided to drag and browse more content. Learners can click and drag the text to select, and trigger select handle module, system will generate a knowledge points list that the knowledge points in it are relevant to selected text.

4.2.2. *Display relevant knowledge points*

The relevant knowledge points list showed in relevant knowledge points list area. Every knowledge point is set as an option, while chosen, select handle module is triggered and get the chapter number of the knowledge point as keyword belongs to by course resource link module, shown in target knowledge point chapter information area. Meanwhile, ontology search module will be triggered when click on any option of relevant knowledge point, a leading and follow-up knowledge points list of the selected knowledge point will be generated.

4.2.3. *Display leading and follow-up knowledge points*

The list generated in function 2 shown in leading and follow-up knowledge area. Click on the knowledge point, and trigger the select handle module, the chapter number in target knowledge point chapter information area change into new message. Click on the chapter number and buttons around it on interface below course content area, change the content in course content area, finally show the content learners want.

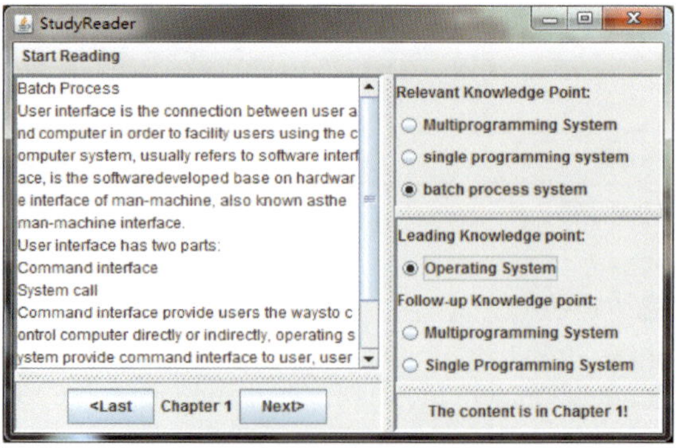

Fig 4. System Interface

As shown in Fig .4, when we select "batch", the relevant knowledge points are "multiprogramming system", "single programming system", "batch process system", they are shown in relevant knowledge points list area. Click "batch process system", the leading knowledge point is "operating system", the follow-up knowledge points are "multiprogramming system", "single programming system", click the "operating system", the message in target knowledge chapter information is "the content is in chapter 1".

5. Conclusions

In this paper, we construct a "network operating system" course domain ontology, states the way to describe leading and follow-up relationships of knowledges in ontology, the way achieve mapping file and system design, and relate course domain ontology to course resources data by setting mapping file.

References

1. Eun Man Choi, Applying Inverted Classroom to Software Engineering Education, International Journal of e-Education, e-Business, e-Management and e-Learning 3, 122(2013).
2. Klemettinen M, Finding interesting rules from large sets of discovered association rules(ACM Press, 1994).
3. T. R. Gruber, A Translation Approach to Portable Ontology Specifications, Knowledge Acquisition 5, 199(1993).
4. Du Xiao-Yong, Li Man, Wang Da-Zhi, Survey on semantic web and ontology, Computer Applications 24, 14(2004).
5. Zhang W, Zhu Q, Research on Construction Methods of Domain Ontology, Library & Information 1, 16(2011).
6. W3C, W3C Semantic Web Activity(2013), http://www.w3.org/2001/sw/.
7. Tim Berners-Lee, Semantic Web (Washington D.C., America, 2000), http://www.w3.org/2000/Talks/1206-xml2k-tbl/slide10-0.html.
8. Shi Yi-Min, Li Guan-Yu, Liu Yu, Overview of ontology in semantic web services, Computer Engineering & Design 23, 5977(2008).
9. T. R. Gruber, Towards Principles for the Design of Ontologies Used for Knowledge Sharing, International Journal of Human Computer Studies 43, 907(1995).
10. Wu Peng-Fei, Yu Sheng-Quan, New progress of apply semantic web in education: linked data perspective, e-Education research 7, 66(2015).
11. Liu Wei, Overview on linked data: concept, Technology and Implementation, Journal of academic libraries 2, 5(2011).

The Efficiency Analysis of the Construction Industry in Linyi City Based on DEA Model

Xiang Liu
School of Architecture, Linyi University,
Linyi, Shandong, China
E-mail: liuxiang0632@126.com

Zong-Ying Li
School of Logistic, Linyi University,
Linyi, Shandong, China
E-mail: zongying_li@126.com

Data envelopment analysis (DEA) is a popular tool to measure the relative efficiency of decision-making units (DMUs). In the paper, the DEA model is applied to measure the relative productive efficiency about construction industry in Linyi city, the application focuses on cross-section data of the construction industry in 12 counties of Linyi city in 2014, accounting for two inputs (labor and operating costs) and one output (income tax). The results indicate that there are great gaps in productivity development level among 12 counties in the construction industry of Linyi. On the basis of the results analysis, some suggestions are provided to improve the performance of the construction industry and promoting the sustainable development of the construction industry between different counties.

Keywords: The Construction Industry; Data Envelopment Analysis; Decision-Making Units; Constant Returns to Scale; Strong Disposable .

1. Introduction

Along with the further advance of urbanization, the construction industry plays a more and more important role in the national economy. During the past few years, the construction industry has been rapid development in Linyi city, but its development model is still in the way of high input and high output, and the disadvantage of low resource utilization is very serious. A large of productivity researches about construction industry are shown in the previous literature [1]–[7], because it is a useful tool to support the design of strategies and government policies towards the improvement of industry performance.

Therefore, it is necessary to measure the production efficiency of construction industry in Linyi city.

We organize the remainder of this paper as follows. Section 2 describes the methodology about Data Envelopment Analysis (DEA) model. The empirical study for construction industry efficiency in Linyi and the discussion of the results are presented in Section 3. The last section concludes the research.

2. Methodology

DEA model [1] was developed by Charnes et al. in 1978, which is a non-parametric, piecewise linear programming method by means of the cross-section. It can be applied to measure the productivity of decision-making units (DMUs) with multiple inputs and multiple outputs.

Assume that there are N (homogeneous) DMUs ,and each DMU produces S outputs through M inputs. Let a $M \times N$ matrix X stands for inputs and an $S \times N$ matrix Y stands for outputs. Then the efficiency score of DMU_0 can be calculated by:

$$\min \ \theta$$
$$s.t. \begin{cases} \sum_{k=1}^{N} x_{ik}\lambda_k \le \theta x_{i0}, i = 1,2,\cdots,M \\ \sum_{k=1}^{N} y_{jk}\lambda_k \ge y_{j0}, j = 1,2,\cdots,S \\ \lambda_k \ge 0,\ k = 1,2,\cdots,N \end{cases} \quad (1)$$

Where x_{ik} and y_{jk} means the ith input and the jth output of DMU_k, and λ_k refers to the weight coefficient of input or output factors of the DMU_k.

The constraint set in model (1) is the corresponding conceptual production technology, which exhibits constant returns of scale(CRS) and has strong disposable inputs and outputs, and it is described as follow:

$$T = \{(x,y): \sum_{k=1}^{N} x_{ik}z_k \le x_i, i = 1,2,\cdots,M$$
$$\sum_{k=1}^{N} y_{jk}z_k \ge y_j, j = 1,2,\cdots,S \quad (2)$$
$$z_k \ge 0,\ k = 1,2,\cdots,N \quad \}$$

Where $x=(x_1,x_2,\cdots,x_M)$ and $y=(y_1,y_2,\cdots,y_S)$ are the vectors of the inputs and the outputs, and T is named the reference technology which consists of all the feasible combinations of the inputs and the outputs.

Model (1) can be transformed with the non-Archimedean infinitesimal (ε), as follows:

$$\min\ \theta-\varepsilon\left(\hat{e}^T S^- + e^T S^+\right)$$

$$s.t.\begin{cases}\sum_{k=1}^{N} x_{ik}\lambda_k + S^- = \theta x_{i0}, i=1,2,\cdots,M \\ \sum_{k=1}^{N} y_{jk}\lambda_k - S^+ = y_{j0}, j=1,2,\cdots,S \\ \lambda_k \ge 0,\ k=1,2,\cdots,N \\ S^- \ge 0, S^+ \ge 0\end{cases} \tag{3}$$

Where S^- and S^+ are the slack variables of the input and output. And other symbols are described as above.

3. Empirical Study

3.1. *Data and DMU*

The paper took the year 2014 as the study period, and we selected 12 counties in Linyi city as DMUs to measure the productive efficiency of the construction industry in these counties. In the paper the level of discrimination is taken into account, we selected two inputs (i.e., labour and operating costs) and one output (income tax). The data used in this study came from Linyi Statistical Yearbook in 2015 [8]. And the descriptive statistics of the data is given in Table 1.

Table 1. Descriptive statistics of input and output variables.

Inputs and outputs	Variable	Unit	Mean	Max	Min	Std.Dev.
Inputs	Labour	person	7500	21352	1352	6266
	Operating costs	10^4yuan	377013.2	2405440	38314	679122.9
Outputs	Income tax	10^4yuan	6270.7	26474	576	9121.2

3.2. *Results analysis*

The radial DEA-based model (3) is applied to calculated the productive efficiency [9]-[10] of the construction industry in 12 counties of Linyi city in 2014 by Matlab software. Table 2 and Table 3 show the obtained efficiency scores, the weights of input and output variables as well as the slacks of the input variables.

The technology efficiency (TE) [1] scores of the construction industry in 12 counties are shown in the first column from Table 2. It is easy to find that only two counties (DMU5-tancheng county and DMU9-pingyi county) whose the TE scores of the construction industry is equal to one, and TE scores of the construction industry in other 10 counties are all less than one. It means that only DMU5 and DMU9 are DEA effective because their TE values are equal to one, and they are the frontier counties. Therefore, DMU5 and DMU9 can be referred to as the peers of other DMUs. On the contrary, other 10 counties are DEA ineffective, they are not the frontier counties, not only their inputs but also their outputs are inefficient. The main causes of inefficiency is that there is less labour demand and less operating costs demand in these 10 counties.

Table 2. Efficiency scores of DMUs and weights of input and output variables.

MU	θ	λ_1	λ_2	λ_3	λ_4	λ_5	λ_6	λ_7	λ_8	λ_9	λ_{10}	λ_{11}	λ_{12}
1	0.805	0.000	0.000	0.000	0.000	0.000	0.000	0.000	0.000	0.923	0.000	0.000	0.000
2	0.854	0.000	0.000	0.000	0.000	1.207	0.000	0.000	0.000	0.166	0.000	0.000	0.000
3	0.300	0.000	0.000	0.000	0.000	0.000	0.000	0.000	0.000	0.022	0.000	0.000	0.000
4	0.584	0.000	0.000	0.000	0.000	0.455	0.000	0.000	0.000	0.128	0.000	0.000	0.000
5	1.000	0.000	0.000	0.000	0.000	1.000	0.000	0.000	0.000	0.000	0.000	0.000	0.000
6	0.822	0.000	0.000	0.000	0.000	3.466	0.000	0.000	0.000	0.000	0.000	0.000	0.000
7	0.515	0.000	0.000	0.000	0.000	0.727	0.000	0.000	0.000	0.000	0.000	0.000	0.000
8	0.463	0.000	0.000	0.000	0.000	1.236	0.000	0.000	0.000	0.000	0.000	0.000	0.000
9	1.000	0.000	0.000	0.000	0.000	0.000	0.000	0.000	0.000	1.000	0.000	0.000	0.000
10	0.705	0.000	0.000	0.000	0.000	3.303	0.000	0.000	0.000	0.000	0.000	0.000	0.000
11	0.692	0.000	0.000	0.000	0.000	1.041	0.000	0.000	0.000	0.000	0.000	0.000	0.000
12	0.365	0.000	0.000	0.000	0.000	0.351	0.000	0.000	0.000	0.032	0.000	0.000	0.000

Table 3. The slacks of input variables.

DMU / Slack	1	2	3	4	5	6	7	8	9	10	11	12
s_1^-	0	0	0	0	0	449	760	309	0	443	2409	0
s_2^-	1113530	0	240.47	0	0	0	0	0	0	0	0	0

Note: The number of DMUs Vs district is:1-lanshan,2- luozhuang,3- hedong,4-yinan,5-tancheng,6-yishui,7-lanling,8-feixian,9-pingyi,10-junan,11-mengyin,12-linshu

In real application, it is very important to talk about the target of a DMU which is DEA inefficient. In the paper, we take DMU1 and DMU4 as example to unfold discussion about this question. DMU4 has TE = 0.584 and its input slacks are equal to zero, so it can possibly reduce the consumption of all the inputs by 41.6% without reducing its output. In addition to this, we also can describe the target of DMU4 by the linear combination of DMU5 and DMU9, where the weights in this linear combination are λ_5 and λ_9 in row 4 of Table 2. Different from DMU4, DMU1 has TE = 0.805 and its slack of the first input is equal to zero, but its slack of the second input is not equal to zero, in order to get its target without reducing output, we must first reduce the usage of all the inputs by 19.5%, then reduce the second input by a further s_2^- units, and the value of s_2^- lies in the first column of Table 3. We can give a similar discussion of the other inefficient DMUs about their targets.

4. Conclusion

It is well known that DEA approach is a popular tool to measure the relative efficiency of DMUs. In the paper, the DEA model is applied to measure the productive efficiency about construction industry in Linyi city, the application focuses on cross-section data of the construction industry in 12 counties of Linyi city in 2014. The results of the empirical application indicate that there are great gaps in productivity development level among 12 counties in the construction industry of Linyi. On the basis of the results analysis, the result of this study provide some significant suggestions that may be useful to decision makers of government and construction organizations in improving the performance of the construction industry and promoting the sustainable development of the construction industry between different counties in Linyi city.

References

1. T. J. Coelli, D.S. Prasada Rao, C. J. O'nnell & G. E. Battese. An Introducition to Efficiency and Productivity Analysis (Second Edition), Spring(2005).
2. A. Charnes, W.W. Cooper, E. Rhodes. Measuring the efficiency of decision making units. European Journal of Operational Research 2, 429-444 (1978).
3. R. Färe, S. Grosskopf, C. A. K. Lovell & C. Pasurka. Multilateral Productivity Comparisons When Some Outputs Are Undesirable: A Non-Parametric Approach. Review of Economics and Statistics 71, 90-98(1989).
4. R.G. Dyson, R. Allen, A.S. Camanho, V.V. Podinovski, C.S. Sarrico & E.A. Shale. Pitfalls and Protocals in DEA. European Journal of Operational Research 132, 429-444 (2001).
5. P. Zhou, K. L. Poh & B. W. Ang. A Non-radial DEA Approach to Measuring Environmental Performance. European Journal of Operational Research 178, 1-9 (2007).
6. X. Xue, Q. Shen, Y. Wang & J. Lu. Measuring the productivity of the construction industry in China by using DEA-based Malmquist productivity indices. J. Constr. Eng. Manage. 134, 64-71(2008).
7. Y. Li, C. Liu. Malmquist indices of total factor productivity changes in the Australian construction industry. Constr. Manag. Econ. 28 (9), 933-945(2010).
8. Statistical Yearbook, 2015. Bureau of Statistics of Linyi, (2015).
9. X.L. Xue, H.Q. Wu, X.L. Zhang, J. Dai & C. Su. Measuring energy consumption efficiency of the construction industry: The case of China. Journal of Cleaner Production 107,509-515 (2015).
10. M. Kapelko, I.M. Horta, A.S. Camanho & A. Oude Lansink. Measurement of input-specific productivity growth with an application to the construction industry in Spain and Portugal. Int. J. Production Economics 166, 64-71 (2015).

Loss effect on the dynamical evolution of rigid holographic bright spatial solitons in a dissipative system

Guang-Yong Zhang†, Mi-Zhuo Zhao, Guo-Zhen Chen and Xiao-Qing Zhang

Department of Physics, China University of Geosciences,
Wuhan, 430074, China

†*E-mail: zhgyong@163.com*

It has been investigated that the crystal loss effect on the dynamical evolutions of rigid holographic bright solitons in photorefractive dissipative systems based on two-wave mixing. Numerical results show that the stability of rigid holographic bright solitons in such systems depends strongly on the crystal loss. The soliton formed at a given crystal loss can propagate stably in the crystal over a distance that is long enough; the input solitary beam can evolve into a stable soliton and propagate in the crystal when the crystal loss drift is quite small, whereas it will not evolve into a stable soliton and its intensity becomes increase or decrease with propagation distance if the loss difference is big enough. A potential application of these properties of rigid holographic solitons in optical attenuators or repeaters is discussed.

Keywords: Optical solitons; Photorefractive; Dissipative system; Holographic; loss.

1. Introduction

PR spatial solitons have become a convenient playground in which to study soliton phenomena, therefore photorefractive (PR) spatial optical solitons have attracted much interest in the last three decades [1]. The mechanism to form PR spatial soliton can be classified into two generic types. The first type is the self-phase-modulation self-focusing mechanism that originates from the spatially nonuniform screening of the applied field [1-3], or from the photovoltaic effects [4-6], or from both simultaneously [7, 8]. The second type is holographic focusing mechanism that arises from a grating in the refractive index induced by interference between the signal and pump beams via two-wave mixing (TWM) [9]. In the later type, the signal and pump beams can simultaneously evolve into holographic solitons in a special case, so the holographic solitons appear as a soliton pair [10,11].

Recently, a new kind of spatial soliton with both the holographic focusing mechanism and the self-phase modulation self-focusing mechanism, named as screening holographic soliton, was proposed by Liu for a PR dissipative system that consists of a biased PR crystal [12]. In this PR dissipative system, the signal beam, i.e. the self-trapping beam, had a one-dimensional soliton-like spatial distribution and could obtain a gain from the pump beam via TWM between the beams, and could evolve into a screening holographic soliton, whereas the pump beam was of a uniform spatial distribution in both transverse dimensions and stronger than the signal beam. The soliton produced in dissipative systems in which soliton solutions result from a double balance, i.e. diffraction is balanced by nonlinearity and loss is balanced by gain. The double balance results in soliton solutions with fixed amplitude and width for fixed values of system parameters [13]. Inasmuch as this, the new kind of screening holographic solitons was also called as rigid screening (RS) soliton [12].

RS solitons depend on double balance, as we known, the crystal loss change, i.e., the change of the absorption coefficients, results in the change of the system parameters, and further affects on the dynamical evolution of the RS soliton. In this paper, we investigate the crystal loss effects on the dynamical evolution of the RS soliton by numerical method under steady condition.

2. The analytical solution of RS solitons

The PR crystal proposed here is SBN with its optical c-axis oriented along the x-axis and the bias field applied in the same direction. A one-dimensional bright solitonlike optical beam as the signal beam is made to propagate in the crystal along the z axis and is permitted to diffract only along the x direction. An optical beam with a uniform spatial distribution in both x and y directions as the pump beam is also made to propagate in the crystal along the z axis and makes a small angle θ in the crystal with signal beam. The two coherent beams couple by PR codirectional degenerate TWM, and then the pump beam can amplify the signal beam. Moreover, let us assume that the signal beam is linearly polarized along the x direction. The pump beam is also linearly polarized light, and its polarized direction makes an angle φ with the x axis.

Let ϕ denote the slowly varying envelopes of the electric-field component of the signal beam. In the slowly-varying approximation, ϕ satisfies the following paraxial wave equation [12]:

$$\frac{1}{2k}\frac{\partial^2\phi}{\partial x^2}+i\frac{\partial\phi}{\partial z}-k_0\frac{n_e^3}{2}\gamma_{33}E_{sc}\phi+(\Gamma_0-\frac{i}{2}\Gamma)\frac{I_P}{I_P+I}\phi+i\frac{1}{2}\alpha_0\phi=0\,, \qquad (1)$$

where $k=n_ek_0$, $k_0=2\pi/\lambda_0$ and λ_0 is the free-space wavelength of the light wave employed, n_e is the extraordinary refractive index, γ_{33} is the electro-optic coefficient, $E_{sc}=E_0(I_\infty+I_p)/(I_p+I)$ is the space charge field induced by the signal beam, E_0 is the bias field strength, I and I_P are the intensities of the signal and pump beam, and $I_\infty=I(x\to\pm,z)$. α_0 is the absorption coefficient of the crystal, and Γ and Γ_0 are the intensity and phase coupling coefficients of the TWM, respectively. Let $U=(2\eta_0I_P/n_e)^{-1/2}\phi$, $|U|^2=I/I_P$, where $I(x,z)=(n_e/2\eta_0)|\phi(x,z)|^2$, $\eta_0=(\mu_0/\varepsilon_0)^{1/2}$. Under the condition $|U|^2=I/I_P\ll 1$, from equation (1) we can obtain a dynamical evolution equation in term U as follows:

$$i\frac{\partial U}{\partial\xi}+\frac{1}{2}\frac{\partial^2U}{\partial s^2}-(P_d+iG)U+(P_d+ig)U|U|^2=0\,, \qquad (2)$$

where $s=x/x_0$, $\xi=z/z_0$, $z_0=kx_0^2$, x_0 is an arbitrary spatial width. $P_d=\beta(1+\rho)-g_0$, $G=g-\alpha$, $\rho=I_\infty/I_p$, $\beta=\chi\gamma_{33}E_0$, $\chi=(k_0x_0)^2n_e^4/2$, $g=z_0\Gamma/2$, $g_0=z_0\Gamma_0$, $\alpha=z_0\alpha_0/2$. For a PR TWM process with a biased field, g and g_0 can be expressed as [14] $g=\delta E_s[E_0^2+E_d(E_d+E_s)]/[(E_d+E_s)^2+E_0^2]$, $g_0=\delta E_0E_s^2/[(E_d+E_s)^2+E_0^2]$, where $E_d=4\pi k_BT\sin\theta/(\lambda_0e)$ is the diffusion field, k_B is the Boltzmann constant, e is the electron charge, T is the absolute temperature, $E_s=eN_A\lambda_0/(4\pi\varepsilon_r\varepsilon_0\sin\theta)$ is saturation field, N_A is the acceptor density, ε_r is the effective dielectric constant, and ε_0 is the permittivity of vacuum. $\delta=\chi\gamma_{eff}$, where γ_{eff} is the effective electro-optic coefficient of the TWM process. For the configuration discussed above, we have $\gamma_{eff}=\bar{n}^3\gamma_{33}\cos\theta\cos(\theta/2)\cos\varphi$, where $\bar{n}=n_e/n_o$, and n_o is the ordinary refractive index [15].

The bright soliton analytical solution of equation (2) was given as follow

$$U(s,\xi)=F\sec h(Bs)\exp\{ib\ln[\sec h(Bs)]\}\exp(-iv\xi) \qquad (3)$$

where $F=[3G/2g]^{1/2}$, $B=[G/b]^{1/2}$, $b=[-3P_b+(9P_b^2+8g^2)^{1/2}]/(2g)$, $P_b=\beta-g_0$ is the chirp parameter, and $v=(b^2-1)G/(2b)+P_b$ denotes the nonlinear shift of the propagation constant.

3. Crystal loss effects on the dynamical evolution of RS solitons

First we investigate the dynamical evolution of the RS soliton determined by equation (3) in the crystal at a given absorption coefficient, i.e., crystal loss. In order to give an actual example, a SBN:60 crystal is used with the following parameters [16]: $n_e = 2.33$, $n_o = 2.36$, $\varepsilon_r = 880$, $\gamma_{33} = 220pm/V$ and $N_A = 1.2\times10^{17}cm^{-3}$ at $\lambda_0 = 0.5\mu m$. Taking $E_0 = 700V/cm$, $x_0 = 25\mu m$, $\alpha_0 = 0.5cm^{-1}$, $\theta = 4^\circ$, and $\varphi = 88^\circ$, when $T = 300K$, these system parameters in equation (2) are $\alpha = 0.46$, $g = 0.51$, $g_0 = 0.38$ and $\beta = 22.4$. Taking these system parameters into equation (3), we obtain a RS bright soliton with the following parameters: $F = 0.39$, $B = 1.82$, $b = 1.54\times10^{-2}$, and $v = 20.4$. This RS soliton possesses an intensity FWHM of $24\mu m$ and the peak value of $I_{\max}/I_p$ is 0.15. The dynamical evolution of such a bright RS soliton can be obtained by numerically solving equation (2) with the above system parameters. As expected, our results confirm that the soliton state remains invariant with propagation distance, as shown in Fig. 1(a).

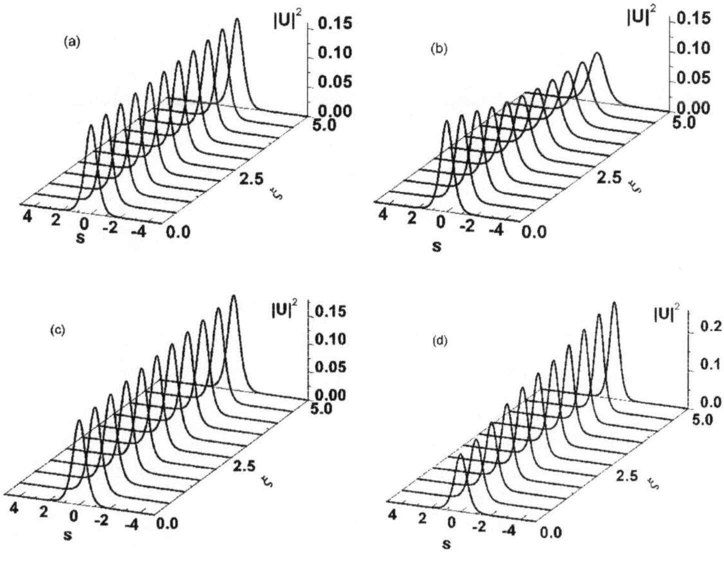

Fig. 1. Crystal loss effects on the dynamical evolution of bright RS solitnos in a dissipative PR system with $\alpha+\Delta\alpha$. (a) $\Delta\alpha = 0$, (b) $\Delta\alpha = 0.046$,(c) $\Delta\alpha = -0.023$ and (d) $\Delta\alpha = -0.046$. The incident beam is a bright RS soliton with $\alpha = 0.46$, $g = 0.51$, $g_0 = 0.38$ and $\beta = 22.4$.

Now we investigated crystal loss, i.e., absorption coefficient effects on the dynamical evolution and the stability of a RS soliton. We firstly determine a bright RS soliton from equation (3) using the above parameters at $\alpha = 0.46$ as the incident solitary beam, and then solve numerically equation (2) under different crystal loss, i.e., with parameter $\alpha = \alpha + \Delta\alpha$. Figs. 1 (b)-(d) give the dynamical evolutions of the incident beam in the crystal at $\Delta\alpha = 0.046, -0.01$, and $\Delta\alpha = -0.046$ respectively. These results show that the crystal loss effects on the evolution and stability of the solitary beam are very obvious. When $\Delta\alpha = 0.046$, the value of α is 0.51, which means the excess crystal loss, thus the signal beam cannot get enough energy from the pump beam to overcome the loss, the intensity of the beam decreases with propagation distance and then the beam is progressively absorbed as it propagates and, consequently, the beam cannot evolve into a bright RS soliton in this case, as shown in Fig. 2 (b). However, when $\Delta\alpha = -0.046$, which corresponds to excess gain case, the beam intensity increases and the beam width decreases with propagation distance, and then the beam cannot evolve into a stable bright RS soliton for a big change of the value α, such as $\Delta\alpha = -0.046$, as shown in Fig. 2 (d). The signal beam can get enough energy from the pump beam and approximately overcome the loss effects, and both the intensity and width of the beam finally reach a saturation value, and the signal beam finally evolves into a stable bright RS soliton for a small change of the crystal loss, such as $\Delta\alpha = -0.023$ as shown in Fig. 2 (c).

The stability of the spatial soliton is very important for the observations and the practical applications of such RS solitons. As considering in this paper, the bright RS soliton can stably propagate in the crystal when the change of the crystal loss is about $-0.023 \sim 0.023$. Therefore, the soliton-driven photonics can be used to realize the stable propagation of the beam when the crystal loss changes slightly. However, when the crystal loss deviates big enough, the beam not only cannot stably propagate, but also increases or decreases with propagation, as shown in Fig.2 (b) and (d). Therefore, the soliton-driven photonics can be made based on these properties of the RS solitons.

4. Conclusion

The crystal loss effects on the dynamical evolution and stability of RS solitons in biased PR crystals with TWM have been investigated. RS solitons are stable to small crystal loss change. When the loss deviation is big enough, however, the peak intensity and width of the beam vary with propagation distance, and the beam cannot evolve into a stable RS soliton. Some novel soliton-driven photonics could be made based on the these properties of RS solitons discussed in this paper.

Acknowledgments

This work is supported by the National Basic Research Program of China(973 Program) (Grant No. 2011CB710606) and the Earth biology fund from State Key Laboratory of Biogeology and Environmental Geology (Grant No. GKZ15Y638).

References

1. M. Segev, G.C. Valley, B. Crosignani, P.D. Porto, A. Yariv, Steady-state spatial soliton in photorefractive material with external field, Phys. Rev. Lett. 73 (1994) 3211-3214.
2. D.N. Christodoulides, M.I. Carvalho, Bright, dark and gray spatial soliton states in photorefractive media, J. Opt. Soc. A m. B 12 (1995) 1628-1633.
3. G. Zhang, J. Liu, C. Wang, H. Zhang, S. Liu, Effect of the dark soliton crystal temperature on the deflection of the bright soliton in a separate bright-dark screening soliton pair, Optik 119 (2008) 303-308.
4. M. Segev, G.C. Valley, M.C. Bashaw, M. Taya, M.M. Fejer, Photovoltaic spatial solitons, J. Opt. Soc. Am. B 14 (1997) 1772-1781.
5. C. F. Hou, Y. Zhang, Y. Y. Jiang, Y. B. Pei, Photovoltaic solitons in two-photon photorefractive materials under open-circuit conditions, Opt. Commun. 273(2007) 544-548.
6. W. She, K. Lee, W. Lee, Observation of two-dimensional bright photovoltaic spatial solitons, Phys. Rev. Lett. 83 (1999) 3183-3185.
7. J. Liu, K. Lu, Screening-photovoltaic spatial solitons in biased photovoltaic photorefractive crystals and their self-deflection, J. Opt. Soc. Am. B 16 (1999) 550-555.
8. S. Konar, S. Jana, S. Shwetanshumala, Incoherently coupled screening photovoltaic spatial solitons in biased photovoltaic photorefractive crystals, Opt. Commun. 273 (2007) 324-333.
9. O. Cohen, T. Carmon, M. Segev, and S. Odoulov, Holographic solitons, Opt. Lett. 27 (2002) 2031–2033.
10. J. Liu, Z. Hao, Coupling and stability of bright holographic soliton pairs, Chin. Phys. 13 (2004)704-711.
11. J. Liu, Existence curves of asymmetric photorefractive holographic spatial optical solitons, Acta Phys. Sin. 53 (2004)3014-3022 (in Chinese).
12. J. Liu, Existence and property of spatial solitons in a photorefractive dissipative system, J. Opt. Soc. Am. B 20 (2003)1732-1738.
13. N. N. Akhmediev, V. V. Afanasjev, J. M. Soto-Crespo, Singularities and special soliton solutions of the cubic-quintic complex Ginzburg-Landau equation, Phys. Rev. E 53 (1996) 1190–1201.
14. P. Yeh, Two-wave mixing in nonlinear media, IEEE J. Quantum Electron. 25 (1989) 484-519.
15. I. Oda., Y. Otani, L. Liu, T. Yoshizawa, Polarization effect for vibration detection using photorefractive two-wave mixing in KNSBN:Cu crystal, Opt. Commun. 148 (1998) 95-100.
16. R. A. Vazqez, M. D. Ewbank, R. R. Neurgaonkar, Photorefractive properties of doped strontium-barium niobate, Opt. Commun. 80 (1991) 253-258.

The Efficiency Analysis of Logistics Industry in Shandong Province Based on VRS DEA Model

Zong-Ying Li

School of Logistic, Linyi University,
Linyi, Shandong, China
E-mail: zongying_li@126.com

Xiang Liu

School of Architecture, Linyi University,
Linyi, Shandong, China
E-mail: liuxiang0632@126.com

In the paper, the VRS DEA model is applied to measure the relative productive efficiency of logistics industry in Shandong province, the application focuses on cross-section data of the logistics industry in 17 cities of Shandong province in 2014, selecting thee inputs and one output. The results indicate that there are great gaps in productivity development level among 17 cities in the logistics industry of Shandong province. On the basis of the results analysis, some suggestions are provided to improve the performance of the logistics industry and promoting the sustainable development of the logistics industry between different cities.

Keywords: The Logistics Industry; Data Envelopment Analysis; Decision-Making Units; Constant Returns to Scale; Variable Returns to Scale; Non-increasing Returns to Scale.

1. Introduction

During the past decade, the logistics industry has been rapid development in China, which plays a more and more important role in the national economy and has become a pillar industry of national economy. In 2014, the added value of the logistics industry was 385.2 billion yuan, accounting for 6.5% of GDP in Shandong province, however, its development model is still in the way of high input and high output, high logistics costs and inefficient operation status are still its basic characteristics. Therefore, it is necessary to measure the operational efficiency [1-5] of logistics industry in Shandong province.

A large number of productivity researchs about various industry are shown in the previous literature [6-11], which are used to support the design of strategies and government policies towards the improvement of industry performance. In the paper, we will study the efficiency of the operational industry in Shandong province by means of Variable Returns to Scale (VRS) DEA model [1].

We organize the remainder sections as follows. Section 2 describes the methodology about DEA model. The empirical study for logistics industry efficiency in Shandong province and the discussion of the results are presented in Section 3. The last section concludes the research.

2. Methodology

Assume that there are N (homogeneous) DMUs, and each DMU can produce S outputs through M inputs. Let a $M \times N$ matrix X stands for inputs matrix and a $S \times N$ matrix Y stands for outputs matrix.

DEA model is a non-parametric, piecewise linear programming method by means of the cross-section, which can be applied to measure the productivity of decision-making units (DMUs) with multiple inputs and multiple outputs. The CCR model and the BCC model are the most popular among the various forms of the DEA models. The CCR model assumes Constant Returns of Scale (CRS), whereas the BCC model is developed under the assumption of VRS.

Then the CRS efficiency score of DMU_0 [1] can be calculated by:

$$\min \ \theta$$

$$s.t.\begin{cases} \sum_{k=1}^{N} x_{ik}\lambda_k \le \theta x_{i0}, i = 1,2,\cdots,M \\ \sum_{k=1}^{N} x_{jk}\lambda_k \ge y_{j0}, j = 1,2,\cdots,S \\ \lambda_k \ge 0,\ k = 1,2,\cdots,N \end{cases} \tag{1}$$

The VRS situations must be taken into account when some DMUs are not operating at an optimal scale. The VRS model [1] can be described by adding the convexity constraint as follows:

$$\min \ \theta$$

$$s.t.\begin{cases} \sum_{k=1}^{N} x_{ik}\lambda_k \le \theta x_{i0}, i=1,2,\cdots,M \\ \sum_{k=1}^{N} y_{jk}\lambda_k \ge y_{j0}, j=1,2,\cdots,S \\ \sum_{k=1}^{N} \lambda_k = 1 \\ \lambda_k \ge 0, k=1,2,\cdots,N \end{cases} \tag{2}$$

And VRS model can be easily modified to account for the problem with Non-increasing Returns to Scale (NIRS) [1] as following form:

$$\min \ \theta$$

$$s.t.\begin{cases} \sum_{k=1}^{N} x_{ik}\lambda_k \le \theta x_{i0}, i=1,2,\cdots,M \\ \sum_{k=1}^{N} y_{jk}\lambda_k \ge y_{j0}, j=1,2,\cdots,S \\ \sum_{k=1}^{N} \lambda_k \le 1 \\ \lambda_k \ge 0, k=1,2,\cdots,N \end{cases} \tag{3}$$

Where x_{ik} and y_{jk} means the *ith* input and the *jth* output of DMU_k, and λ_k refers to the weight coefficient of input or output factors of the DMU_k .

Based on the above model (1) to model (3), we have the following results. The CRS technical efficiency (CRS TE) can be described as the product of the VRS technical efficiency (VRS TE) and the scale efficiency (SE). For a scale inefficient DMU, the increasing returns to scale (IRS) exit for it when the VRS TE is unequal to the NIRS technical efficiency (NIRS TE), and the DMU is on the decreasing returns to scale (DRS) portion of the VRS frontier while the VRS TE is equal to the NIRS TE.

3. Empirical Study

3.1. *Data and DMU*

The year 2014 was taken as the study period in the paper, and we selected 17 cities of Shandong province as DMUs to measure the productive efficiency of the Logistics industry in these cities. In the paper, when some rules of thumb [5] are taken into account, we selected thee inputs (i.e., Logistics related industry total investment in fixed assets, Transportation, warehousing and postal service workers and Expressway and class I to IV Highways) and one output (Freight Turnover). The data used in this study came from Shandong Statistical Yearbook in 2015 [12]. And the descriptive statistics of the data is given in Table 1.

3.2. *Results analysis*

The radial input-oriented DEA model (1) to model (3) are applied to calculated the TE [6-11] of the logistics industry of 17 cities in Shandong province in 2014 by Matlab software. Table 2 shows us the input-oriented CRS, VRS and NIRS DEA results of the logistics industry as well as scale efficiency and returns-to-scale (RTS) categories (increasing returns to scale, constant returns to scale and decreasing returns to scale).

Table 1. Descriptive statistics of input and output variables.

Inputs /outputs	Variable	Unit	Mean	Max	Min	Std.Dev.
	X_1	10^8yuan	247	638.9	34.1	152.8
Inputs	X_2	10^4persons	3.2	7.5	0.5	2.1
	X_3	km	15202.5	26779	4223	6761.6
Outputs	Y	10^6 ton-km	40816.5	118762	8852	28908.4

Note: X_1 - Logistics related industry total investment in fixed assets, X_2 - Transportation, warehousing and postal service workers, X_3 - Expressway and class I to IV Highways, Y - Freight Turnover.

Table 2. Input-Oriented CRS, VRS and NIRS DEA Results.

DMUS	CRS TE	VRS TE	SE	NIRS TE	RTS
1	0.6536	0.7039	0.9285	0.6536	IRS
2	1.0000	1.0000	1.0000	1.0000	-
3	0.8176	0.9034	0.9050	0.8176	IRS
4	0.4208	0.6271	0.6710	0.4208	IRS
5	0.8773	1.0000	0.8773	0.8773	IRS
6	0.5744	0.5780	0.9938	0.5780	DRS
7	0.4951	0.5147	0.9619	0.4951	IRS
8	0.8216	0.8447	0.9727	0.8216	IRS
9	0.2167	0.3347	0.6474	0.2167	IRS
10	1.0000	1.0000	1.0000	1.0000	-
11	0.6193	0.7399	0.8370	0.6193	IRS
12	0.6559	1.0000	0.6559	0.6559	IRS
13	1.0000	1.0000	1.0000	1.0000	-

14	0.2928	0.3425	0.8549	0.2928	IRS
15	0.9277	0.9730	0.9534	0.9277	IRS
16	0.7881	0.8454	0.9322	0.7881	IRS
17	0.6929	0.9082	0.7629	0.6929	IRS
mean	0.6973	0.7833	0.8797	0.6975	

Note: The number of DMUs Vs city is: 1-Jinan, 2-Tsingtao, 3-Zibo, 4-Zaozhuang, 5-Dongying, 6-Yantai, 7-Weifang, 8-Jining, 9-Taian, 10-Weihai, 11-Rizhao, 12-Laiwu, 13-Linyi, 14-Dezhou, 15-Liaocheng, 16-Binzhou, 17-Heze

It is easy to find that only three cities (DMU2-Tsingtao city,DMU10-Weihai City and DMU13-Linyi City) are efficient when CRS is assumed, whose TE scores is equal to one, so they are operating at an optimal scale. However, when VRS is assumed, there are five efficient cities (DMU2-Tsingtao city, DMU5-Dongying City, DMU10-Weihai City, DMU12-Laiwu City and DMU13-Linyi City). And the rest of other 12 cities are inefficient under both CRS and VRS technologies, among them, DMU9 (Taian city) is the city of the least VRS TE score, its score of VRS TE is equal to 0.3347. In case that VRS TE is low, a DMU can increase its efficiency score through the proper combination of the input and output variables.

In terms of RTS, among 14 CRS inefficient cities, only DMU6 (Yantai city) is on the DRS portion of the VRS frontier, whose SE score can be increased through the appropriate decreasing change in scale. The rest of other 13 CRS inefficient cities are on the IRS status, and the appropriate increasing change in scale can increase their SE score.

4. Conclusion

It is well known that DEA approach is a popular tool to measure the relative efficiency of DMUs. In the paper, the input-oriented VRS DEA model is applied to measure the operational efficiency about logistics industry in Shandong province, the application focuses on cross-section data of the logistics industry in 17 cities of Shandong province in 2014. The results of the empirical application indicate that there are great gaps in productivity development level among 17 cities in the logistics industry of Shandong province. On the basis of the results analysis, some significant suggestions is put forward which may be useful to decision makers of government and industry organizations in improving the performance of the logistics industry and promoting the sustainable development of the logistics industry among different cities in Shandong province.

References

1. T. J. Coelli, D.S. Prasada Rao, C. J. O'nnell & G. E. Battese. An Introducition to Efficiency and Productivity Analysis (Second Edition), Spring(2005).
2. A. Charnes, W.W. Cooper, E. Rhodes. Measuring the efficiency of decision making units. European Journal of Operational Research 2, 429-444 (1978).
3. R.D. Banker, A. Charnes & W.W. cooper. Some Models for Estimating Technical and Scale Inefficiencies in Data Envelopment Analysis. Manag. Sci. 30,1078-1092(1984).
4. R. Färe, S. Grosskopf, C. A. K. Lovell & C. Pasurka. Multilateral Productivity Comparisons When Some Outputs Are Undesirable: A Non-Parametric Approach. Review of Economics and Statistics 71, 90-98(1989).
5. R.G. Dyson, R. Allen, A.S. Camanho, V.V. Podinovski, C.S. Sarrico & E.A. Shale. Pitfalls and Protocals in DEA. European Journal of Operational Research 132, 429-444 (2001).
6. P. Zhou, K. L. Poh & B. W. Ang. A Non-radial DEA Approach to Measuring Environmental Performance. European Journal of Operational Research 178, 1-9 (2007).
7. X. Xue, Q. Shen, Y. Wang & J. Lu. Measuring the productivity of the construction industry in China by using DEA-based Malmquist productivity indices. J. Constr. Eng. Manage. 134, 64-71(2008).
8. Y. Li, C. Liu. Malmquist indices of total factor productivity changes in the Australian construction industry. Constr. Manag. Econ. 28 (9), 933-945(2010).
9. Y.H. Ahn & H. Min. Evaluating the multi-period operating efficiency of international airports using data envelopment analysis and the Malmquist productivity index. Journal of Air Transport Management 39,12-22(2014).
10. X.L. Xue, H.Q. Wu, X.L. Zhang, J. Dai & C. Su. Measuring energy consumption efficiency of the construction industry: the case of China. Journal of Cleaner Production 107,509-515 (2015).
11. M. Kapelko, I.M. Horta, A.S. Camanho & A. O. Lansink. Measurement of input-specific productivity growth with an application to the construction industry in Spain and Portugal. Int. J. Production Economics 166, 64-71 (2015).
12. Statistical Yearbook, 2015. Bureau of Statistics of Shandong province, (2015).

Study on International Electricity Market Regulation Models and China's Practice

Huang Qing*

State Grid Energy Research Institute,

Beijing, China

†*E-mail: huangqing@sgeri.sgcc.com.cn*

Abstract: This paper reviews and analyzes international experiences in electricity market regulation, and with full regard to China electricity sector future reform trends, starts from electricity market reforms, constructs scientific regulation system, analyzes its effects and puts forward regulation policy suggestions. It is suggested that the regulatory institutions redefine regulation goals and strategies in regard to various aspects and problems, establish comprehensive before-process, in-process and after-process electric power regulation procedures. It is recommended that the supervised power companies actively adapt to government regulation, establish a dedicated regulation service department.

Keywords: Electricity market, regulation, international experiences, regulation model, regulation system.

1. Analysis of international electricity market regulation models

Electricity market regulation main tasks include ensuring the stability and safety of power supply and promoting effective resource allocation. Electricity supervisory agencies carry out electricity market monitoring and management according to relevant laws and regulations, with respect toward market rules. The objective of such regulation is to maintain market order, ensure the stability and safety of power supply, fully utilize the basic function of resource allocation, and promote effective market transformation and healthy development of the electricity supply sector.

Before the market reforms of the electricity sector, in the majority of countries electricity regulation was among the executive functions of the government. As the market reforms of the electricity industry grow faster, market competitiveness mechanisms are starting to play its roles, and the necessity and importance of power market regulation become more and more evident. Therefore, a method generally adopted by a number of countries is to move all or some electricity regulation functions from the government to specially set up institutions, responsible for implementing electricity regulation. According to the practice in the developed Western countries and neighboring developing countries, there are two classical models of electricity supply regulation:

1.1. *Independent regulation model*

In this model, a regulatory agency is independent from the government, concentrates all supervisory functions and has relatively strong authority and neutrality. On the one hand, a regulatory agency is independent from the government bodies and exercises its power to the extent allowed by law, therefore it is characterized with exclusive authority and high neutrality. On the other hand, it also is independent from the supervised and maintains proper distance to the supervised, investors and consumers for maintaining its neutral and impartial nature. Due to different relations between central and local regulatory agencies, this kind of regulation model can be further divided into two types. One is vertical regulation model, when all regulation is carried out by a unified, national regulatory agency and a number of branch offices. For example, countries such as the UK, Argentina, New Zealand established only one national electric power supply supervisory institution, and, according to the actual demand for electric power regulation, branch offices have been set up in various regions. The other type is a hierarchical regulation approach, usually adopted by federal states such as the US or Australia.

1.2. *Non-independent regulation model*

This model is characterized by the unification of government and supervisory institutions, employed by Japan and some civil law states in Europe before 1998 (such as France or Germany). Due to political, economical, marketization degree differences, concrete methods adapted by various countries are diverse. For example, Japan and Korea, are both East Asian countries which governments like to interfere in business operations in the process of economic development, therefore they both have not established a supervisory institution independent from the government. At the same time, there are relatively big differences between history, culture and economic development of these two countries, so even though they have both established supervisory institutions overseen by the government, these institutions have distinct institutional autonomies.

2. Survey of Chinese electric power regulation reform trends

2.1. *Status quo*

The main objective of government regulation of electric power sector is to make sure the regulated companies have reasonable gains and profit margin and that they satisfy the society's demand for power at the lowest possible cost (Shi Liangping, Liu Xiaoqian, 2007). Even though in 2005 the State Council promulgated the "Electricity Regulation Ordinance", defining the institutions, responsibilities, methods, legal liabilities etc. of electricity regulation, this ordinance still is the product of compromise and balancing among the interests of government departments and does not fully embody international trends of electricity regulation reforms. In free market conditions, market access regulation and price regulation are the core competencies of electricity regulatory institution. And this ordinance has not clearly assigned those competencies to supervisory institutions, which makes it hard for such institutions to regulate and maintain balance among the national interests and the interests of the consumers, investors and electrical companies, and to implement effective management.

During the 18th National Congress, it was said that "efforts shall be made to remove market obstacles, increase the effectiveness and fairness of resource allocation; fair, open and transparent market rules need to be established". The future goals of electricity regulation are to transform the present, dispersed, primary, non-standardized market into a unified, national one, to remove market obstacles, create a standardized market platform and increase the effectiveness and fairness of resource allocation. Supervised content shall also become richer, better suited to actual development needs and demands of market reforms. Afterwards regulation shall be transformed into a combination of beforehand and afterwards, power transmission tariff shall change from purchase and sales gap to independent power distribution tariff, non-standardized market access shall be standardized. Electric power regulation system shall be further perfected, competencies and regulatory scope division among supervisory institutions shall be further clarified and adjusted.

2.2. *Trends*

The future trends of China's electric power market regulation are as follows:

(1) Price, costs and investment regulation shall become further standardized and strict. More attention shall be paid to clear up the interrelationship among fuel price, grid access price, power transmission-distribution price and sales price; power grid sales overall price shall still be supervised based on cost plus profit method, in order to encourage power grid companies to improve their

basic power grid infrastructure, while making further improvements of tariff regulation mechanism. While the overall cost plus profit price regulation method shall be continued, power grid costs shall be supervised more strictly, regulatory agencies shall require electrical companies to define the cost scope in a more clear, standardized way, non-confidential business-related costs shall be more transparent.

(2) Regulation of fair power grid access shall be more impartial and transparent. More attention shall be paid to information transparency, so as to ensure that power generating companies have fair, equal access to power grid. Regulation of renewable power and grid operations shall increase. Following the swift development of clean energy such as wind power, solar power, supervisory methods will focus on encouraging electrical companies to increase wind power, solar power and other renewable power generation access and absorption capacity, promote clean energy and power grid coordinated development.

(3) Further improvement of general service regulatory mechanisms. In the future, general service responsibility subject and enforcement subject shall be further clarified, a rational, sustainable general service financial compensation system shall be developed, and general service object and quality standards shall be defined.

(4) Electricity regulation laws and regulations system shall be gradually established and further improved. In the future, "Electricity Law" shall be corrected and perfected, the basis of existing laws shall be thoroughly revised, the laws and regulations on the subject in charge of electricity regulation shall be further improved. In the end a multi-layered electricity regulatory system, consisting of basic laws, subject regulations and professional rules shall be established, so as to enable the regulatory regulation to be done according to the law.

3. China's electricity market regulation content

3.1. *Regulation of investment and transmission-distribution price*

(1) Improved investment regulation. Improved regulation of electrical companies' investment by, for example, systematic, normalized reporting of power transmission assets utilization, and further expanding of power grids of all voltage levels. Strict regulation of effective assets used as an input of pricing, further improving of effective asset approval procedures and management methods.

(2) Steady pushing forward electricity transmission-distribution prices reform. Gradual establishment of an independent, rational electricity transmission-distribution pricing system according to costs plus profit method,

steady carrying out of electricity tariff pilot projects, improved inter-provincial electricity transmission and distribution pricing systems.

(3) Increased cost accounting. Introduction of electricity transmission-distribution pricing cost regulation policy, standardized cost accounting methods, expanded scope of electricity transmission-distribution cost accounting pilot tests.

3.2. *Regulation of power quality*

Basing on multiple attributes of energy products, combining power quality definition and third-party power quality regulation, with regard to customer benefits and social benefits, in Chinese market environment, power quality covers three aspects: stability, energy quality and customer service quality. Customer service includes general customer service and basic electricity service.

3.3. *Regulation of electricity general service*

(1) The development and improvement of electricity general service laws and regulations system. By improving regulations relevant to electricity general services and adjusting, enriching and perfecting according to actual situation, the legal applicability and operability of laws and regulations can be ensured.

(2) Gradual improvement of electricity general service mechanisms. Clearly set general service goals, standards, content, implementation methods, compensation mechanisms, general service responsibility subject and enforcement subject, in order to provide reliable basis for general service regulation.

(3) Establishment of electricity general service compensation mechanisms. Cost compensation provided for companies implementing general service, stable general service funding source.

3.4. *Regulation of energy-saving and emission reduction*

According to related policies of the State Council and National Energy Administration, China's regulation of energy-saving and emission reduction mainly includes the following aspects:

(1) Implementation of energy-saving, emission reduction technological projects and key sectors low-carbon technology innovation demonstration projects.

(2) Standardization and improvement of electricity tariff policy implementation, revoking of preferential electricity tariff policies illegally offered by local governments in favor of high energy consumption companies.

(3) Implementation of power demand side management method.

3.5. *Regulation of information disclosure*

(1) Intensification of information disclosure on electricity transactions and scheduling. Strengthening of the regulation of information disclosure and publishing of electricity market transactions. Implementation of electricity scheduling operations mode reporting system, increase of information sharing between power grid and power plants, promotion of new energy synchronization.

(2) Stress put on information transparency during power grid access equality regulation, in order to make sure power grid is in accordance with national energy policies and includes all power generating companies in a fair and impartial way.

(3) Improvement of costs transparency. While continuing the overall cost plus profit price regulation model, require related companies to prepare more clear, standardized cost definitions; non-confidential business-related costs shall be more transparent

4. Relevant suggestions

It is suggested that the regulatory institutions redefine regulation goals and strategies in regard to various aspects and problems, establish comprehensive before-process, in-process and after-process electric power regulation procedures. It is recommended to establish a regulatory mechanism consisting of regulatory performance assessments, hearings, public disclosure and regulation, at the same time electricity market must be cultivated, laws and regulations need to be revised, market environment must be improved and attention must be paid to competency and professionalism level of electricity regulation.

It is recommended that the supervised power companies actively change their ideas, adapt to government regulation, establish a dedicated regulation service department and change from passive "supervised body" into proactive "servicing regulation body". Improve lawful enterprise governance, standardized operations, provide regulation services. Establish a regulation service working mechanism, set up work team used to government regulation in advance, carry out inter-departmental coordination of regulatory work, solve any unexpected major supervisory issues. After that, create a distinct regulation service department, consisting of professionals in fields such as finance, audit, law, operations; improve compliance and normativeness.

References

1. Shi Liangping, Liu Xiaoqian, China's Power Industry Regulatory Efficiency Analysis, Financial Research, 2007/07.
2. Huang Liang Jin, He Lijun, Xiao Daibo, Generation Of Electricity Regulatory Reform Of China's Internet Governance Perspective, New Vision, 2010/02.
3. (US) Sally Hunt. Making Competition Work in Electricity [M]. Beijing: China Economic Publishing House. 2004.
4. (US) Daniel F. Spulber. Regulation and Markets [M]. Shanghai: Shanghai People's Publishing House . 2008.
5. Zhu Chengzhang, Yang Mingzhou, Huang Yuansheng. Power Industry Regulation and Market Supervision [M]. Beijing: China Electric Power Press, 2003.
6. Li Guilin, Xu Hua'an. UK Electricity Reform and Inspirations for China [J]. Economic Research Group . 2007, H.
7. Long Shengping, Zhang Lirong, Zhang Fang, Zhang Yan, Ge Shelun. Analysis and Inspirations on Western Electricity Regulatory System [J]. 2009,7 Reform and opening.
8. Ji Limin. System of Electric Power Regulation System Change Analysis [D]. Shandong University. 2007, 10.
9. Chai Liping. China's Electricity Regulatory Legal System [D]. 2008. 04 Fudan University.
10. Yang Kun, Sun Yaowei, Liang Zhihong Electricity Market and Its Target Modes [M]. Beijing: China Electric Power Press (2008).

Building and Application of Project Management Maturity Model for Military Equipment Medium-Maintenance

Yong-Dong Zhang, Zheng-Ping Shu and Zhong-Guang Li
Department of Graduate Management, Equipment Academy
Beijng, Huairou, China
†*E-mail: laodazyd@163.com*

Abstract: The project management maturity model for military equipment medium-maintenance is built and applied to the evaluation of equipment maintenance organizational maturity grade, which help to provide improved process and enhance the managing competence of military equipment organization. This paper defines the knowledge scopes of military equipment maintenance project management, analyzes the four management courses of medium-maintenance, and the maturity grade and evaluation index system is also set up, based on which the project management maturity model for military equipment medium-maintenance is found. This paper also probes into the application of project management maturity model for military equipment medium-maintenance grounded on fuzzy integrated evaluation. The research indicates that the project management maturity model for military equipment medium-maintenance is the effective tool for competence improvement, and provide guidance for the military equipment maintenance construction.

Keywords: equipment; medium-maintenance; project management maturity model; analytic hierarchy process; fuzzy comprehensive evaluation.

1. Introduction

Maintenance includes the total technique and managing activity which keeps, resume and improve the prescriptive technic state of equipment in employ and storage time[1], and produces great effect on the battle effectiveness growth. In information era, the advanced technology and complex structure of equipment leads to high maintenance cost, long maintenance period and big maintenance risk, this demands that the management level of equipment maintenance should improve rapidly. The improvement of management level should apply the advanced logos and technology of modern project management. Project management maturity model provides a frame for the evaluation and improvement of organization management level, and become hotspot concerned by many research institutions from the 1990s[2]. Until now, Project management maturity model has acquired successful application in many fields. The application of project management maturity model to equipment medium-maintenance will help to evaluate the management maturity, provide concrete plan for sustaining improvement. At the same time, some problem such as "how is the managing level", "what's the defect" and "what's the direction" has been understood by some equipment medium-maintenance management organization.

2. Knowledge scopes of equipment medium maintenance project

The 115 knowledge modules are defined in china project management knowledge system, and the project management knowledge from 9 fields is introduced[3]. The 6 project management knowledge system is more important to the equipment medium-maintenance, they are scope management, quality management, resource management, time management, risk management and flow management, whose frame is shown in Fig.1:

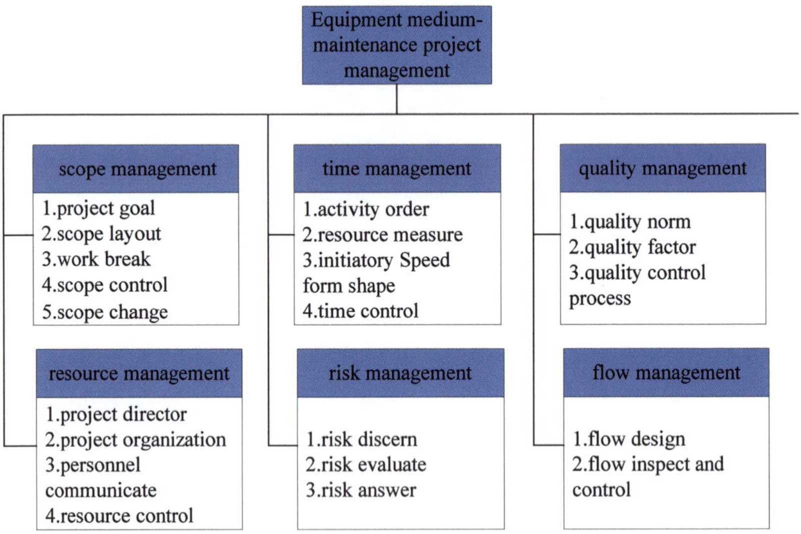

Fig.1 equipment medium-maintenance project management knowledge field frame

The scope management of equipment medium-maintenance includes goal definition, scope ascertaining, scope verify, and change control; the time management comprises activity definition, activity order, resource estimating, schedule workout and time control; the quality management is composed of quality managing plan and quality control; the resource management include human resource, material and outlay management; the risk management makes u of risk identify, risk evaluation and risk answer; the flow management includes flow design, flow optimization and flow inspect.

3. Building of equipment medium-maintenance project management maturity

3.1. *Equipment medium-maintenance project management process*

The period of equipment medium-maintenance project management can be divided into four process, the startup process, layout process, execution process and finishing process, which is shown in Fig.2.

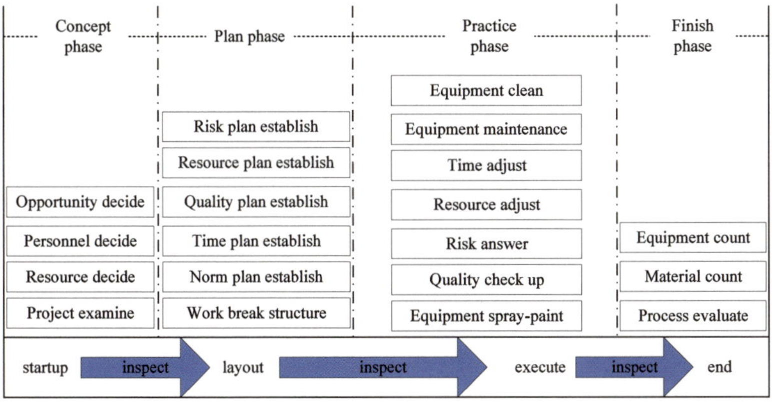

Fig.2 Equipment medium-maintenance management period

3.2. *Grade and character of equipment medium-maintenance management maturity*

In this paper, the project management maturity of equipment medium-maintenance is separated into five grades, there are disordered grade, groping grade, normative grade, controllable grade and improved grade, whose character is described in Table 1.

Table 1. Project management maturity grade character of equipment medium-maintenance

maturity grade	difference	Character
disordered grade	unorderly management	more optional process; disordered process to availably control; attach elementary importance to project management
groping grade	standardization management	management experience is repetitious used; process management depends more on repetitious; the project process

		is defined and regulation is formed
normative grade	canonical process management	project management process is understood and approved
controllable grade	subtle management	project acquires more perceive; the project process can be quantitative basically
improved grade	project strategic management	Normative project strategic management

3.3. *Evaluation index system for project management maturity of equipment medium-maintenance*

Based on the project management knowledge scope of equipment medium-maintenance and expert's opinion, the evaluation index system with four layers is confirmed through two choice, as is shown in Table 2.

The weight of each index is confirmed by analytic hierarchy process[4]. With regard to rule layer, it's weight vector is supposed as $w=\{w_1,w_2,w_3,w_4\}$, during which $w_i\,(i=1,2,3,4)$ denotes the weight of the i index in rule layer. After the analysis of expert's opinion, the estimating matrix A is set up.

$$A=\begin{pmatrix} 1 & 2 & 5 & 3 \\ 1/2 & 1 & 3 & 1/2 \\ 1/5 & 1/3 & 1 & 1/4 \\ 1/3 & 2 & 4 & 1 \end{pmatrix}$$

The geometric average of all element in each row is calculated and calculate processed, and the weight vector is acquired. The weight vector $w=\{0.144,0.212,0.491,0.153\}$ of rule layer is get with use of formula one and formula two.

$$\overline{w_j}=\left(\prod_{i=1}^{n} a_{ij}\right)^{\frac{1}{n}},\ j=1,2,...,n. \qquad (1)$$

$$w_j=\overline{w_j}\Big/\sum_{j=1}^{n}\overline{w_j},\ j=1,2,...,n. \qquad (2)$$

In the same way, the weight of demand layer and index layer is obtained, and the weight of goal layer is defined as one.

Table 2. Evaluation index for project management maturity of equipment medium-maintenance

Goal layer	Rule layer	Demand layer	Index layer
Project management maturity of equipment medium maintenance	startup process management	Project concept	Demonstration capacity, decision-making capacity, scope demonstrate shaping capacity
		Project startup	Organizing startup meeting capacity, flow establishing capacity, project organization building capacity
	plan process management	WBS	Confirming content capacity, establishing task list capacity, computing time demand capacity, computing resource demand capacity, computing cost demand capacity
		Shape plan	Establishing norm plan capacity, Establishing initial plan capacity, Establishing resource distribution plan capacity, Establishing quality plan capacity, Establishing risk plan capacity
	execution process management	Progress control	Progress information collecting capacity, progress comparing capacity, progress plan adjusting capacity
		Resource control	Personnel communication capacity, personnel conflict solving capacity, turnover budget capacity, equipment collecting capacity, equipment distributing capacity
		Quality control	Quality checkup capacity, quality comparing capacity, warp correcting capacity
		Risk control	Risk tell capacity, risk evaluation capacity, risk answer capacity
	ending process management	examination	Quality examination capacity, file examination capacity
		final accounts	Information handling capacity, establishing final accounts capacity
		evaluation	Self-evaluating capacity, benefit evaluating capacity

3.4. *Equipment medium-maintenance project management maturity model*

Till now, there are dozens of project management maturity models issued at home and abroad, the representative four of which is as follow: Capability Maturity Model Integration (CMMI)[5]; Kerzner Project Management Maturity Model (K-PMMM)[6]; Organizational Project Management Maturity Model (OPM3)[7]； Portfolio, Programme and Project Management Maturity Model (P3M3)[8].Han-chen Jiang and Mao-shan Qiang compared with the four model[9] and concluded that OPM3 is more logistic to learn and understand, and the index of OPM3 is frondose to apply. The project management maturity model of equipment medium-maintenance is shown in Fig. 3.

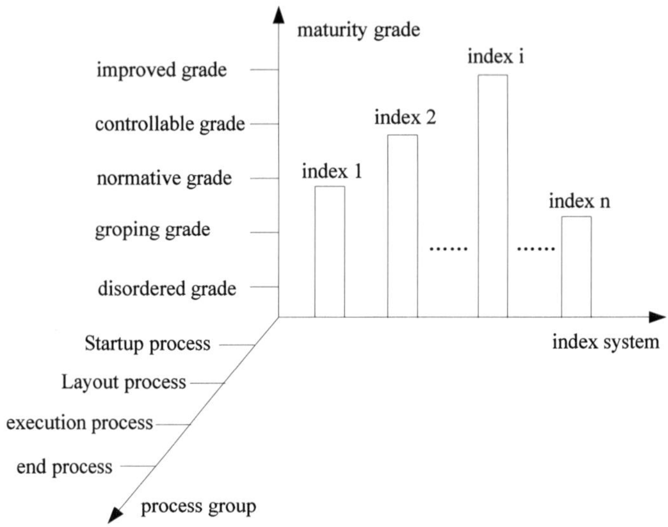

Fig. 3 project management maturity model of equipment medium-maintenance

4. Application of equipment medium-maintenance project management maturity model

4.1. *Evaluation preparation*

There are ten experts from equipment maintenance research and teaching to constitute a marking group, to whom ten pieces of survey questionnaire are provided and all are reclaimed. At the same time, the result of survey questionnaire is quantified with subject function.

4.2. *Evaluation of equipment medium-maintenance project management maturity*

In this paper, the fuzzy comprehensive evaluation[10] is applied to evaluated the maturity grade.

4.2.1. *Evaluation factor aggregation and grade*

$U=\{U_1,U_2,U_3,U_4\}$ is supposed as the first layer factor aggregation of project management maturity, $U_i=\{U_{i1},U_{i2},\cdots,U_{ij}\}$ is as the second layer, and $U_{ij}=\{U_{ij1},U_{ij2},\cdots,U_{ijk}\}$ is as the third layer factor aggregation, during which U,U_i,U_{ij},U_{ijk} respectively stand for the goal layer, the norm layer, the demand layer and the index layer. The evaluation grade aggregation is as follow:

$v=\{v_1,v_2,v_3,v_4,v_5\}$ ={disordered grade, groping grade, normative grade, controllable grade, improved grade}.

4.2.2. *Fuzzy comprehensive evaluation*

Firstly, each factor in U_{ij} is evaluated and the transformation matrix for U_{ij} is acquired.

$$R_{ij}=\begin{bmatrix} r_{ij1}^1 & \cdots & r_{ij1}^k \\ \vdots & \ddots & \vdots \\ r_{ijm}^1 & \cdots & r_{ijm}^k \end{bmatrix}$$

variable k in the transformation matrix is the progression of v, r_{ijm}^j denotes the subject degree of U_{ijm} belonging to v_j. element $r_{ijm}^k=d_{ijm}^k/d$, d denotes the number of the participant experts, d_{ijm}^j denotes the number of experts who gives v_j to factor m.

Secondly, $B_{ij}=w_{ij}\bullet R_{ij}$ is get through the secondary fuzzy evaluation to U_{ij}, after the unitary dealing with vector B_{ij}, the evaluation matrix $B_i'=\{B_{i1}',B_{i2}',\cdots,B_{ik}'\}^T$ of U_{ij} attained. The norm grade's evaluation matrix $B'=\{B_1',B_2',\cdots,B_k'\}^T$ is obtained in the same way.

At last, the comprehensive evaluation vector and project management maturity value are calculated with formula three and four.

$$Z=w\bullet B' \tag{3}$$

$$C=Z\bullet F^T \tag{4}$$

In the former formulas, Z is the comprehensive evaluation vector, and F is the row vector of equipment medium-maintenance project management maturity grade, $F=\{1,2,3,4,5\}$, respectively corresponding to disordered grade, groping grade, normative grade, controllable grade, and improved grade. The comprehensive evaluation value of the troop's vehicle medium-maintenance project management is 1.9671.

4.2.3. *Maturity grade decision*

The maturity comprehensive evaluation value C is corresponding with the management maturity grade of equipment medium-maintenance. When C is between zero and one, the maturity grade is disordered grade, and between one and two is groping grade, between two and three is normative grade, between three and four is controllable grade, and between four and five is improved grade. From the former evaluation, the comprehensive evaluation value is 1.9671, the maturity grade of the troop's vehicle equipment medium-maintenance project management is groping grade.

4.3. *Improvement of vehicle equipment medium-maintenance project management*

The maturity grades provides basis for the improvement of project management. The improvement of vehicle equipment medium-maintenance project management go through seven steps: preparation, maturity grade evaluation, finding problem, defining the improved content, establishing plan, execution and evaluation after improvement.

References

1. Yao-Hui Zhang, Equipment Maintenance Technology, Beijing, National Defence Industry Press, 2008:4-6.
2. T. Kent Crawford, Project Management Maturity Model, Beijing, China Machine Press, 2008:6-9.
3. China Project Management Research Committee, China Project Management Knowledge System(C-PMBOK2006), Beijing, Electronic Industry Press, 2006:12-14.
4. T.L.Saaty, Shu-Bai Xu, Analytic Hierarchy Process, Beijing, Coal Industry Press, 1988:35-40.
5. Chrissis M B, Konrad M, Shrum S, CMMI: Guidelines for Process Intergration and Product Improvement. 2nd Ed. Boston:Addison-Wesley Professional, 2006:5-8.
6. Harold Kerzner, Strategic planning for Project Management Using: Project Management Maturity Model, New York NY, 2001.
7. PMI, Organizational Project Management Maturity Model Program Plan, OPM3 Program Plan Ver.l.3, 2000.9.
8. Xiao-Wei Yan, Fei Gao, Research on P3M3 in the Promotion of Project Management Capacity, Management Observer, 2012(28):3-4.

9. Han-Chen Jiang, Mao-Shan Qiang, Compare of Four Project Management Maturity Model, Project Management Techniques, 2013(11):17-22.
10. Ya-Jun Guo, Comprehensive Evaluation Theory and Method, Science Press, 2002:40-46.

The Algorithm of the Regional Centroid Indoor Positioning Based on RSSI Filter

Ji-Hua Ye, Qing-Ping Lan, Jun Chen

College of Computer Information Engineering, Jiangxi Normal University, Nanchang, 330022, China

†E-mail: lanqingping2013@163.com，754054327@qq.com

RSSI(received signal strength indication) algorithm which based on ranging is widely used, but it is easily impacted by environment. It is difficult for the RSSI algorithm to position in the complex indoor environment, in order to solve the problem, an algorithm of the regional centroid indoor positioning based on Kalman filter, Gaussion filter and Statistics median filter (KGSM filter) was proposed. In the filtering algorithm, Kalman filter was used to smooth the random error firstly; then the Gauss filter was used to remove the significant error; finally, the Statistical median filter was used to further smooth the random error and suppress the relatively significant errors. After calculating the distance by the logarithmic-distance loss model, four different circumstances that the three circles intersect at one point, intersect in an area, two areas and three regions were discussed by the algorithm of trilateral. Finally, the estimated position was got by the algorithm of the regional centroid .The result of the experiment show that, KGSM filter is better than the traditional filtering in the complex indoor environment, the accuracy of the RSSI is improved by the KGSM filter, and the accuracy is kept between 0.7 meter to 1 meter, after the regional targeting, improving the accuracy and stability of the indoor positioning.

Keywords: Received Signal Strength Indication; indoor positioning; filter methods; regional centroid algorithm; the error of the positioning.

1. Introduction

In the real life, GPS is widely used, but in the indoor environment, due to the structure of the block, GPS unable to carry out. In order to solve the above indoor positioning problem, people have tried a variety of methods.

At present, method which commonly used in indoor positioning can be divided into two categories, ranging and non-ranging[1], the ranging algorithm including TOA[2], TDOA[3], RSSI[4],AOA[5], and so on; non-ranging algorithm including APIT[6],convex programming method, Centroid method[7], DV-Hop[8].

RSSI algorithm which based on ranging is widely used, but it is easily impacted by environment, therefore, people often process it with filter[9]. The commonly used RSSI filtering methods are mean filter, Gaussian filter, median filter, Kalman filter and their fusion. The reference[10] has improved the Gaussian filter, make it more consistent with the theoretical model and the filtering effect is better. The reference[11] has combined Kalman filter with median filter, the effect of coarse RSSI value on the results is reduced effectively. Inspired by the above related research, in this paper, a new method called KGSM(Kalman, Gaussian and Statistics Median) filter which combine Kalman filter, Gaussian filter with Statistical Median filter, the new method improves the accuracy, reduces the localization error by region centroid algorithm.

2. KGSM filter optimizes RSSI values

2.1. *Analysis of traditional filtering method*

In indoor positioning, due to the unstable factors: personnel moving, signal reflection, the interference of electronic instruments itself, and so on, there is a large error in the received RSSI value, in order to improve its accuracy, we often process it with filter. Median filter can effectively overcome the volatility of the interference caused by accidental factor, but it is not ideal for the signal filter with strong impulse interference; Gauss filter can solve the problem of RSSI which easy to be disturbed and poor stability in the actual test, but it is not good to deal with the effect of the shadow effect and the interference of the electronic instrument; Kalman filter can remove the additive noise superimposed on the signal, smooth significant error, but there are problems that errors can't be eliminated, and if the end of a few measurements have a big deviation, it will have a significant impact on the results.

2.2. *KGSM filter*

In this paper, a hybrid filtering method which called KGSM filter is adopted, it combines Kalman filter, Gaussian filter with improved statistical median filter. Its specific algorithm is as follows.

1) At a certain time period, collect RSSI signal for times, then deal the sampling value with Kalman filter. After that, the gross error signal turn to smooth, the noise made by the instrument has been reduced. However, due to

the existence of inertia accumulation of Kalman filter which means the greater the sampling value fluctuations at this moment, the greater the filter value fluctuation at the next moment, Kalman filter also has a great error after smoothing.

2) Collect every sampling value corresponding to the kalman filter, then value it with Gaussian filter, leave the middle of about 70 percent of the data. The advantages of this method is cutting of some Kalman filter which deviating from the true value greatly, reduce the end of a few RSSI sampling value fluctuation influence on filtering result.

3) Since the experiment is in a relatively complex indoor environment, the results error still exist after Gaussian filtering value. In order to further improve the RSSI value accuracy and so as not to affect the regional area centroid algorithm, after Gaussian filtering value, we process it with Statistical median filter, the detail steps as below:

a) Sorting the RSSI signal from small to large after Gaussian filtering value, then get the results, leave the median value to median filter.

b) Suppose the weight of the i th value is w_i, the threshold value T is the mean square value of each signal value and their mean value, when the square deviation is greater than the threshold, the weight decided by the mean square deviation, else decided by the threshold. This design has a better filtering effect in the situation which Gauss filter still has a lot of deviation from the true value.

$$\begin{cases} \omega_i = \dfrac{1}{1+\max\{T,(RSSI_i - RSSI_{mid})^2\}} \Bigg/ \displaystyle\sum_{i=1}^{N} \dfrac{1}{1+\max\{T,(RSSI_i - RSSI_{mid})^2\}} \\ T = \displaystyle\sum_{i=1}^{N} (RSSI_i - RSSI_{mid})^2 \Big/ N \end{cases} \tag{1}$$

c) Multiply Gaussian filtering value by its weighting coefficient, get the final Statistical median filtering value, that is KGSM filter:

$$RSSI_{KGSM} = \sum_{i=1}^{N} \omega_i * RSSI_i \tag{2}$$

3. Localization algorithm base on regional centroid

We can obtain $RSSI_{KGSM}$ by KGSM filter and the corresponding distance by logarithmic loss model. We decorate three AP: AP_1 , AP_2 , AP_3 in room, their coordinates are (x_1,y_1) , (x_2,y_2) , (x_3,y_3), the location is $D(x,y)$, distance between $D(x,y)$ and the three AP is d_i $(i=1,2,3)$, use AP_1 , AP_2 , AP_3 as the center of a circle, d_1,d_2,d_3 as radius and draw three circles, due to the fact that the three circles may intersect in a variety of different situations, we carried out a classification discussion as follows.

1) In an ideal situation, the three circles intersect at a point, as shown in Fig 1:

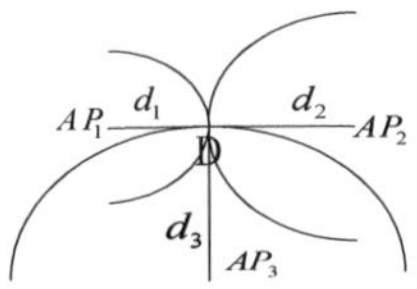

Fig.1. Intersecting at a point

The computational formula is:

$$\begin{cases}(x-x_1)^2+(y-y_1)^2=d_1^2\\(x-x_2)^2+(y-y_2)^2=d_2^2\\(x-x_3)^2+(y-y_3)^2=d_3^2\end{cases} \tag{3}$$

Obtain the coordinates of Geo-Location D by formula 3:

$$\begin{pmatrix}x\\y\end{pmatrix}=\begin{pmatrix}2(x_1-x_3) & 2(y_1-y_3)\\2(x_2-x_3) & 2(y_2-y_3)\end{pmatrix}^{-1}\begin{pmatrix}x_1^2-x_3^2+y_1^2-y_3^2+d_3^2-d_1^2\\x_1^2-x_3^2+y_2^2-y_3^2+d_3^2-d_2^2\end{pmatrix} \tag{4}$$

2) When the three circles intersect in one region, there are two cases, as shown in Fig 2.

Fig.2. Intersecting in one region

Their computational formulas is respectively as Eq (5) and Eq (6):

$$\begin{cases}(x-x_1)^2+(y-y_1)^2=d_1^2\\(x-x_2)^2+(y-y_2)^2=d_2^2\end{cases} \tag{5}$$

$$\begin{cases}(x-x_1)^2+(y-y_1)^2=d_1^2\\(x-x_2)^2+(y-y_2)^2=d_2^2\\(x-x_3)^2+(y-y_3)^2=d_3^2\end{cases} \tag{6}$$

The intersection area is determined by Eq (5) or Eq (6), assuming that intersection area is consist of N points (x_i, y_i), $i \in [1, N]$, use the centroid to show the estimated position as below:

$$D(\hat{x}, \hat{y}) = \left(\frac{\sum_{i=1}^{N} x_i}{N}, \frac{\sum_{i=1}^{N} y_i}{N}\right) \tag{7}$$

3) When they are intersect in two regions, show as Fig 3:

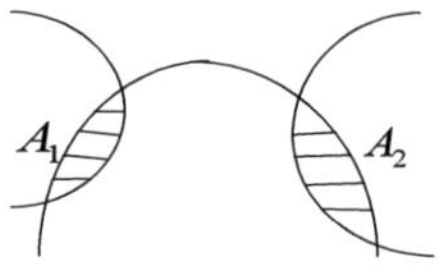

Fig.3. Intersecting in two regions

Calculation formula of regional A_1:

$$\begin{cases}(x-x_1)^2+(y-y_1)^2=d_1^2\\(x-x_3)^2+(y-y_3)^2=d_3^2\end{cases} \tag{8}$$

Calculation formula of regional A_2:

$$\begin{cases}(x-x_2)^2+(y-y_2)^2=d_2^2\\(x-x_3)^2+(y-y_3)^2=d_3^2\end{cases} \tag{9}$$

The intersection region A_1, A_2 is determined by Eq.(8) and Eq.(9) respectively, each region is consist of N points $(x_i, y_i), i \in [1, N]$, and N' points $(x_i', y_i'), i \in [1, N']$ respectively, combine the centroid algorithm, we can obtain the estimated position as below:

$$D(\hat{x},\hat{y})=\frac{1}{2}(\frac{\sum_{i=1}^{N}x_i}{N}+\frac{\sum_{i=1}^{N'}x_i^{'}}{N^{'}},\frac{\sum_{i=1}^{N}y_i}{N}+\frac{\sum_{i=1}^{N'}y_i^{'}}{N^{'}}) \tag{10}$$

4) When they are intersect in three regions, as shown in Fig 4:

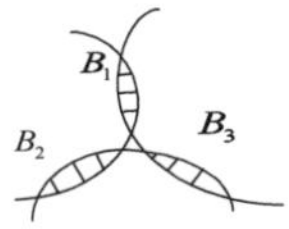

Fig.4. Intersecting in three regions

Calculation formula of regional B_1:

$$\begin{cases}(x-x_1)^2+(y-y_1)^2=d_1^2\\(x-x_2)^2+(y-y_2)^2=d_2^2\end{cases} \tag{11}$$

Calculation formula of regional B_2:

$$\begin{cases}(x-x_1)^2+(y-y_1)^2=d_1^2\\(x-x_3)^2+(y-y_3)^2=d_3^2\end{cases} \tag{12}$$

Calculation formula of regional B_3:

$$\begin{cases}(x-x_2)^2+(y-y_2)^2=d_2^2\\(x-x_3)^2+(y-y_3)^2=d_3^2\end{cases} \tag{13}$$

The intersection region B_1,B_2,B_3 is determined by Eq.(11), Eq.(12), Eq.(13) respectively, and each region is consist by N points (x_i,y_i), $i\in[1,N]$, $N^{'}$ points $(x_i^{'},y_i^{'})$, $i\in[1,N^{'}]$,and $N^{''}$ points $(x_i^{''},y_i^{''})$, $i\in[1,N^{''}]$ respectively, combine the centroid algorithm, we can obtain the estimated position as below:

$$D(\hat{x},\hat{y})=\frac{1}{3}(\frac{\sum_{i=1}^{N}x_i}{N}+\frac{\sum_{i=1}^{N'}x_i^{'}}{N^{'}}+\frac{\sum_{i=1}^{N''}x_i^{''}}{N^{''}},\frac{\sum_{i=1}^{N}y_i}{N}+\frac{\sum_{i=1}^{N'}y_i^{'}}{N^{'}}+\frac{\sum_{i=1}^{N''}y_i^{''}}{N^{''}}) \tag{14}$$

4. Experimental results and analysis

In the indoor location, the burst interferences such as personal moving may cause RSSI value deviates from the real value largely; random interferences such as signal reflection, electronic instrument itself may cause sampling value fluctuates up and down around the true value. However, there may be a large number of unexpected interference and random interference in the relatively complex indoor environment.

In order to verify that the algorithm has good accuracy and stability in the complex indoor environment, design the experiment environment as below: the indoor size is 10 meters by 10 meters; due to the room exist personal moving, layout 3 Aps in the room; in location position, we use mobile with sampling frequency of 5 Hz to collect signal, after a period of continuous time, each of the AP collects 200 RSSI value. Change the location position, repeat the experiment for 10 times, we can obtain 10 different sets of data.

Using RSSI sampling standard deviation to measure the filtering performance, position error to measure the positioning performance. The formula of RSSI sampling standard deviation σ_s show as follows:

$$\sigma_S = \sqrt{\frac{1}{N-1}\sum_{i=1}^{N}(RSSI_i - RSSI)^2} \qquad (15)$$

Where RSSI is real value, $RSSI_i$ is sampling filter value.

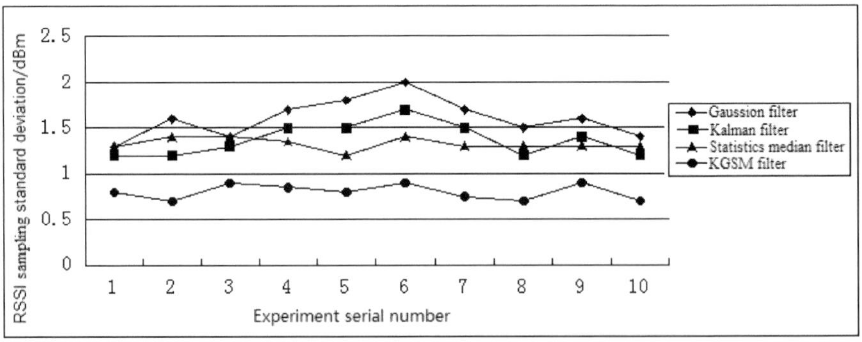

Fig. 5. RSSI sample standard deviation contrast figure

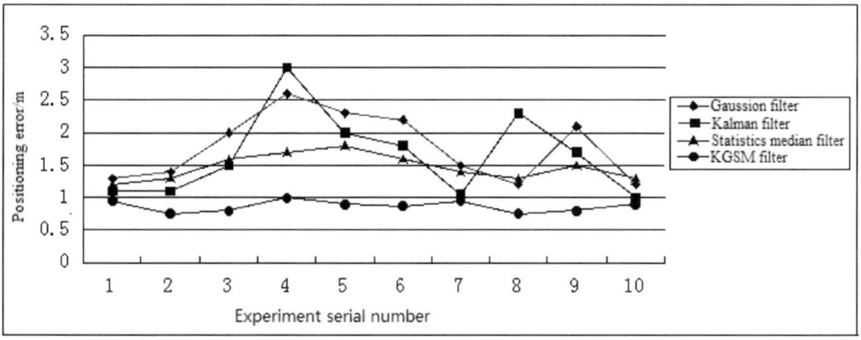

Fig.6. The positioning error comparison chart

Fig 5 compared Gaussian filter, Kalman filter, Statistical median filter and KGSM filter the effect of RSSI value filtering. Fig 6 compared the filtering algorithm of positioning error base on the four filtering algorithms. From Fig 5 and Fig 6, we can find that:

1) In the Complex indoor environment, due to the signal was interference long-term, the Sampling value after Gaussian filtering value, signal fluctuation large, it shows as sample standard deviation of the Gaussian filter are large on the whole, positioning error base on Gaussian filter are also large on the whole.

2) From the serial number 1, 2, 3, 8, 10 in Fig 5, we can find that, in the case that the significant error caused by the burst interference is not large, the Kalman filter model is better than the statistical mean value model; from the else serial number in Fig 5, in the situation that the significant error is large, due to the Kalman filter can only smooth significant error, its performance of removing significant error is worse than statistical median model. From the serial number 6, 8 in Fig 6, we can find that, when the end of a few samples values fluctuate considerably, Kalman filter positioning effect will be affected majorly.

3) The sample standard deviation of KGSM filter is the smallest on the whole, the positioning error is also the smallest and the most stable, the error range is 0.7~1.0 meter. it proves that KGSM filter base on RSSI can improve the precision and stability of the region centroid localization algorithm

From what has been discussed above：KGSM filter firstly smooth the error caused by all kinds of interference though Kalman filter, then remove the significant error after Kalman filtering by Gaussian filter valuing, at last, through statistical median filter to further smooth the random error, restrain relative significant error, improving the precision of the RSSI value, make the error in the range of 0.7~1.0 meter stably after region centroid localization, effectively improve the positioning accuracy and stability.

5. Conclusion

In this paper, a new filtering method called KGSM filter is proposed based on the analysis of various filtering methods in the complex indoor environment, after applying the logarithmic loss model to transform the filtered values into distance, we discuss four kinds of situations that three circles intersecting, finally, we obtain the estimated position by the region centroid algorithm. Experimental results show that: in the complex indoor environment, KGSM filter is better than the traditional filtering, and the position error is smaller, more stable.

Although the region centroid localization algorithm based on KGSM filter can improve the positioning accuracy, there are still some limitations, we can make improvement from below several directions in the future:1) KGSM filter increase the algorithm complexity, the next work is finding a kind/a few kinds of filter which can improve the RSSI value accuracy and reduce the algorithm complexity at the same time; 2) we only obtain an estimate of the location by region centroid algorithm, we need to find a positioning method which can make value more accurate and positioning accuracy is higher.

Acknowledgments

This work was supported by the Nation Natural Science Foundation of China (NSFC).

Reference

1. Nasipuri A, LI K. A directionality based location discovery scheme for wireless sensor networks, In Proceedings of ACM International Workshop on Wireless Sensor Networks and Application. Atlanta, USA,2002: 105-111.
2. Li X., and Pahlavan K., "Super-resolution TOA estimation with diversity for indoor geolocation." IEEE Transactions on Wireless Communications 3.1 (2004) : 224-234.
3. Luthy K , Grant E , and Chenderson T, "Leveraging RSSI for Robotic Repair of Disconnected Wireless Sensor Networks." International Conference on Robotics and Automation (2007)
4. Shen G., Zetik R., and Sthoma R., "Performance comparison of TOA and TDOA based location estimation algorithms in LOS environment." Workshop on Positioning Navigation and Communication (2008).
5. Zeng F, Yu M , Zou C, et, al. "An Improved Point-in-Triangulation Localization Algorithm Based on Cosine Theorem." (2012).

6. Zhang Jianping, Yu Min, Zhang Ke, et al, An improved weighted triangle centroid localization algorithm of APIT for wireless sensor networks, 2012 International Conference on Computer and Information Science, Safety Engineering, Wuhan, China: IE-EE, 2012:18-21.
7. Wang J, et, al. "Weighted Centroid Algorithm for Estimating Primary User Location: Theoretical Analysis and Distributed Implementation." Clinical Orthopaedics and Related Research (2010).
8. Dengyin Z , and Guodong C., "A Union Node Localization Algorithm Based on RSSI and DV-Hop for WSNs." (2012).
9. Huang Ying Ying, The technology of the three dimensional positioning in Wi-Fi, Beijing University of Posts and Telecommunications, 2012
10. Chen Li, Wang Xuedong, Sun Jingjing, The algorithm of the indoor wireless positioning based on the improved Gaussian filter, The Electric Mation,2014, 36(3):31-33.
11. Liu Jun, The researching and designing of the system of the WSNs positioning based on Kalman filter, The Central China Normal University, 2014.

Acquisition and Research of the Soil Building Environmental Data Based on the Sensor Network*

Yuan Zhang†, Chao-Ming Wang, Fan Zhang, Zhi-Yong Zhang and Xue-Wen Wang

School of Information Science and Technology, Northwest University,

Xi'an, 710127, China

†*E-mail: zhangyuan@nwu.edu.cn,*

www.nwu.edu.cn

Sensor network system has been commercialized in the environmental monitoring of soil buildings. However, there had problems of insufficient data analysis model and network structure for soil monitoring of the building. The objectives of this research were: (i) the design of the sensor network system, install and acquire the soil building environmental data and (ii) multi-position temperature and moisture data acquisition and analysis. According to the requirements of data acquisition and analysis, we design tree structure sensor network, sensor and communication protocol of the data acquisition. Download the data and combine the simulation experiment, analyzing and modeling according to the experiment data . By analyzing the temperature and moisture data model according to the simulated rainfall experiment , we find that single position moisture content changing with time are basically identical with theoretical relations. It is proved that the design of this system meet the demands of the analysis of the temperature and moisture monitoring for soil building, and it has a high reliability.

Keywords: Sensor networks; temperature and moisture; data acquisition; tree-like network structure; soil buildings.

1. Introduction

Many ancient soil buildings need to be protected in China, for the soil buildings could be destructed by various factors in natural. Therefore, it is of great significance of monitoring these factors for the protection and inheritance

* This work is supported by the Science Foundation of Northwest University (13NW39).

† Work partially supported by the World Cultural Heritage Risk Control Key Technology Research and Demonstration Project.

of these ancient soil buildings. By using the sensor network to monitor soil environment such as temperature, moisture, as the significant monitoring data, it can be used to analyze the relationships between the environmental factors and the soil buildings[1].

The environmental monitoring system based on wireless sensor network can achieve a long-term wide range of data acquisition and recording on the soil buildings, but the monitoring position inside the building is less and the data sampling speed is low[2]. In order to search the effects of environmental factors inside the buildings, we need to install the sensor inside the buildings to investigate the effects of environmental factors to it. We cannot realize it with the wireless sensor network (WSN), so we need to design the structure of sensor networks and node communication protocols to achieve the multi-position data acquisition[3].

According to the needs of the soil building protection, this paper studies the topology structure of sensor network and it's communication protocols. By using RS - 485 network build a set of tree structure sensor network, we optimize the network protocol, and realize modular and structure of the system design scheme. We can verify the reliability of system by analyzing the simulated rainfall experimental data model.

2. Materials and methods

According to the distribution of the soil buildings, we locate the experiment site in the North Town Station which is 5 kilometers north from Yulin city of Shaanxi Province. Most of the soil in this regions is silt soil and a few of the soil is silty clay. Soil moisture content is small in this area, which is about 2%[4]. The solid wall in the experiment are both 3 meters long, 1.5 meters wide and 2 meters high. We place the temperature sensor and moisture sensor a layer after layer during making the solid wall.

2.1. *The environmental monitoring system*

The whole system is mainly composed of temperature moisture sensors, sensor controller, embedded data acquisition boards, network connection and power supply.

2.1.1. *Principle and structure*

We use the DS18B20 temperature sensor and soil moisture sensors based on the principle of time domain transmission (TDT) as the temperature and moisture acquisition module, each set of sensor using a sensor controller. Embedded data acquisition boards adopts ARM processor, it can provide user data query and download service through the Web server[5]. Temperature and moisture data (including time, sensors address, etc.) are stored in SQLite database[6]. It can form a tree network through network links and a sensor controller on the embedded data acquisition board. Embedded data acquisition boards access every sensor controller via a serial port, and read the acquired data. Switching power supply power supply for the ARM processor and wireless router.

2.1.2. *Design and discuss*

In the view of the experimental requirements, the design has many changes: (i) real-time sampling of more than 80 groups of temperature and moisture sensor's data and (ii) what kind of network structure should be adopted to realize the rapid and stable data transmission.

In order to meet the requirement of distance transmission characteristic and stable transmission, we choose the RS-485 serial bus communication mode and use tree network topology .

RS-485 serial bus use the balance transmission and differential reception, so it has the ability of restraining common mode interference. And it's bus transceiver equipment has a high sensitivity, it can detect low voltage of 200mv, so the transmission signals can be restored in kilometers distance. RS-485 serial bus uses half duplex working mode, there is only one position in the sending state during the same time, as a result, the sending circuit must be controlled by enabling signal. It is very convenient for the RS-485 serial bus for multi-position connection, so we can save a lot of signal lines[7]. Serial communication use the polling mode: we acquire data through each set of temperature and moisture sensor , from the first to the last sensor, to finish the Q&A process of each set of sensors, so we use the polling mode rather than the interrupting mode[8].

Studies have shown that if we consider the obstacles and interference conditions, the data packet loss rate and response time achieve the best under the wire transmission condition[9]. The tree network topology has root node and branch node which is suitable for the branch management and system control ,also it is easy to extend and fault isolation; Bus topology structure does not have real time function; The net structure cannot be applied in the wire transmission.

2.2. *The data acquisition*

According to the meteorological monitoring data, from 1971 to 2000,North Town Station of Yulin City's annual average rainfall is 365.7 mm . July and August are the rainy season , can be as high as 178.2 mm which often appears larger storm rainfall phenomenon. Rainfall process for different location of the temperature and moisture value influence factors including buildings rammed earth thermal conductivity and the water permeability, and size of the rainfall, etc. We use the accelerated testing experiment simulate the rainfall to obtain the temperature and moisture data of the soil body through the sensor group, then we can get the distribution of water content in the wall.

The experimental soil is taken from North Town Station of Yulin City in Shaanxi Province. The wall was built by stamping earth method in the selected place. In the North Town Station, we build two rammed earth wall for experiment, which is 3 meters long, 1.5 meters wide, 2 meters high. The two walls are in east-west and north-south direction respectively, they are not connected, and the wall slope is set to 80 degrees (table 1).

Table 1. Analog wall parameter table

Top surface			Facade	Specific parameters			
form	top: bottom	slope	height: Bottom width	height	Bottom width	Top width	length
Flat-top	0.8	80°	1.3	2m	1.5m	1.2m	3m

Temperature and moisture acquisition module use the DS18B20 temperature sensor and it is based on the principle of time domain transmission (TDT) of soil moisture sensors which is miniaturization, low power and low moisture . A single set of sensor is a cylinder of 7.5 cm long, 8 mm diameter. In order to reflect the soil environment parameters and monitor the changes of internal environment parameters of soil under different depth, temperature and moisture sensors are installed inside the wall. So in the ramming earth wall, temperature and moisture sensors were laid in wall body at the same time. The placement of sensors are shown in Fig.1. A rammed earth wall has 43 sensors of temperature and moisture. The distribution is: the line distribution in the vertical center and horizontal distribution on the surface. The distribution in another rammed earth wall is the same.

Considering the soil permeability and heat transfer characteristics, we do not adopt the method of uniform distribution. Generally speaking, it is considered unsaturated rainfall infiltration into the surface can be divided into three stages: constantly absorbing water, saturated water and stable infiltration in three stages. During this, the infiltration rate is a nonlinear change from big to small. The artificial rainfall simulation use the water gravity for supply, which is the spray type automatic rainfall simulation system. Water nozzle is 6 meters high, pinhole are distributed evenly, and the rainfall uniformity coefficient is more than 0.8. The experiment begin in in the August 22th, 2015.We use 1.0 mm/min rainfall intensity, and the rainfall time is 4h. In the process of simulation of artificial rainfall, we use Handheld terminals to access database with wireless, and check the sensor data changes of the temperature and moisture in real time.

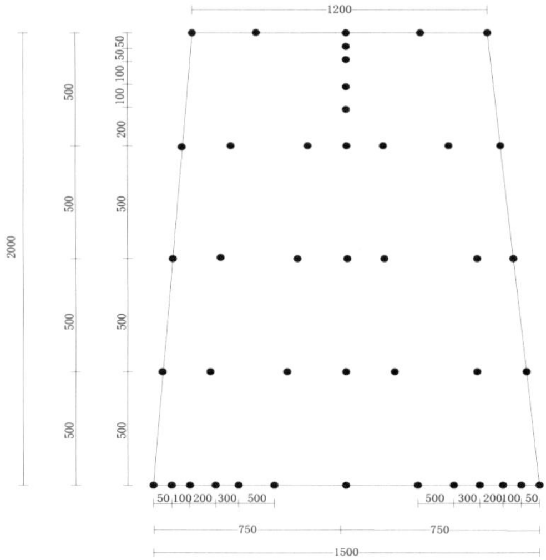

Fig 1. Stationing profile of rammed earth wall :the distance of top center between the position (temperature and moisture sensors) and the surface is 2cm , 5 cm, to 10 cm, 20 cm, 30 cm, 50 cm, 100 cm, 150 cm, 200 cm successively; The elevation of the wall is packed with sensors from 2 cm, 20 cm, 50 cm of the surface. The distance of each vertical layer is 50 cm. The distance of grounding connecting layer sensor placement is 2 cm, 5 cm, to 10 cm, 20 cm, 30 cm, 50 cm to the elevation surface.

3. Results and discussion

To analyze the process of rainfall infiltration based on the changes of the water content at any position in soil building. The process of rainfall infiltration have a total of four stages, respectively is: start with stable, a sharp rise in the water content, water content increased slowly and achieve stability at last[10]. The water content changes with time can be used for analyzing the relationship between soil water infiltration capacity variation with time. The data of the sensor which is sampled every 10 minutes was acquired in the center of the vertical direction and distance from the surface 5 cm ,and from 16:03 on October 22, 2015 to 14:53 on October 23, 2015(Fig.2). The simulation rainfall experiments was end at 15:23 on October 22, 2015.Within about 40 minutes after the time, the water content was stable at 11.25% in the position. With continuous infiltration water into the position, the water content was increasing. The water content increased from 11.25% to 18.46% over the 18 hours , which increased 7.21% in total. the position has remained steady at 18.46% of moisture content finally. It is found that the change trend of the water content at the position can reflect the relationship between the soil water infiltration capacity changes with time (Fig.3). There is a little difference in the process of a sharp rise in the water content with the theory studies complexion and the water content will decline after it stabilize for about 2 hours at 18.46%, considering the evaporation and infiltration of gravity condition result from the environmental factors and the influence of sensor error.

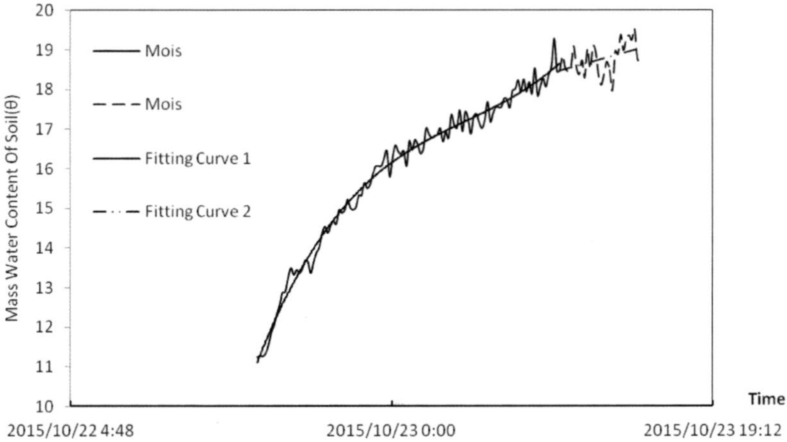

Fig 2. This is the water content changes with time

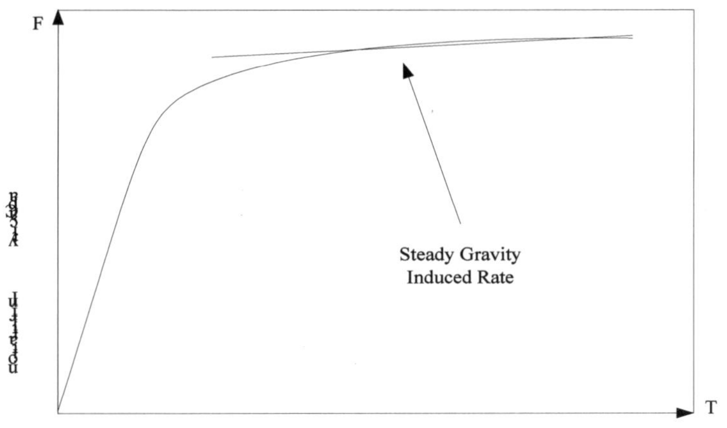

Fig 3. This is th soil water infiltration capacity variation with time[11]

4. Summary and conclusions

The temperature and moisture data acquired by tree structure sensor network system based on RS-485 network has a high reliability. This system can measure the water content changes of the soil building in real time, stability.

Data model is set up in line with the relationship model of the water content with time. In the process of data transmission in the experiment, the data packet loss rate is less than five over ten thousand. This system has a great application prospect in multi-position temperature and moisture acquisition, short distance transmission, remote monitoring, etc. There is a certain error in the experimental study of the model of the water content changes with time in the position due to manual operation and the influence of environmental factors.

Acknowledgments

This research was found by the Science Foundation of Northwest University(13NW39). Work partially supported by the World Cultural Heritage Risk Control Key Technology Research and Demonstration Project.

References

1. Sun Man-li, Research status and development of the conservation of earthen sites, Sciences of Conservation and Archaeology **4**,19(2007).
2. Liu Q, Huang X H, and Leng S P. , Deployment strategy of wireless sensor networks for Internet of Things, China Communications **8**,111(2011).
3. Tamoghna Ojha, Sudip Misra, Narendra Singh Raghuwanshi, Wireless sensor networks for agriculture: The state-of-the-art in practice and future challenges, Computers and Electronics in Agriculture **118**,66(2015).
4. Zhao Fan, Yao Xue, Sun Man li, Preliminary analysis of the destructive effect of rain on the Ming Dynasty Great Wall in Yuyang borough, Yulin, Shaanxi Province, Sciences of Conservation and Archaeology **26**,1(2014).
5. Guan Hua-qing, Transplantation and Verification of Embedded Web Server Boa, Computer Technology and Development **16**,128(2016).
6. Li Jun Feng, Application of SQLite in Embedded Middle ware System., Computer and Modernization. **5**,50(2010).
7. Li Cheng, Wang Peng, Ding Tianhuai, Chen Ken, Geng Lizhong, RS-485 bus-based high-speed serial remote data transmissions, Journal of Tsinghua University (Science and Technology) **5**,15(2009).
8. Wang Hai-Xiang, Chen Mei-jun, Design of Serial Port Expansion Based on ARM9, Computer and Modernization. **12**,25(2008).
9. Man Sha, Yang Hui-xian, Peng You, Wang Xu-si, Design of embedded wireless smart home gateway based on ARM9, Journal of Computer Applications **30**,2541(2010).
10. Dershowitz W.S, New three dimensional model for flow in fractured rock, Int.Assoc.Hydrogeol. **17**,441(1985)
11. Richards L A. , Capillary conduction of liquids in porous mediums, Physics. **1**, 318(1931).

Research on Influencing Factors of Employees' Acceptance of MOOC Training

Shuai Wang, Pin-Hua Dong and Shuai Shao,

School of Government, Beijing Normal University,

Beijing, 100875, China

†E-mail: imsean1029@gmail.com

The acceptance intention of MOOC plays a significant role in the efficiency of MOOC learning. Based on the Technology Acceptance Model (TAM), a new model concerning the influencing factors of employees' acceptance of MOOC training is established, which is tested by 300 participants who use MOOC as a routine method of training. It concludes that technical features-perceived flexibility, perceived interactivity and perceived resource advantage, the organizational feature-organizational support and individual features-individual creativity and perceived self-efficacy can influence the acceptance intention indirectly through individual perception factors. Perceived easiness can influence the acceptance intention both directly and indirectly through perceived usefulness. According to these conclusions, several suggestions are provided to take better advantage of MOOC training.

Keywords: MOOC; Training; Technology Acceptance Model; Perceived Usefulness; Acceptance.

1. Introduction

MOOC, or Massive Online Open Course, is an advanced technology based on Open Educational Resources[1]. It is considered to be an accessible education materials opening for learners regardless of cultures, occupations as well as other backgrounds.

At the very beginning MOOC is mainly employed in higher education. Teachers in the university uploaded digital materials for learners who would like to attend the course but are unable to appear in the classroom[2]. Gradually enterprises come to know the great advantages of MOOC on employment training. Compared with traditional way of training, MOOC is more flexible, cheaper and has richer resources[3].

However, its research in enterprise training is relatively rare, most of which is about the attitude towards MOOC training (Alexandria Walton Radford & Jessica Robles, 2014; Stephanie Castellano, 2014; Li Xin'e, 2014), the analysis of enterprise MOOC training cases at home and abroad (Zhang Sanbao, 2014), exploring of MOOC training system (Guo Yijun, 2015; Huang Yongping, 2015), MOOC training effect (David M. Savino, 2014; Lv Guochang, 2015) and so on.

Few studies using empirical method to explore influencing factors of MOOC training acceptance. As users of MOOC platform, employees' acceptance decides the development trend of MOOC, so research of their acceptance towards MOOC has guiding significance and practical value in enterprise training[4].

2. Literature Review

In recent years, foreign scholars' studies of the acceptance of new information systems have increased steadily, most of which were based on the theory of Technology Acceptance Model (TAM) put forward by Davis to study external influencing factors of consistent use of an information system.[5] TAM depicts the external factors that affect the actual use of information systems and reflects the logical structure of the influencing factors, making TAM one of the most significant models of adoption decisions of information technology. As shown in Figure 1, perceived usefulness and perceived ease of use are the two key factors that affect the acceptance of new technologies. Perceived usefulness means the extent to which users think they will improve their work performance when they adopt a certain system. Perceived ease of use means the extent to which users think they will save their energy when they adopt a certain system[6].

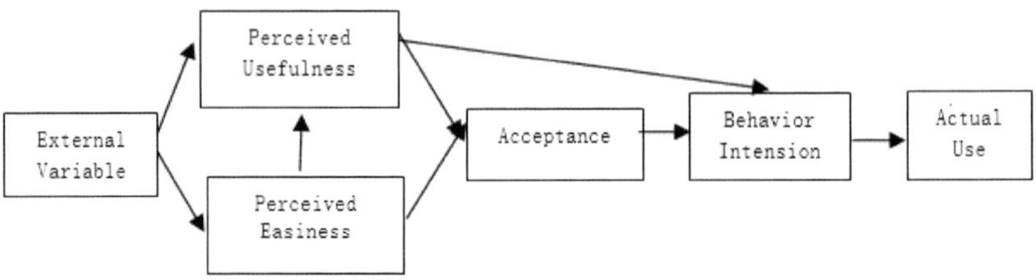

Fig.1. Technology Acceptance Model (TAM)

This paper bases the study on the current situation of MOOC training of employees in China and puts forward external variables that affect acceptance of MOOC training from three aspects, MOOC system, organization and individual, and then construct the factor model of employees' acceptance of MOOC training.

3. Research Model and Hypotheses

On the basis of Technology Acceptance Model, we build the influencing factors model of employees' acceptance of MOOC training, which is showed in Figure 2. Our hypotheses are as following.

3.1. *Variables in TAM*

According to TAM , perceived usefulness is the main influencing factor of users' intention to accept information technology and perceived easiness affects perceived usefulness directly or indirectly, thus it will affect users' accept intention[7]. Perceived usefulness is to extent to which users think MOOC training system can help them improve the work performance. Perceived easiness is the ease of use MOOC training system[8]. The following hypotheses are proposed.

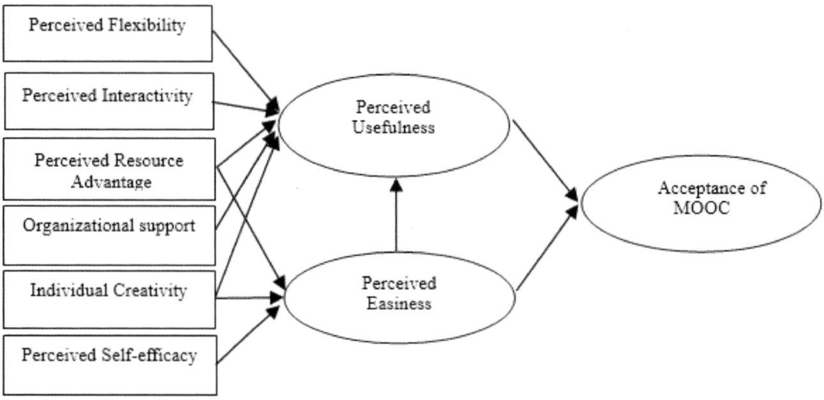

Fig.2. Influencing factors model of employees' acceptance of MOOC training

H1: Perceived usefulness positively affects users' acceptance.

H2: Perceived easiness positively affects users' acceptance.

H3: Perceived easiness positively affects perceived usefulness.

3.2. *Technical feature*

Perceived interactivity is users' perception of interoperability with platform or other users[9]. Thus the following hypothesis is proposed.

H4: Perceived interactivity positively affects perceived easiness.

In this paper perceived flexibility is defined as the degree of MOOC training's flexibility. Users can use MOOC whenever and wherever they want and choose the learning material according to themselves, they feel MOOC can improve the effectiveness of the training. So the hypothesis can be drawn.

H5: Perceived flexibility positively affects perceived usefulness.

Based on planned behavior theory, the more resources people required to support a particular behavior, the stronger control of action they will make, then the higher his behavior intention is (Azjien, 1985). So we think that:

H6: Perceived resource advantage positively affects perceived usefulness.

H7: Perceived resource advantage positively affects perceived easiness.

3.3. *Organizational feature*

Organizational support is employees' feeling of recognition of their contribution and whether their company cares about their interests and happiness or not. Rewarding both on work itself and working environment can enhance the effect of MOOC training[10]. Thus we assume that:

H8: Organizational support positively affects perceived usefulness.

3.4. *Individual feature*

Individual creativity is defined as whether people want to try new information technology. Previous studies show that Individual creativity influences on perceived usefulness and perceived ease of use. So the following assumptions are proposed.

H9: Individual creativity positively affects perceived easiness.

H10: Individual creativity positively affects perceived usefulness.

Perceived self-efficacy means the level of confidence in one's ability to some behavior in a given situation. So the hypothesis can be drawn.

H11: Perceived self-efficacy positively affects perceived easiness.

4. Data Collection and Analysis

4.1. *Questionnaire design and data collection*

The formal questionnaire includes 9 aspects, 40 items totally. Respondents were required to rate every statement ranging from 1 point to 5 points according to their real experience. The higher the score was, the more agreeable they felt. The questionnaire had been revised for several times before formal investigation to meet the criteria of validity and reliability. 300 copies of questionnaire were delivered and finally 224 copies were reclaimed as valid ones.

4.2. *Reliability analysis and validity analysis*

Cronbach's α is regarded as a good indicator to evaluate reliability. In this research Cronbach's α of 9 constructs are all above 0.7 (most of them are above 0.8), indicating that the model has adequate reliability (Table 1).

Exploratory factor analysis is applied to verify construct validity, in which Kaiser-Meyer-Olkin (KMO) and factor loadings are employed. KMO of these 9 constructs are mostly above 0.7, while factor loadings of each item are all above 0.7, indicating that the model has sufficient construct validity.

4.3. *Structural equation model*

11 hypotheses are tested by SEM with AMOS 21.0 according to the theoretical model in Figure 3 after constructing the measurement model. The

results are provided in Table 2. All the paths are significant at the level of 0.05, indicating that the hypotheses are well established.

The final Influencing factors model of employees' acceptance of MOOC training and the standardized path coefficient is shown in figure 3.

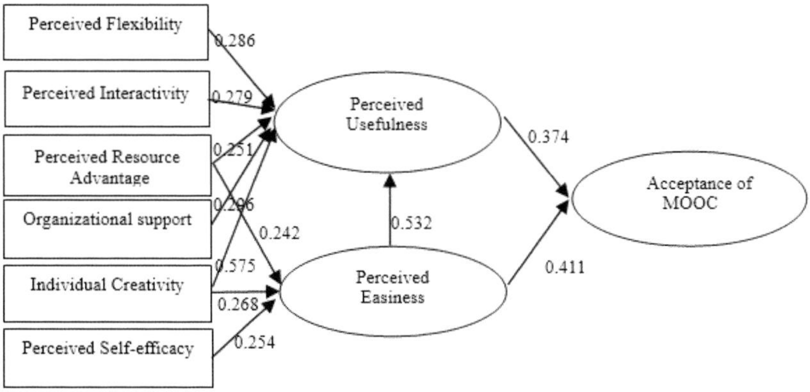

Fig.3. the standardized path coefficient

5. Discussion and conclusion

5.1. *Test of variables in TAM*

This paper uses the approach of empirical study to test interactional relation between PU, PE and Acceptance of TAM, and proves the applicability of TAM model in the study of employees' acceptance of MOOC training. Hypothesis 1 and 2 are confirmed. Direct path coefficients of PU and PE are 0.374 and 0.411, which suggests that employees are willing to accept MOOC training when they feel that adopting MOOC training will help them accomplish training and improve their work performance or it is easy to adopt MOOC training. In addition, PE has an indirect influence on acceptance by PU. Hypothesis 3 is confirmed.

5.2. *The influence of external variables*

Technical feature: PF, PI and PRA will indirectly affect users' acceptance of MOOC training by PU, and the extent of influences decrease in turn. Hypothesis 4, 5 and 6 are confirmed. And perceived resource advantage also will indirectly affect the acceptance by PE. Hypothesis 7 is confirmed. Organizational feature: Organizational support (OS) has a positive impact on Perceived Usefulness (PU) of users, and the path coefficient is 0.296. It suggests that perceived usefulness of employees will be promoted when companies offer corresponding support, such as abundant learning resources, publicity of MOOC training and related incentive mechanism. Hypothesis 8 is confirmed. Individual feature: Individual creativity and perceived self-efficacy have positive impacts on perceived ease of use, the path coefficient of which are 0.268 and 0.254. Also, individual creativity has a significant positive impact on perceived usefulness. The path coefficient is 0.575. Hypothesis 9, 10 and 11 are confirmed.

Table 1. Reliability and Validity

Factors	Cronbach's Alpha	KMO	Loading
Perceived Interactivity	.876	0.737	0.806
Perceived Flexibility	.874	0.679	0.799
Perceived Resource Advantage	.784	0.699	0.705
Organizational Support	.924	0.900	0.688
Individual Creativity	.907	0.831	0.733
Perceived Self-efficacy	.914	0.808	0.747
Perceived Usefulness	.912	0.821	0.792
Perceived Easiness	.876	0.791	0.747
Acceptance Intention	.931	0.769	0.833

Table 2. Model Hypothesis Testing Results

Path	Coefficient	Standard Coefficient	Critical Ratio	Significance	Result
H1	0.397	0.374	3.227	***	√
H2	0.372	0.411	3.796	***	√
H3	0.429	0.532	5.143	***	√
H4	0.231	0.286	3.548	***	√
H5	0.159	0.279	3.447	***	√
H6	0.234	0.251	2.944	0.004	√
H7	0.155	0.242	2.907	0.004	√
H8	0.324	0.296	3.395	***	√
H9	0.452	0.575	5.547	***	√
H10	0.163	0.268	3.294	0.001	√
H11	0.185	0.254	3.276	0.001	√

5.3. *Suggestions on the introducing of MOOC training*

Combining the conclusions of this paper, we provide some suggestions on the introducing of MOOC training to employees. First, before the introducing of MOOC training, companies should promote and advocate MOOC training to make employees familiar with MOOC and reduce resistance. Second, during the MOOC training, companies should offer good technical support and reduce the complexity and difficulty of MOOC training system to increase employees' perception ease of use. Third, companies should create a cultural atmosphere of innovation and develop self-efficacy and innovativeness of employees to increase employees' perception ease of use. Last but not least, companies should also establish MOOC training department to provide specialized support for employee training and motivate employees to accept MOOC training.

5.4. *Limitations*

Despite several findings of this paper, there are also several limitations. The selection of external variables is based on previous studies and very representative, but it's not comprehensive and detailed. In addition, the sample of the study is limited to employees in China, the conclusions may be different with other countries. Also, the small sample size of 300 respondents may not well reveal the acceptance of MOOC training.

Reference

[1] Jian-Rong Liu. Theoretical & Quantitative Research on Individual & Organizational Influence on Corporate Training Effectiveness[D]. Shanghai: East China Normal University, 2005.

[2] McGraw, M. (2013). Measuring the Impact of MOOCs, Human Resource Executive Online, January 22, found at www.hreonline.com/HRE, accessed September 10, 2013.

[3] Stephanie Castellano. (2014). MOOCs in the Workplace.

[4] Zhao-Hong Liao. Look from MOOCs Global Trend Enterprise University for the Future Development Trend[J]. China's Remote Education, 2013.

[5] Ceng Liu, Min Liu. The Revelation of Massive Open Online Courses for Collectivization Enterprise Education Training Work[J]. China Academic Journal Electronic Publishing House, 2014(8):17.

[6] Long-Zhen Hou. MOOC Application Analysis In the Enterprise Staff Training[A]. Journal Of Guangzhou Open University, 2014.

[7] Rogers E M. Diffusion of innovations [M]. Simon and Schuster， 2010.

[8] Fang Dong. The Classic of Transmission Flow Study- Talk About Rogers Diffusion of Innovation[J]. Journal of Southwest University(Social Science Edition), 2010(4):191-192.

[9] Davis F D et al.. User Acceptance of Computer Technology： A Comparison of Two Theoretical Models[J]. Management Science， 1989.

[10] Jie-Yi Wei et al. Research on Influencing Factors of Staff s' Intention to Accept E - Learning[J]. World Sci-Tech R&D, 2014(03)： 317-323+335.

Decomposition and Modeling of Equipment Support Mission*

Tian Zhao†, Wei Zhang, Yao-Ping Lv and Xian-Jun Pan

Academy of Equipment,

Beijing, China

†E-mail: ianworkmail@sina.com

As a key subject of equipment support mission planning, the relationship of equipment support mission decomposition and equipment support mission modeling is going forward one by one. We intergrade qualitative decomposition with quantitative modeling. Above all, starting at the essentials and principles analysis of mission decomposition, basing on synthetic micro-analytic approach, we design the decomposition method of equipment support mission. Then, basing on formalization model of equipment support mission, according to PERT Chart, Hexagonal Grid, L-R Trapezoidal Fuzzy Number etc. we analyze several key problems of equipment support mission modeling from three dimensions, which is time, space and information.

Keywords: Equipment Support; Mission Planning; Mission Decomposition; Mission Modeling; Synthetic Micro-analytic Approach.

1. Introduction

Decomposition and modeling of equipment support mission is a process that decompose vague, complex and un-executable equipment support missions to concrete, simple and executable equipment support missions, and model those decomposed equipment support missions. Mission decomposition adopt qualitative analysis, mission modeling adopt quantitative analysis.

2. Mission Decomposition

2.1. *Essential Analysis*

Essential of equipment support mission is the indispensable fundamental factors for keeping equipment support mission working and existing[1]. It

* This work is supported by 15GJ005-039.

† Work partially supported by grant 2-4570.5 of the Swiss National Science Foundation.

include of objective, organization and rule. A objective is the working standard and outcome description for equipment support force realizing the particular goals. A organization is the equipment support substances for executing particular equipment support mission. A rule is a statement telling equipment support force what they should do in order to achieve equipment support mission.

2.2. *Principle Analysis*

- **Dependence**. Equipment support mission have a dependency relationship to combat mission[2]. The style, scale and degree of combat mission determines the style, scale and degree of equipment support mission.
- **Hierarchy**. The structure of equipment support mission system is a pyramid-shaped. The level of mission is inversely proportional to the performability of mission.
- **Independence**. It means that the result of the equipment support mission decomposition should be independent, both in function and time.
- **Integrity**. the results of equipment support missions decomposition should ensure the integrity of the original equipment support missions, both in content and function.

2.3. *Method Analysis*

2.3.1. *Method Principle*

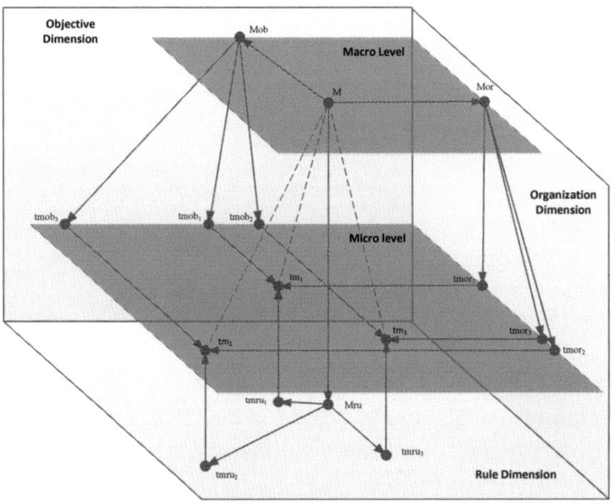

Fig. 1. Equipment support mission decomposition model based on SMA

Micro-analytic Approach Synthetic (SMA) [3] is a basic method theory for complex system analysis, which is proposed by Ying-zhi Ouyang. The method of equipment support mission decomposition, which is based on Micro-analytic Approach Synthetic, integrate the macro level and micro level of equipment support mission. It establish the analysis mechanism from macro to micro and from micro to macro. Generally speaking, the method is "two levels and three dimensions" (Figure 1). "Two levels" refer to the equipment support missions in the macro level and micro level. "Three dimensions" refer to the equipment support missions attribute separately from objective, organization, rule of three dimensions.

2.3.2. *Method Process*

The method process mainly consists of two parts, "three-dimension" micro analysis and "three dimension" matching verification (Figure 2).

Fig. 2. Decomposition process of equipment support mission

- **"Three-dimensional" micro analysis**. It is using the SMA to analyze the relationship between complex missions and simple missions from three aspects of objective, organization and rule. The basic process and method is similar. For example, MAS of objective. First of all, at the macro level to confirm the objective attribute *Mob* for equipment support complex mission *M*. Secondly, according to confirm the objective assemblage *(tmob1,tmob2,tmob3)* for simple mission assemblage *(tm1,tm2,tm3)* in micro level. Finally, the objective attribute *Mob* could restrict simple mission assemblage *(tm1,tm2,tm3)* in macro level, and the objective assemblage *(tmob1,tmob2,tmob3)* could support complex mission *M* in micro level.

- **"Three dimension" matching check**. It is a feedback mechanism. Using the iterative computation, checking out the match degree between objective assemblage *(tmob1,tmob2,tmob3),* organization assemblage *(tmor1,tmor2,tmor3)* and rule assemblage *(tmru1,tmru2,tmru3).* Then, we will keep adjusting relevant parameters until the match degree in an acceptable scope.

3. Mission Modeling

3.1. *Formalization Model*

The formalization model[4] of equipment support mission could accurately express the properties of equipment support missions. According to the characteristics of equipment support mission and the influence of operation on equipment support, the three dimensions of time, space and information[5] can be used to describe equipment support mission.

$$\begin{aligned} &T =< TID, TName, TTime, TSpace, TInformation > \\ &TTime =< Tbegin, Tduration, Tend > \\ &TInformation =< TResource, TAbility, TAmount, \ldots > \end{aligned} \tag{1}$$

Where *T* is equipment support mission, *TID* is *T'* ID, *TName* is *T'* name, *TTime* is *T'* time assemblage, *Tbegin* is *T'* start time, *Tduration* is *T'* duration time, *Tend* is *T'* end time, *TSpace* is *T'* space assemblage, *TInformation* is *T'* information assemblage, *Tresource* is *T'* resource assemblage, *Tability* is *T'* ability assemblage, *Tamount* is *T'* amount assemblage.

3.2. *Time Model*

The PERT Chart[6] could be used in time model for equipment support mission. The key points of PERT Chart is predicting for mission duration time. The accumulation of equipment support work practice is very important in prediction, especially in data accumulation, analysis and verification, which is from the daily equipment support work, equipment support training and so on. According to the Probability Theory[7], the duration of equipment support mission is subject to normal distribution. Duration prediction model of equipment support mission is using three time prediction method, as in Eq.(2):

$$ET_i = \alpha Ta_i + \beta Tb_i + \gamma Tc_i$$
$$\sigma T_i^2 = (\beta Tb_i - \alpha Ta_i)^2 \quad (2)$$
$$s.t. \alpha + \beta + \gamma = 1$$

Where ET_i is the mathematical expectation of mission i' duration, σT_i is the variance of mission i' duration, Ta_i is the most optimistic estimate of mission i' duration, Tb_i is the most pessimistic estimate of mission i' duration, Tc_i is the most probably estimate of mission i' duration. α is the coefficient of the most optimistic estimate, β is the coefficient of the most pessimistic estimate, γ is the coefficient of the most probably estimate.

3.3. *Space Model*

Because of special geometric structure and information advantage, the terrain quantitative model, that based on Hexagonal Grid[8], has become an important method in military topography. The procedure of terrain quantitative model is:

- **Step one**. By determining the length of Hexagonal Grid to define the basic grid size.
- **Step two**. Establishing the terrain vector coordinate system. The hole mission area can be represented by hexagonal grid of $m \times n$.
- **Step three**. Terrain attribute description.
- **Step four**. Hexagonal Grid quantitative model.

```
Struct Hexi
{ Int m_i ;   //Row-coordinate of Hexagonal Grid
  Int n_i ;   // Column-coordinate of Hexagonal Grid
  Float elevationi;   // Altitude of Hexagonal Grid
  Int typei;   // Attribute of Hexagonal Grid  …}
```

3.4. *Information Model*

In order to quantify the vague information for equipment support mission, we establishing the information model by L-R trapezoidal fuzzy number[9].

According to the definition of trapezoidal fuzzy number, the A is a trapezoidal fuzzy set on the domain U. α, m, n, β are the fuzzy value of fuzzy set A.

$$A = (\alpha, m, n, \beta) \mid \alpha, m, n, \beta \in U; \alpha \le m \le n \le \beta \quad (3)$$

Turning fuzzy set A to L-R fuzzy set *AL-R*.

$$A_{L-R} = (m, n, \gamma, \delta) \mid \gamma = m - \alpha, \delta = \beta - n \quad (4)$$

Numerical calculation formula of L-R fuzzy set is:

$$A_{1L-R} + A_{2L-R} = (m_1 + m_2, n_1 + n_2, \gamma_1 + \gamma_2, \delta_1 + \delta_2)$$
$$A_{1L-R} \cdot A_{2L-R} = (m_1 \cdot m_2, n_1 \cdot n_2, m_2\gamma_1 + m_1\gamma_2 - \gamma_1\gamma_2, n_2\delta_1 + n_1\delta_2 - \delta_1\delta_2)$$
$$A_{1L-R} / A_{2L-R} = (\frac{m_1}{n_2}, \frac{n_1}{m_2}, \frac{m_1\delta_2 + n_2\gamma_1}{n_2(n_2 + \delta_2)}, \frac{n_1\gamma_2 + m_2\delta_1}{m_2(m_2 + \gamma_2)}) \quad (5)$$
$$s.t. A_{1L-R} = (m_1, n_1, \gamma_1, \delta_1), A_{2L-R} = (m_2, n_2, \gamma_2, \delta_2)$$

Mathematical expectation of L-R fuzzy set *AL-R* is:

$$\mathrm{E}(A_{L-R}) = {m+n+\gamma+\delta}/{4} \quad (6)$$

4. Conclusions

Equipment support mission decomposition and modeling is the basic step of equipment support mission planning[10]. According to the theory of Micro-analytic Approach Synthetic, PERT Chart, Hexagonal Grid, L-R Trapezoidal Fuzzy Number etc. we intergrade qualitative decomposition with quantitative modeling. In further study, we will focus on data analysis and running framework for equipment support mission decomposition and modeling.

References

1. Hon-bao Mao, UAV Mission Planning, 1st ed. (National Defence Industry Press, Reading, 2015).
2. Miu Xu-dong and Wang Yong-chun, Mission Planning Theory and Application for Air Defense of Naval Vessel Formation, 1st ed. (National Defence Industry Press, Reading, 2013).
3. Ouyang Ying-zhi, Foundations of Complex-System Theories, 1st ed. (Shanghai Technological Education Publishing House, Reading, 2002).
4. Tan Yue-jin, Principle of System Engineering, 6st ed. (Science Press, Reading, 2015).
5. Liu Xing-tang, The Theory, Method and Technique for Complex System Modeling, 4th ed. (Science Press, Reading, 2011) .
6. Liang Chun-mei, Military Operation Research, 1st ed. (National Defence Industry Press, Reading, 2013) .
7. Sheng Ju, Probability Theory and Mathematical Statistics, 12th ed. (Higher Education Press, Reading, 2005) .
8. Zhou Xian-zhong, Aided Decision-Making Techniques in Command and Control System, 1st ed. (National Defence Industry Press, Reading, 2012) .
9. Hou Fu-jun, Fuzzy Preference Relation and Decision, 1st ed. (Beijing Institute of Technology Press, Reading, 2009) .
10. G.M. Levchuk, Y.N. Levchuk, Jie Luo, K.R. Pattipati, D.L. Kleinman. Normative Design of Organizations-Part I: Mission Planning. IEEE Transactions on Systems, Man, and Cybernetics, (2002).

Motion Simulation of Tracked Robot Under the Virtual Reality Environment†

Ma Jiu-he†1, Chen Zhi-wei2, Sun Hai-tao1, Deng Hui-yong2

1. Dept. of Guns Engineering, Ordinance Engineering College, Shijiazhuang 050003, China; 2. Dept. of Vehicle and Electrical Engineering, Ordnance Engineering College, Shijiazhuang 050003, China

Abstract: The motion simulation of tracked robot needs a physical model as real as possible. In view of the current situation: the existing models are mostly very complex, without experimental verification, and the majority is two-dimensional model, in order to develop a three-dimensional motion simulation platform, the simulation system including modules, such as remote control, communication protocols, motion simulation, terrain generation and three-dimensional display, is designed, the idea of dividing the motion simulation into two steps, plane motion and matching terrain, is presented, in order to achieve the three-dimensional motion simulation of the tracked robot on uneven ground.

Keywords: Tracked Robot; Virtual Reality; Motion Simulation.

1. Introduction

Creating visual simulation platform of the robot provide a flexible research tool for robot control method, path planning and remote operation. The motion performance simulation of robot in the computer, usually needs to know robot pose of each discrete time in the terrain. The general approach is: one order differential component of pose parameters of the robot in the terrain environment is obtained by the kinematics model (or kinetic), the pose parameters are obtained by integration; and then this result as an input parameter is supplied to the three-dimensional display, finally, the interaction between the robot and the terrain environment is simulated [1]. The robot visual simulation platform is mainly related to kinematics or dynamic calculation, wheel-ground interaction, pose three-dimensional display and human-computer interaction [2]. Many dynamic models have been developed for large tracked vehicles for both planar motion [3] and uneven terrain [4]. Some have modeled

† Work partially supported by grant 2-4570.5 of the Swiss National Science Foundation.

the weight distribution under the tracks as uniformly distributed [5] or not [6-9]. Li Yan [10] proposed a tracked robot model suitable for controller design, Sun Feng chun [11] makes a kinematics and dynamic analysis for tracked vehicles steering law on the slopes using the space vector method in the math mechanics, Ma Xing Guo [12] establishes the dynamic model of high-speed tracked vehicle suspension system interaction with the ground based on a multi-body dynamics theory, Huang Si ji [13] establishes a precise tracked ground mobile robot dynamic model using the Lagrange robot dynamics method and incomplete dynamic Routh equations, Wang Liang xi [14] analyses tracked vehicle steering kinematics and uniform velocity steering dynamics.

In this paper, the motion of tracked robot is divided into plane motion and matching terrain, the results of the two aspects are combined in order to calculate the pose of the robot in the terrain in real time, and the slip and skid is taken into account in the process of calculating the actual track speed, the inertia of the tracked robot is also integrated into matching the terrain.

2. Architecture of simulation system

Motion simulation module is responsible for receiving and executing commands from the remote control module, robot simulation, and running parameters feedback to the remote control module. As shown in Figure 1, the remote control module directly sends track speed to motion simulation module, the module does not directly use the track speed for motion simulation, but according to the track speed computes theoretical speed, considering the slip and skid impact, ultimately the actual track speed is achieved.

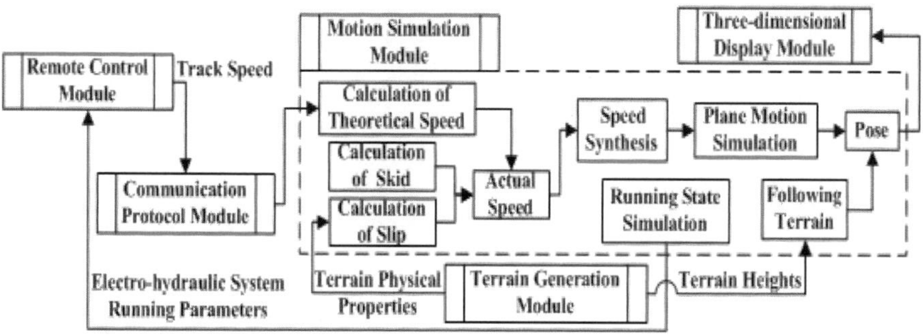

Figure 1 System architecture and simulation flowchart

3. Calculation of track speed

To calculate the amount of displacement and rotation of the robot, the speed of both tracks must be first calculated, in order to ensure realism, the slip and skid impact can not be ignored.

3.1 Calculation of the theoretical track speed

The calculation procedure is shown as follows:

(1) First, the possible maximum speed and the minimum speed of the track are calculated:

$$v_{\max} = v_0 + a \times t \tag{1}$$

$$v_{\min} = v_0 - a \times t \tag{2}$$

In the formula v_0 is the current track speed, $a = F/m$ is the maximum acceleration, t is the time elapsed since the last calculation.

(2) The robot receives data frame from the remote control terminal, which includes a forward or backward speed v_r (the speed ranges between 0-32) of the left and right track, based on this speed the theoretical speed of the track is calculated:

$$v_t = v_r / 32 \times v_{\max} \tag{3}$$

If the theoretical speed ranges between the possible minimum and maximum speeds, it was chosen as the new speed. Otherwise, the minimum or maximum speed is chosen.

3.2 Slip

On the slopes, if ground friction is not enough to offset the weight component of the robot parallel to the slope direction, the robot will slip, as shown in Figure 2:

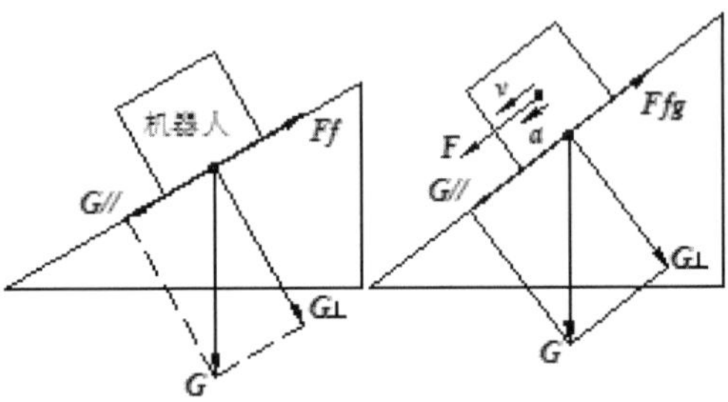

Figure 2 slip

Weight G is the force prompting the robot down, it can be decomposed into two components, respectively, parallel and perpendicular to the ground. Friction F_f is in the opposite direction of G //, equals to the $G \perp$ multiplied by the coefficient of friction μ. The friction coefficient is divided into the coefficient of static friction μ_a and the coefficient of sliding friction μ_g. As long as the force G // does not exceed F_f, there is no slip and μ_a is used to calculate F_{fa}. If at any time G // becomes larger than F_{fa}, then the robot starts to slip. From that moment on μ_g is used to calculate F_{fg} (sliding friction).

The robot acceleration along the slope direction is equal to the vector sum of F_{fg} and G // divided by the mass, because of separately considering each side of the tracks, the mass quality here is half the mass of the robot. If G // is greater than F_{fg}, the direction of the acceleration is down along the slope, and vice versa the acceleration direction is in the opposite (that is, sliding is slowing down).

3.3 skid

Skid is calculated as a factor that slows the movement of the robot. It is originally integrated into the slip. But the calculations are too complex and little improvement is provided. So a simpler model is finally adopted.

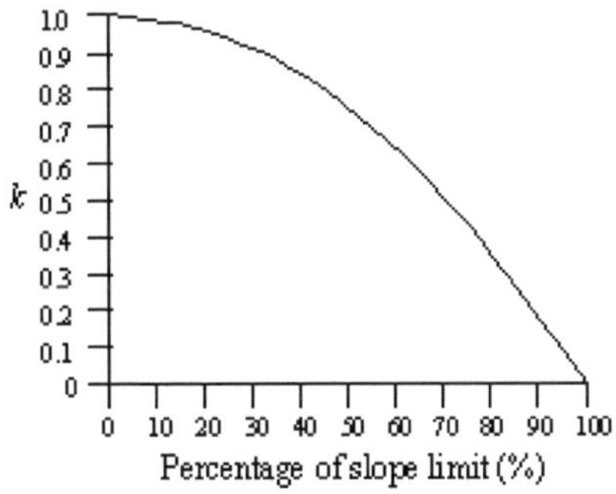

Figure 3. Interpolation of the factor k

Track skid is represented by a factor k, by which multiplies the speed of the track. This factor is equal to 1 for a horizontal ground, and 0 for a slope corresponding to the boundary sliding friction coefficient (slope for which force F is equal to the force of sliding friction). Intermediate values are interpolated. It is tested that a quadratic interpolation can give more realistic results.

4. Combination of the speed of the tracks

Even with equal speeds, the two tracks do not necessarily have the same total speed, taking into account the slip and skid. So a system is needed to transform these two different speeds in the displacement and rotation of the entire robot.

First, the track speed is decomposed into two components, one parallel to the forward direction of the robot, and the other perpendicular, to the left (or right). The overall lateral speed of the robot will be the average of the two lateral speeds.

For speed in forward or backward, there are several cases.

First case: as shown in Figure 4, both tracks will go forward, but with possibly different speeds. For forward (or backward) speed of the robot, the smaller of the two is taken, the left in Figure 4. Then the difference of the two speeds is calculated, which is called ΔV and used to calculate the rotation.

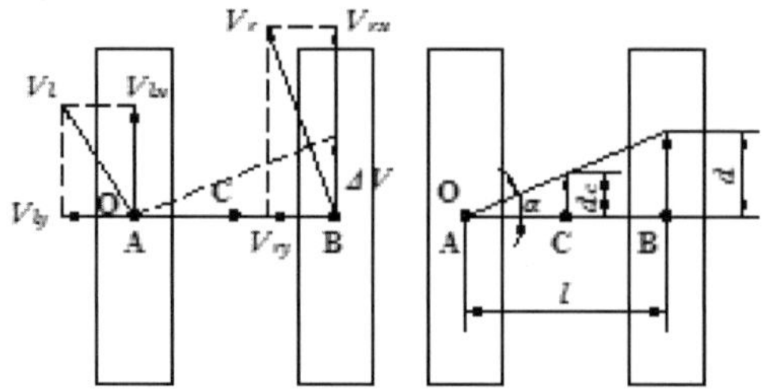

Figure 4. Tracks with the same direction of speeds

In the second case: as shown in Figure 5, the tracks are not just in the same direction, and they may have a different speed or not.

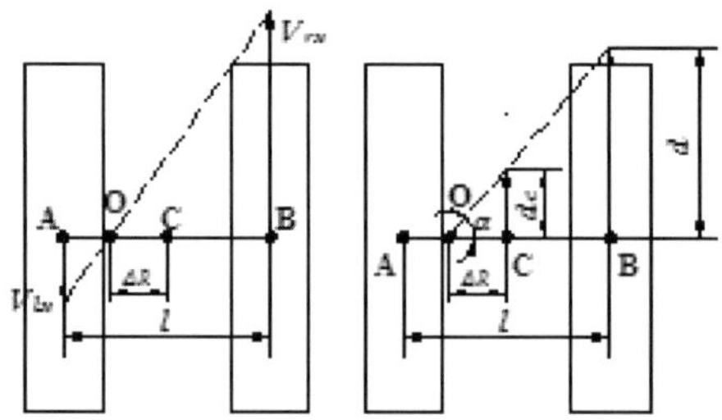

Figure 5. Tracks with speeds in the opposite direction

The distance ΔR from the center of rotation to the center of the robot is calculated, by the following formula, where $\vec{V}_1$ and $\vec{V}_2$ are the rates of the two tracks in vector form and l is the width of the robot:

$$\Delta R = \frac{\left|\vec{V}_1 + \vec{V}_2\right|}{\left|\vec{V}_1\right| + \left|\vec{V}_2\right|} \cdot \frac{l}{2} \tag{4}$$

Then the same method mentioned above is applied, that is to say that the distance d is divided by the perimeter of the circle whose radius is the half-width of the robot added with ΔR. Similarly, the progress distance of the robot is equal to the sine of the angle α multiplied by ΔR.

5. Conclusions

In fact, in the course of the development, trying to create a physical model as real as possible, in addition to the above considerations, also the subsidence of robot body is simulated through the introduction of the sedimentation coefficient, based on Java3D technology the problem of collision test is solved, also the track trace is simulated. Nevertheless, some simulation of the physical effects is given up or the simulation was not real, such as the impact of moment of inertia of the robot. But after the actual application, the researchers reflect the simulation platform provides an effective simulation tool for research on control method, path planning and remote operation of the tracked robot.

References

[1] Liu Ji, Wang Yuechao, Zhou Chuan. Method of Eliminating Wheel-terrain Interaction Errors in Lunar Rover Simulation. Journal of System Simulation, 2008, Vol. 20 No. 14: 3733-3737, 3741.

[2] Peng Xingwen, Ju Hehua, Cui Pingyuan et al. Research on Key Technology of Visual Simulation System of Lunar Rover Dynamic. Journal of System Simulation, 2008, Vol. 20 Suppl.: 355-358.

[3] Martinez, J., Mandow, A., Morales, J., Pedraza, S. and Garcia-Cerezo, A. Approximating kinematics for tracked mobile robots. *The International Journal of Robotics Research*, 2005, Vol. 24, No. 10, pp.867–878.

[4] Murakami, H., Watanabe, K. and Kitano, M. A mathematical model for spatial motion of tracked vehicles on soft ground. *Journal of Terramechanics*, 1992, Vol. 29, No. 1, pp.71–81.

[5] Shiller, Z., Serate, W. and Hua, M. Trajectory planning of tracked vehicles. *Proceedings of the IEEE Control Systems Magazine*, 1993, Vol. 3, pp.796–801.

[6] Wong, J. and Chiang, C. A general theory for skid steering of tracked vehicles on firm ground. *Institution of Mechanical Engineers*, 2001, Vol. 215, pp.343–355.

[7] Wong, J. An improved method for predicting tracked vehicle performance. *Journal of Terramechanics*, 1984, Vol. 21, No. 1, pp.35–43.

[8] Wong, J. *Theory of Ground Vehicles*, 3rd ed.. New York: John Wiley & Sons, Inc., 2001.

[9] Wong, J. and Huang, W. Evaluation of the effects of design features on tracked vehicle mobility using an advanced computer simulation model. *Int. J. Heavy Vehicle Systems*, 2005, Vol. 12, No. 4, pp.344–365.

[10] Li yan, Yang Xiangdong, Chen Ken. Dynamics model and feedback control of tracked robots. J Tsinghua Univ(Sci & Tech), 2006, Vo1.46, No.8: 1377-1380.

[11] Sun Fengchun, Shi Qinglu, Guo Fen et al. Characteristics of Dynamics of Tracked Vehicles Steering on Ramp. China Mechanical Engineering, 2007, Vol. 18, No. 22: 2766-2771.

[12] Ma Xingguo, Wu Zhaodong, You Xiaomei. Dynamical simulation on suspension system for high speed tracked armored vehicles. Chinese journal of construction machinery, 2009, Vol. 7, No. 1: 36-40.

[13] Huang Siji, Ou Yi, Tao Weijun, et al. Dynamics modeling analysis for tracked mobile robot. Journal of Mechanical & Electrical Engineering, 2011, Vo1.28, No.3: 280-285.

Research on Social Media Marketing Strategy of Small and Micro Enterprises Under the Background of "New Normal" of Economy

Fang Gao, Xin Zhao
School of Business, Jianghan University, Wuhan, Caidian, China
E-mail: gaofangwh@163.com, 342321534@qq.com

With the rapid development of the Internet, social media growing prosperity, under the background of "new normal" of economy, how to use social networking and other Internet collaboration platform and media to carry out marketing activities, public relations, maintain and develop customer service, which posed a severe test to China's small and micro enterprises. The growth of small and micro enterprises in our country is very difficult, but at the same time, it is accompanied by a lot of opportunities. Based on the analysis of the current development of small and micro enterprises in our country and current situation of social media marketing, this paper discusses the social media marketing strategy of small and micro enterprises under background of the "new normal" of economy.

Keywords: China's new normal; Small and micro enterprises; Social media; Social network; Marketing strategy.

1. Introduction

At present, the "new normal" of China's economy has been widely recognized in the economic field. Under this background, with more and more personalized and diversity of consumer lifestyle, the traditional way of enterprise marketing is facing challenges, especially for small and micro enterprises. In addition, with the progress of science and technology, the scale effect is no longer play a decisive role as before, which has provided a lot of opportunities for the growth of small and micro enterprises. With the rapid development of the Internet, how to use social networking and other Internet collaboration platform to carry out marketing activities, which posed a severe test to China's small and micro enterprises. Under the background of "new normal" of economy, how to seize the marketing opportunity to achieve a major breakthrough in small and micro enterprises, which is worth exploring.

2. Theoretical basis

2.1. *The development of small and micro enterprises under the background of "new normal" of economy*

Lian (2014) [6] through the observation of the development trend of China's economy, summarizes the six characteristics of the "new normal": GDP growth in high-speed; minor fluctuations in economic growth; employment balance; low inflation; economic operation of locality financial risk; limited stimulus. Zhang (2015) [9] makes a more detailed definition of the meaning of the China's new normal: the average annual GDP growth of around 7%; show the vitality of reform; ranking of industrial structure is tertiary, secondary and primary; rapid development of emerging industries; economic development tends to be balanced and diversified; the general price level to maintain an average of about 2%. Zhu (2015) [10] pointed out that, under the background of "new normal", to promote the development of small and micro enterprises has important strategic significance, not only can increase employment, improve people's livelihood, but also to promote enterprise restructuring, promote technology innovation.

Chen (2015) [1] etc. explored the unfavorable factors which hinder the growth of small and micro enterprises under the background of "new normal" of economy, including congenitally deficient, its weak; lower control and gradually rising production cost; lack of funds; obvious difficulties on human resources management.

2.2. *Social media marketing for small and micro enterprises*

Constantinides (2014) [2] put forward two possible social media marketing strategy. One is a passive approach, focused on the use of customer reflect and market intelligence in the social media field. The other is a positive approach, social media as a channel for marketing and handling relationships, as personalized products and tools, and finally develop as a cooperative platform.

Kilgour etc. (2015) [5] study enterprises how to effectively use the social media channels, they believe that the key of successful communication strategy is to deliver the message to the target audience, and realize the interaction between the customer and enterprise. Dahnil etc. (2014) [3] study the factors that drive the small and medium enterprises and organizations to adopt social media marketing. And Jones etc. (2015) [4] study the impact of Internet and social media on small business. In view of the actual situation of small enterprises, puts forward the measures and suggestions for the use of network and social media to promote the development of enterprises. Wang (2015) [7], Wang (2015) [8] etc. analysis on the marketing situation of our country small and micro enterprises, aiming at the existing problems, and put forward the specific countermeasures and solutions.

3. The current development status of China's small and micro enterprises

3.1. *Analysis of the current difficulties faced by small and micro enterprises*

3.1.1. *Production costs rose too fast*

In recent years, the price level rising, small and micro enterprises due to its own limitations show more powerless for rising raw materials, human capital and financial costs, which set up obstacles to the realization of business survival and development. With operating costs rising, it is difficult to earn high efficiency, even caught in a vicious cycle of loss. With the decreased business volume and overall increase in operating costs, small and micro enterprises face dilemma.

3.1.2. *Financing difficulties, lack of funds*

As the basic component of the market economy, the basic point of the national economy, the small and micro enterprises play a limited role in resource allocation process. Plus its smaller capital investment, less number of capital access path, easily volatile floating capital and this series of factors will push the small and micro enterprises into the crack of large enterprises, monopoly enterprises and listed companies.

3.1.3. *Difficult employment*

Although the number of graduating students is huge, but their career planning chaos, unable to position themselves correctly, cannot be rational view of the job. Even accept the training of enterprises, they will leave for various reasons, liquidity is too large. Small and micro enterprises due to the low assets, lower wages and benefits provided, which lead to a big difficulty in recruiting employees and maintaining staff stability.

3.2. *Analysis of the development opportunities of China's small and micro enterprises*

3.2.1. *Intensified government support*

Under the background of "new normal" of economy, to speed up the development of small and micro enterprises, the government has issued a lot of policy support, in order to achieve the purpose of reducing taxes and reducing the financing difficulties. In addition, the State Council has released more operational and targeted "New Ten National Notices", aimed at providing policy support and assistance to the coordinated and stable development of small and micro enterprises from three important aspects of taxation, spending and monetary.

3.2.2. *Residents consumption upgrade*

Basic consumption in the proportion of total consumption is low, the changes in consumption structure shows that the consumers' purpose from the survival of switch to the enjoyment, people spent on information, tourism, culture, health care costs finest gradually increased. In addition, thanks to the rapid progress of network technology, online shopping has been accepted by more people, the consumer gradually transferred from the line to online. County and urban residents also bring great opportunities for the market on the basis of accelerating urbanization in China.

3.2.3. *Personalized demand is universal*

With the continuous improvement and development of China's market economy, the consumption demand of residents gradually changed from the universal consumption to the pursuit of personality and individual characteristics. The future is individuality, the pursuit of personality show era, sensitive to changes in market demand for personalized and reflect them accordingly, is the key to determine whether to occupy the market dominance. Small and micro enterprises need to do is to be able to deeply understand the needs of users, so as to achieve the transition from low-end services to high-end services.

4. Current situation of social media marketing of small and micro enterprises in China

4.1. *Lack of understanding of social media marketing*

Social media is born with technological advancements, most of the small and micro business operators subject to their own culture and education, their thoughts easily rigid, still cannot accept the advanced enterprise management mode. Although currently part of small and micro enterprises groping at the plan and carry out social media marketing campaign, overall, the operators are still one-sided understanding of social media marketing, unable to face, there is no clear understanding that the social media marketing used properly can make enterprises occupy more virtual market share.

4.2. *Uneven quality of personnel*

The actual situation is that small and micro enterprises existing staff lack of social media marketing knowledge and skills, and therefore the enterprises are facing recruitment difficulties and hard to keep staff. Small and micro enterprises currently in order to better carry out the work require its employees to understand the business environment, rapid input and presided over the work. However, even if there are people with such ability, it is difficult to stay for a long time for their own development. The end result is a vicious circle of talent.

4.3. *Lack of effective social media marketing activities*

At present, China's small and micro enterprises are scrambling to establish the official website, but they did not think carefully about the future of their own, realistic, rarely enrich the web content, web site information cannot be updated timely leading to a disconnect with the users. Meanwhile, the enterprise cannot be positioned correctly, nor efficient implementation of the marketing plan, resulting in the information of enterprises unable targeted delivery to the user, finally did not achieve the first expected marketing results.

5. Social media marketing strategy of small and micro enterprises under the background of "new normal" of economy

5.1. *Strengthen the importance of social media marketing*

Social media marketing is a derivative of scientific and technological progress, and if its unique advantages used properly, will produce a comprehensive promotion for small and micro enterprises. With the social media gradually penetrated into our lives, the words and deeds manner produced a silent change and people began to recognize the new forms of consumption.

5.2. *Build high cohesion and creativity of social media marketing team*

Small and micro enterprises need to increase training of staff about social media marketing related knowledge and skills, establish perfect on-the-job personnel training plan. Establish a multifaceted system management, institutional manage marketing staff, so that the behavior of employees more in line with the institutional norms. Through the incentive measures, to achieve the combination of short and long effects, personal motivation and team motivation, so as to build a high cohesion and loyalty social media marketing team.

5.3. *Integration of social media marketing strategy based on enterprise characteristics*

5.3.1. *Select appropriate social media platforms and target customers*

In general, the crowd gathered is different from the different social media. Therefore, small and micro enterprises need to accurately select its market and target customers. Only accurate positioning, enterprises can choose the most appropriate social media marketing platform and more efficient planning and development.

5.3.2. *Focus on product information publish and update*

In general, social media users will share and discuss for the information they are interested in at any time. Therefore, the release of information can not be arbitrary. Exhibiting form and content of information will influence the audience's perception of a certain extent. If the information is published properly, it will cover the advantages of enterprises, to enlarge the disadvantage of the enterprise, and even cause the user's aversion. Once this happen, companies need to spend energy to recover.

References

1. Chen Y. Research on the Development Difficulties and Countermeasures of Small Micro Enterprise under the"New Normal"in China[J].Value Engineering, 2015, 20: 168-170.
2. Constantinides E. Foundations of Social Media Marketing [J]. Procedia-Social and Behavioral Sciences, 2014, (148): 40-57.
3. Dahnil M.I., Marzuki K.M., Langgat J., Fabeil N.F. Factors Influencing SMEs Adoption of Social Media Marketing[J]. Procedia-Social and Behavioral Sciences, 2014, (148): 119-126.
4. Jones N., Borgman R., Ulusoy E. Impact of Social Media on Small Businesses [J]. Journal of Small Business and Enterprise Development, 2015, 22(4): 611-632.
5. Kilgour M., Sasser S.L., Larke R. The Social Media Transformation Process: Curating Content into Strategy [J]. Corporate Communications: An International Journal, 2015, 20(3): 490-495.
6. Lian P. The Analysis of New Normalcy of China Economic Operation[J]. Scientific Development, 2014, 8: 5-8.
7. Wang G.H. Marketing status and strategy of small and micro enterprises [J]. Modern Marketing, 2015, 7: 58-59.
8. Wang L., Liu G. Mobile Internet Marketing Strategies for SMEs[J]. Business Economy, 2015, 7: 96-98.
9. Zhang Z.C. The New Normal of Chinese Economy and New Trend of Its Sustainable Developmen [J]. Henan Science, 2015, 1: 91-98.
10. Zhu Y.S. On the Marketing Transformation of Micro and Small Businesses under China's Economic New Normal[J]. Journal of Hunan Business College, 2015, 4: 64-68.

Section 2

Optics and Electronics Engineering

A Novel Optical Fiber Transmission System Based on Dynamic Payload Clock Recovery Technology

Jiang-Yi Qin†, Zhi-Ping Huang, Jun-Yu Wei, Zhi Dong and Zhen Zuo

College of Mechatronic Engineering and Automation, National University of Defense Technology, Changsha 410073, Hunan Province, P. R. China

†E-mail: qjyacmilan@163.com

A novel optical fiber video transmission system based on dynamic payload clock recovery technology is proposed to achieve high reliability transmission performances. In this transmission system, a dynamic payload clock recovery algorithm is introduced to detect the payload clock change automatically. Compared to traditional optical fiber video transmission system, the proposed designation combines the external Digital Phase Locked Loop (DPLL) and the Field Programmable Gate Array (FPGA) PLL to recover the payload clock accurately. Another improvement is analyzing the Camera Link protocol and encoding the video signal with additional information for the clock recovery algorithm. Finally, the performance of the designed system is demonstrated by the simulations and experiment results in actual video transmission engineering projects. The result has proved the system can operate at a rate of up to 7.14Gbps and the transmission link is stable with no error for more than 72 hours.

Keywords: Optical Fiber Video Transmission; Dynamic Payload; Clock Recovery; Channel Encoding; Camera Link.

1. Introduction

As the development of modern industrial technology and video imaging technology, the digital video is applied more and more widely. The high speed camera utilizes Camera Link cable to transmit video data and control signal with the Camera Link cable. The maximum transmission rate is up to 7.14Gbps in the Camera Link full mode. However， the standard transmission distance is only 3 meters and not over 10 meters. Additionally, the speed of transmission and the transmission distance is inversely proportional relationship. Therefore, high speed digital video transmission technology is the difficult problem in application. Optical communication with large capacity, strong anti-interference ability, good secrecy performance, transmission distance and so the advantage is very suitable for the transmission of the video signal [1-4]. In the optical fiber video transmission system, the payload is the high speed video signal data and the payload clock is the pixel clock. In some special application, the camera running mode will be changed dynamically which will make the pixel clock changed. Even in the same camera running mode, the pixel clock will be changed dynamically. Therefore, how to recover the pixel clock accurately is one of the most important factors for recovering the original video signal. If the recovered pixel clock is different from the original pixel clock, in the receiver, the video signal recovery will be disordered and the video signal transmission will be interrupted. For this reason, the optical fiber video transmission system must recover the dynamical pixel clock of high speed video signal automatically and use the clock to recover the original video signal accurately.

2. System model

The schematic diagram of an optical fiber video transmission system is shown in Fig. 1.

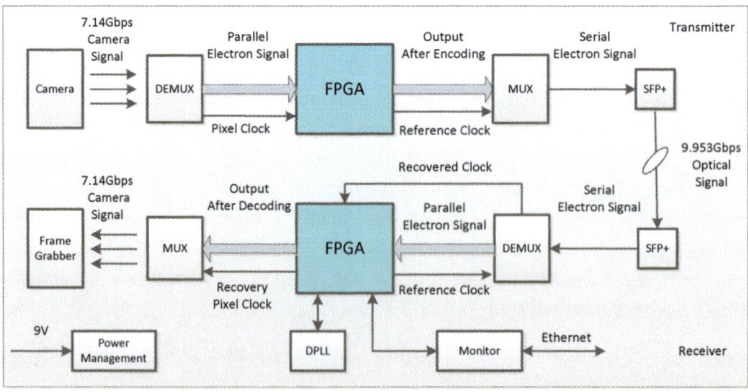

Fig. 1. Schematic diagram of the optical fiber video transmission system

The FPGA is the processing and controlling center [5]. The video signal is generated by a high speed industry camera which has a digital Camera Link Full interface. In Full mode, the data rate of video signal is up to 7.14Gbps. Therefore, in the transmitter of the system, a demux device converts the high speed video signal to 84 channels parallel-signal including pixel clock to fit the FPGA receiving. The FPGA will encode the video signal according to Camera Link protocol. The details of the encoding format come in the following passage. After encoding, the signal will implement parallel-serial conversion in a mux device and be transmitted through optical fiber. In the receiver, a SFP+ is used to receive high speed optical signal, and a demux device converts the high speed serial electron signal to 16 channels parallel-signal to fit the FPGA receiving firstly. Then, the demux device can recover an optical signal transmission clock which can be used to recover the pixel clock and receive the optical signal. Next, the FPGA will decode the signal and recover the pixel clock by an external DPLL. Finally, the video signal will be restored by the recovered pixel clock and send to a frame grabber through a mux device

In order to ensure the optical fiber video transmission system in a normal operation, the system needs a proper power management and a monitor model. After these processes, the frame grabber can use the high speed video signal inerrably. Furthermore, the optical fiber video transmission system can transmit camera control signal at the same time. However, this function will not be introduced here.

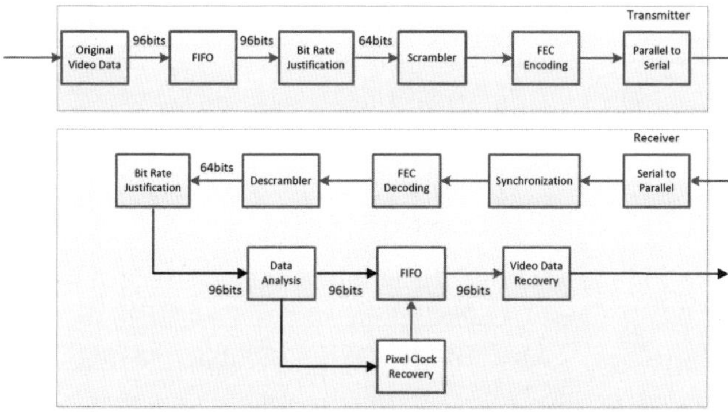

Fig. 2. Data stream of the software in FPGA

The data stream in the FPAG is shown in Fig. 2. In the transmitter, the original video signal data and camera control signal constitute a 96bits data. First-In-First-Out memories (FIFOs) with 96bits in port and 96bits out port cache the data and implement the bit rate justification. The maximum rate of camera pixel clock is up to 85 MHz and the optical signal transmission clock is 155.52 MHz, the bit rate justification is necessary for the optical fiber video transmission system. Then, a scrambler is used to enhance the efficiency of clock and data recovery. What is more, Forward Error-Correction (FEC) techniques are used to reduce the errors which occur in transmissions [6-8]. Through parallel-serial convention, the data is sent to next model for optical transmission. In the receiver, the FPGA should synchronize the receiving data firstly. During the synchronization, a clock is recovered by the demux device and the clock has the same frequency with the optical signal transmission clock.

According to the encode format, FPGA implement the FEC decoding and descrambling. Then, the system uses the clock which is recovered by the demux device to count an unstable 8 KHz period for the DPLL and send 84bits data to FIFO. The DPLL will return a stable clock to FPGA, and FPGA can use the clock to recover pixel clock. The recovered pixel clock will be used to read the data from FIFO. At last, according to the Camera Link protocol, FPGA restored the original video signal and send the data to next model.

3. Dynamic payload clock recover algorithm

Pixel clock recovery is the most important part of high speed optical fiber video transmission. The performance of the recovered pixel clock will affect the quality and distance of transmission. In this part, we will propose an algorithm of dynamic payload clock recovery in detail.

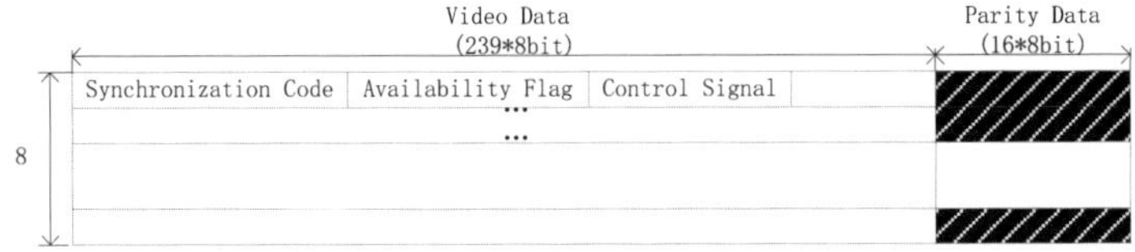

Fig. 3. Frame format of optical signal

Firstly, we will introduce the frame format of optical signal which is shown in Fig. 3. The whole frame is composed by 8 rows. Each row is a Reed-Solomon (RS) error-correction codeword [9-10]. According to the transmission requirements, we adapt RS (255,239) codes to control the errors occurred during transmission. Each codeword is composed of 239 information bytes and 16 parity bytes, which can correct at most 8 random error bytes. The frame header is composed of 128bits data and the first 64bits data is used for synchronization. As previous mention, the data rate is different between the video signal and optical signal. Therefore, the next 64bits data are used to indicate the available number of video data in the frame and transmit camera control signal. In the receiver, we will detect the frame header to judge whether these data are available.

In the system, we use an external DPLL to obtain a stable clock which is related to the pixel clock, and then utilize the FPGA PLL to recover the pixel clock. The DPLL require an 8 KHz period as the input reference. The available number in the frame header is used to count the 8 KHz period. In this system, the frequency multiplication factor of the DPLL is 3125. The frequency multiplication factor of FPGA PLL is N/25 [11-12]. Therefore, the detailed steps can be described as follows:

(1) Generate a local 8 KHz clock by FPGA PLL.

(2) Utilize the recovered optical signal transmission clock to count the available number in a single clock cycle of the local 8 KHz period and note the available number as A.

(3) Look up the parameter in graph which is proximate to A. And the parameter is noted as 125*M.

(4) Utilize the recovered optical signal transmission clock to generate another 8 KHz period for DPLL. If the data is available, utilize the recovered optical signal transmission clock to count (A-1)/2 for high signal, and the next (A+1)/2 for low signal. Suppose the A is even number. The high and low signals constitute the single clock cycle of the 8 KHz period which is sent to DPLL. Note the 8 KHz period as f8k.

(5) FPGA receives the stable clock comes from DPLL and multiplies the clock by N/25. Therefore, the recovered pixel clock (fpixel) can be expressed as follow.

$$f_{pixel} = f_{8k} * 3125 * N / 25 \tag{1}$$

Consequently, when the pixel clock is changed, the system can detect the change and search the fpixel automatically as the steps from (1) to (5).

Most DPLL have the input timing requirement for the 8 KHz period, the DPLL of the system require the input period is from 7.8125 KHz to 8.2644 KHz. The relation can be expressed as follow.

$$7.8125 < f_{pixel} / 125M < 8.2644 \tag{2}$$

According to the Camera Link protocol, the fpixel is range from 20MHz to 85MHz. The relationship between M value and fpixel is shown as follow.

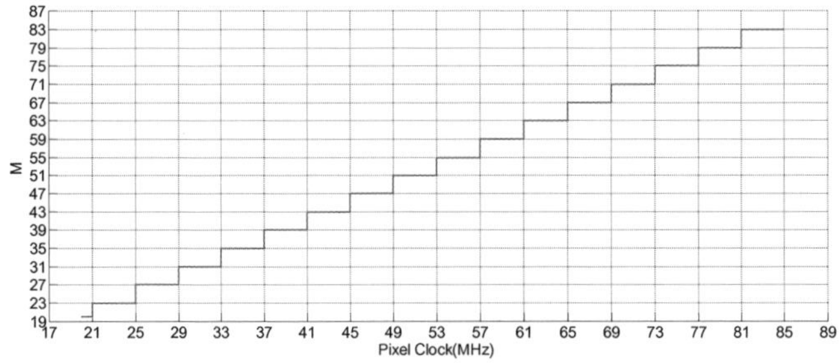

Fig. 4. The graph of M value and f_{pixel}

In the receiver, if the recovered pixel clock is faster than the original pixel clock, the FIFO will be empty. If the recovered pixel is slower than the original pixel clock, the FIFO will overflow. Therefore, when the FIFO is overflow or empty, it means that the recovered pixel clock is wrong or the original pixel clock is changed. What is more, the time that FIFO become overflow or empty is depend on the recovered pixel clock. If the recovered pixel clock is close to the original pixel clock, the FIFO will overflow or empty slower. However, if it is much different from the original one, the FIFO will overflow or empty faster.

Therefore, when the FIFO is overflow or empty, system should reset the parameters and the models firstly. Then recover the pixel clock as the steps from (1) to (5).

The relationship between the amount of data stored in FIFO, M value and running time is shown as follows.

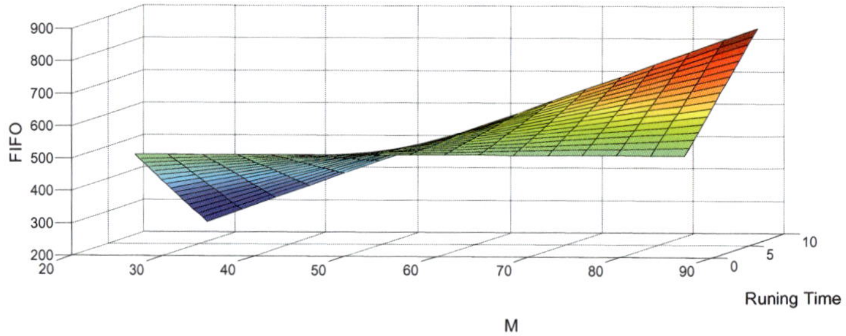

Fig. 5. Ideal simulative relationship between FIFO, M value and running time

In this simulation, the original pixel clock is 55MHz, the deep of the FIFO is 1024 words and the wide of the FIFO is 84bits. From Fig. 5, we can see that the M value near 55, the amount of data stored in FIFO is stable. However, when the M value is much bigger or smaller than 55, the amount of data stored in FIFO will decrease or increase obviously.

4. Experiment results

In this section we present the experiment in an actual optical fiber video transmission system. In this experiment, we use Optronis CL600*2 camera to generate original video signal. The camera is running in the full and free running mode. The Camera link output is Full 8x8bits. Pixel resolution is 1280*1024 and the frame rate is 500 fps. Moreover, the pixel clock is 66 MHz and the optical signal transmission distance is 1 kilometers. In order to verify the validity and stability of the algorithm, we use the test images of CL600*2 camera for our experiment. Test images are generated in the camera FPGA, independent of the image sensor. We can use test images to check the transmission path from the camera to the frame grabber. The ramp test image outputs a constant pattern with increasing grey level from the left to the right side and the single grayscale test image outputs only one certain grayscale. In the receiver, system recover the original video signal based on the principles descript in section 2~3. Applying the dynamic payload clock recover algorithm, we get the M value is 67. Meanwhile we have recorded the grayscale distribution of ramp test image and single grayscale test image as Fig. 6 and Fig. 7 have shown.

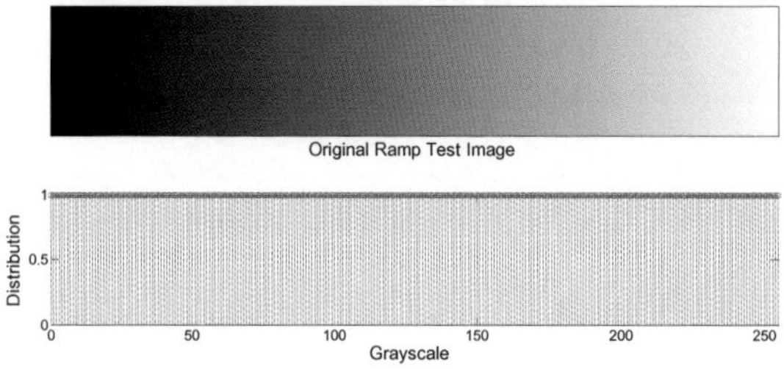

Fig. 6. Experiment results of ramp test image

Fig. 6 shows the original ramp test image in the transmitter and the grayscale distribution in the receiver. Because in the ramp test image, the grayscale sequence contains every possible grayscale that is repeated for every row and these grayscale have the same distribution. Therefore, in the receiver, the grayscale distribution we recorded indicates that every grayscale has the same distribution and it means that the system receives and recovers the original ramp test image accurately. Fig. 7 shows the original single grayscale test image in the transmitter and the grayscale distribution in the receiver. In this experiment, the grayscale value of the single grayscale test image in transmitter is 100. There is only one grayscale which is 100 appearing in the single grayscale test image distribution chart and it indicates that the system receives and recovers the original single grayscale test image accurately.

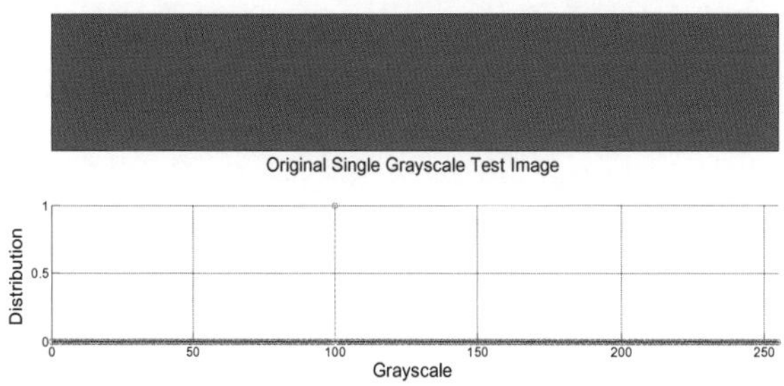

Fig. 7. Experiment results of single grayscale test image

We use the single grayscale test image to test the Image Bit-Error-Rate (IBER) of the optical fiber video transmission system. If the other grayscale is not equal to 0 in the distribution chart, the system has IBER. The IBER experiment results show that the system has not appeared any IBER more than

72 hours. It is a long time for most applications. And the M value is adjusted to 67 in more than 1000 consecutive tests.

5. Conclusion

A design of optical fiber video transmission system based on a novel dynamic payload clock recover technology is proposed in this paper. Simulation and experiment results on actual optical fiber video transmission system show the effectiveness of the proposed system and technology. What is more, the design is useful and significant for other domains which need the optical fiber transmission.

References

1. Jing Zhou, Zhiping Huang, Chunwu Liu, Shaojing Su, Design of adaptive dispersion compensation and demodulation system for 40Gbps RZ-DQPSK optical fiber communication receivers, Optik 125 (2014) 370-373.
2. Carlos Eduardo Maffini Santos, Eduardo Parente Ribeiro, Carlos Marcelo Pedroso, The application of neural networks to improve the quality of experience of video transmission over IP networks, Engineering Applications of Artificial Intelligence 27 (2014) 137-147.
3. Danping Ren, Hui Li, Yuefeng Ji, Demonstration of QoS-aware wireless protection scheme for video service in fiber-wireless access network Optik 124 (2013) 1827-1831.
4. A. Reaz, V. Ramamurthi, S. Sarkar, D. Ghosal, S. Dixit, B. Mukherjee, CaDAR: an efficient routing algorithm for a wireless-optical broadband access network (WOBAN), J. Opt. Commun. Netw. 1 (2009) 392-403.
5. Zhou Jing , Huang Zhiping, Liu Chunwu, Su Shaojing, Zhang Yimeng, A resource-saving decoding scheme of the orthogonally concatenated codes for optical transport networks in FPGA, Optik 125 (2014) 1174-1177.
6. B.P. Smith, A. Farhood, A. Hunt, F.R. Kschischang, J. Lodge, Staircase codes: FEC for 100 Gbps OTN, J. Lightwave. Technol. 30 (2012) 110-117.
7. ITU-T G.975.1, Forward error correction for high bit rate DWDM submarine systems, 2004.
8. C. Xie, Y. Zhao, Z. Xiao, et al., FEC for high speed optical transmission, in: Proc. SPIE Int. Soc. Opt. Eng., Shanghai, China, 2011.
9. T. Mizuochi, Y. Konishi, Y. Miyata, et al., Experimental demonstration of concatenated LDPC and RS codes by FPGAs emulation, IEEE Photon. Technol. Lett.21 (2009) 1302-1304.
10. Zhou Jing, Huang Zhiping, Su Shaojing, Blind frame synchronization of Reed-Solomon coded optical transmission systems, Optik 124 (2013) 998-1002
11. Junming Ye, Guangxiang Zhou, Haiyan Liu, Design and Research of Improved Digital Phase-Locked Loop Based on FPGA, Procedia Engineering 29 (2012) 3327-3332.
12. Jianguo Yuan, Qingping He, Yunxia Hu, Hao Li, Lin Wang, Wenlong Liu, Ya Xie, Optimization analysis on the bandwidth algorithm for the phase-locked loop (PLL) in OUT, Optik 123 (2012) 1681-1685.

The Research of Signal Power Quality Disturbance Identification and Classification Based on the GA-ELM

Bo Sang[†] and Hong-Wen Liu

Department of Vehicle and Electrical Engineering, Ordnance Engineering College, Shijiazhuang, China

[†]*E-mail: 726932384@qq.com*

Power quality disturbance identification and classification are important for power quality management, for which a new method is proposed to deal with power quality disturbance identification and classification based on GA-ELM (Genetic Algorithm-Extreme Learning Machine). Decompose the disturbance signals with wavelet for ten layers, it is proposed that using standard deviation and mean of the energy difference, and the ratio of them as the feature vectors. In addition, the root mean of disturbance signals and normal signals is calculated as a supplement in order to reduce the dimension of the importing vectors. The Input Layer and the Hidden Layer Connection Weights and Hidden Layer Neuron Threshold which are set randomly in ELM can influence classification accuracy. So it is proposed that ELM training error is used as the fitness function of GA to optimize Input Layer and the Hidden Layer Connection Weights and Hidden Layer Neuron Threshold to enhance the accuracy of classification and also maintain high classification speed. Simulation results demonstrate that this method can accurately and effectively identify seven common disturbance types.

Keywords: Power Quality; Wavelet Transform; Feature Vector; Classification; ELM; GA..

1. Introduction

Power quality is put forward higher requirements with the extensive use of impact and nonlinear loads. So power quality monitoring is becoming necessary in order to ensure safety and stable operation of power system. Power quality disturbances classification plays an important role in the power quality monitoring. At present the main steps of power quality disturbances classification are feature extraction and classifier design. Feature extraction is the signal processing essentially, the main method are Short-time Fourier Transform, Wavelet Transform, S-Transform, HHT, Mathematical Morphology, e.g.[1]-[5], which can be used to analyze the non-stationary signal power quality disturbances effectively because of the Time-Frequency domain characteristics. However, these methods also have some disadvantages like complicated calculation, decline of classification rate and other issues. Classifier design are currently used Neural Networks, Support Vector Machines, Fuzzy Systems and other intelligent algorithms[3],[5]-[7]. These intelligent algorithms can be achieved relatively high level in the classification accuracy, but they usually spend a lot of time training because of large number of samples. So classification speed cannot be achieved the requirements.

In this paper, input vectors dimension are reduced through the effective extraction of feature vectors and ELM[8]-[9] is used as the classifier design. It is proposed that using GA to optimize Input Layer and the Hidden Layer Connection Weights which is set ω and Hidden Layer Neuron Threshold which is set b in ELM, which can improve classification accuracy while maintaining a high classification speed. Simulation results show that this method can accurately and effectively identify seven common disturbance types.

2. The Basic Principles of GA-ELM

GA[11] consists three key parts: selection operation, crossover operation and mutation operation. GA optimizes the parameters of a group of series coded with the chosen fitness function to achieve selection, crossover and mutation operations following the rule of "survival of the fittest evolution", which can preserve the better individual and eliminate the worse individual until the best individual is chosen.

Cross Validation (CV) is used to select ω and b which have a great influence on the ELM classification accuracy. Firstly original samples are divided into Group K, with one group as test set, the rest of the group as training set. Then the average of these final classification accuracy of the K model is chosen as the performance of classifier. ELM training error is used as the fitness function of GA to optimize ω and b. Fitness function expression as in Eq. (1).

$$f_{cv} = \frac{1}{K}\sum_{i=1}^{K} g_i \tag{1}$$

Where: f_{cv} is average precision under CV; K=5; g_i is the classification accuracy of each sample.

The purpose of selection operation is to choose the better individual from the old to the new population groups with a certain probability. In this paper, roulette method is used in selection operation. Set the ratio of fitness as a selection tactic and the selection probability of the *jth* classification accuracy rate is p_j as in Eq. (2).

$$P_j = \frac{af_{cvj}}{\sum_{m=1}^{N} h_m} \tag{2}$$

Where: f_{cvj} is the *jth* fitness value of the classification result; α is coefficient; a larger f_{cvj} means a higher classification accuracy, individual with larger f_{cvj} can be saved more easily; N is the number of population, where N = 20.

The purpose of crossover operation is to exchange the two individual chromosomes randomly and get a new outstanding individual, operation as in Eq. (3).

$$\begin{cases} r_{mk} = r_{mk}(1-s) + r_{nk} \\ r_{nk} = r_{nk}(1-s) + r_{mk} \end{cases} \tag{3}$$

Where: *s* is a random number, crossover probability is 0.3; r_{mk} is the *mth* chromosome on the *kth* bit; r_{nk} is the *nth chromosome* on the *kth* bit.

Mutation operation is to choose an individual *m* and mutate the *nth* gene r_{mn} of the individual chromosome in order to get a new outstanding individual, operation as in Eq. (4).

$$r_{mn} = \begin{cases} r_{mn} + (r_{mn} - r_{\max}) \cdot h(x) & g > 0.5 \\ r_{mn} + (r_{\min} - r_{mn}) \cdot h(x) & g \le 0.5 \end{cases} \tag{4}$$

Where: $r_{\max}$, $r_{\min}$ are respectively upper bound, lower bound of gene r_{mn} ;$h(x)=b(1-x/X_{max})^2$,*b* is a random number, *x* and X_{max} are respectively current number of iterations and the maximum number of iterations; X_{max}=50; mutation probability is 0.1. GA-ELM specific program flow is shown in Figure 1:

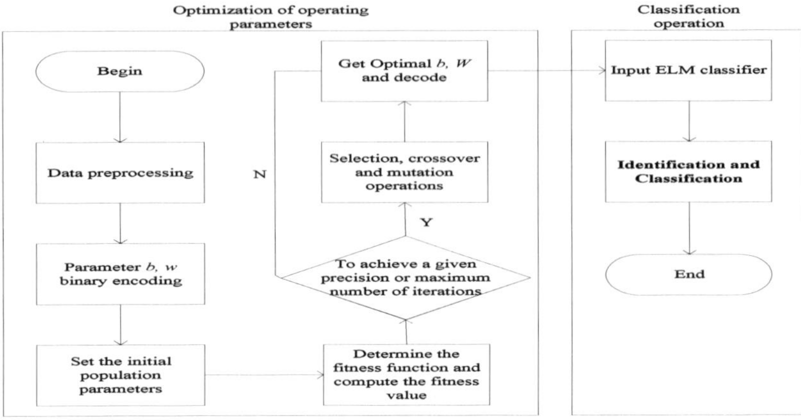

Fig. 1 GA-ELM optimization classification flowchart

3. Simulation and Analysis

3.1. *Feature Vector Extraction*

Seven common disturbance signal: swell, sag, interruption, harmonics, transient pulse, transient oscillations, flicker. These disturbance signals were added noise of 30dB considering noise pollution in actual disturbance signals, mathematical model is shown Literature [11]. Decompose the disturbance signals with 'db4'wavelet for ten layers, then normalize the energy difference in order to improve the difference of disturbance signals. Normalized energy difference waveform were shown in Figure 2.

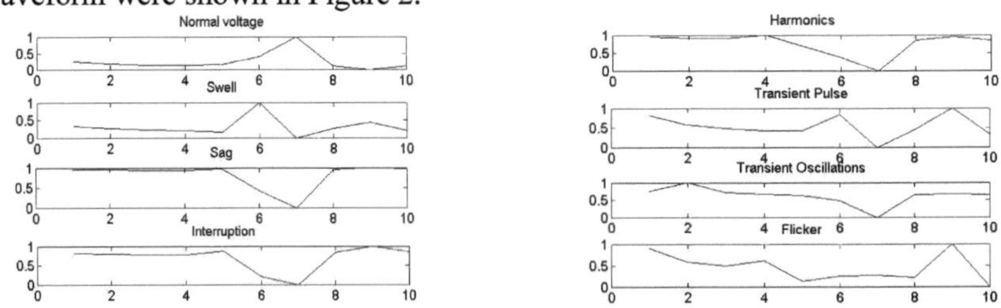

Fig. 2 Energy difference of power quality signals waveform

It is proposed that using standard deviation of the energy difference ΔE_{std} and mean of the energy difference ΔE_{ave} and the ratio of them ε and the root mean of disturbance signals and normal signals R as the feature vectors, R as in Eq.(5).

$$R = \sqrt{\frac{1}{N}\sum_{i=1}^{N}\left[s(t)-x(t)\right]^2} \tag{5}$$

Where: N is the number of sampling points; $s(t)$ is disturbance signal; $x(t)$ is normal signal. So the final feature vector are $E=[\Delta E_{ave}, \Delta E_{std}, \varepsilon, R]$.

3.2. *Algorithm Performance Comparison*

Analysis disturbance signals in 10 frequency cycles, the sampling frequency is 3200Hz. Each disturbance signal takes 200 samples. The first 150 are training samples and the other are test samples. 30 training samples(240 samples for 8 kinds of signals) and 10 test samples(80 samples for 8 kinds of signals) were selected randomly in each test. The test results will be different because training samples and testing samples were randomly selected in each test. The test will be run 50 times and a result of test is shown in Figure 3.

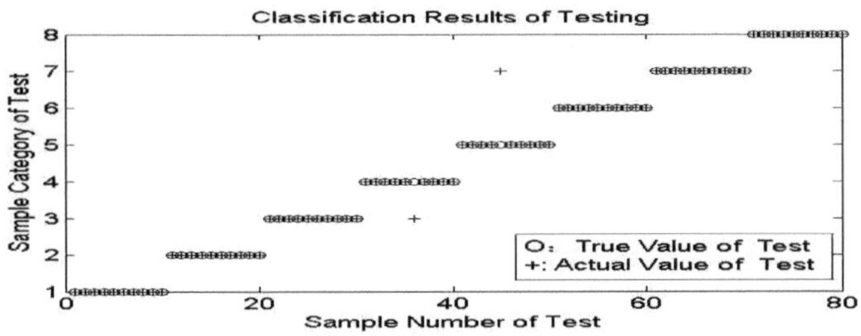

Fig.3 Results of test classification

From this test, there are only two classification errors in 80 testing samples. Calculate the average classification accuracy and running time as the final results after running 50 times and compare with BPNN(BP Neural Network)[5] and ELM without optimizing the parameters. The results were shown in Table 1.

Table 1 Comparison results of different classification methods

Method	Average Training Time /s	Average Test Time /s	Average Classification Accuracy /%
BPNN	9.7524	3.3621	95.75
ELM	6.6456	2.3011	96.40
GA-ELM	6.7363	2.3529	97.65

4. Conclusion

Grid information can be feedback effectively through identification and classification of power quality disturbances, which is good to analyze and evaluate the power system operation. In this paper, the feature vectors of signal disturbances were effectively extract by wavelet transform and the feature vector dimension were reduced. In addition, parameters of ELM were optimized by GA. After simulation, the method proposed in this paper which has correct classification rate, less training samples and less training time can be better used in power quality disturbance recognition online.

Acknowledgment

This work is supported by China National Science Foundation NO. 51307184.

References

1. Tang Qiu, Wang Yaonan, Guo Siyu, Jiang Feng. Power Quality Disturbance Classification Based on S Transform and Fourier Transform[J]Journal of Hunan University, 2009，35(4):38-42.
2. Wang Linghong, Chen Xuechang. Power Quality Disturbances Identification Based on Double-density Dual-tree Discrete Wavelet Transform[J]. Electron Measurement and Instrument，2012,49(8):18-21.
3. Lu Ganyun, Cheng Haozhong, Zheng Jinju et al. Power quality disturbances detection and identification based on S transform and multilay SVMs[J]. Transactions of China Electro technical Society, 2006, 21(1):121-126.
4. Zhong D J, Li Q,Zhang J F, et al. Improving the Accuracy of Single-ended Transient Fault Locations Using Mathematical Morphology[A]. In : IEEE /CSEE International Conference on Power System Technology [C].Kunming(China):2002.788-792.
5. Kanitpany W, Premrude S. Power Quality Problem Classification Using Wavelet Transformation and Artificial Neural Networks[C].Power Systems Conference and Exposition, 2004. IEEE PES.10-13 Oct. 2004(3):1496-1501.
6. Chen Huafeng，Zhang Gexiang，Power Quality Disturbance Identification Using Decision Tree and Support Vector Machine［J］.Power System Technology, 2013, 37 (5)：1272-1278.
7. Zhang Hong，Lei Zhiguo，Jin Yue, Tan Wanyu. Research of Power Quality Disturbance Classification Based On SVM［J］Electrical Measurement & Instrumentation, 2014,51(23):69-72.
8. Huang G B, Zhu Q Y,SIEW C K. Extreme Learning Machine: Theory and Applications[J]. Neurocomputing, 2006, 70:489-501.
9. Huang G B, Lei C K., Siew C K. Univeral Approximation Using Incremental Constructive Feedforward Networks with Random Hidden Nodes [J].IEEE Transactions on Neural Networks, 2006, 17 (4):879-892.
10. Pedram Ghamisi, Jon Atli Benediktsson. Feature Selection Based on Hybridization GA and PSO[J].IEEE Geoscience and Remote Sensing Letters,2015,12(2):309-314.
11. Fang Guozhi. Study and Implementation of Transient Power Quality Detection Method[D], Harbin Science and Technology University 2013:10-14.

Research on Protection Configuration Optimization of Power System

Xian Li

State Grid Henan Economic Research Institute,

Zhengzhou, Henan, 450000, China

E-mail: lixian_lix@sina.com

Yubin Mao, Xinju Guo and Jinli Yu

Henan Electric Power Survey & Design Institute

Zhengzhou, Henan, 450000, China

In this paper, safe, reliable, economic and effective protection configuration plan are formulated, aiming at electrical railroad traction station power supply project. It is from the perspectives of electrical railroad traction station power supply protection characteristics, protection configuration and the relationship between power grid stability and traction station main transformer security. Meanwhile, simplification and optimization plan are proposed.

Keywords: Electrical railroad traction station; Power supply; Power grid stability; Main transformer security; Protection configuration.

1. Introduction

Power supply to electrical railroad traction stations is involved in two industries of electric power and railroad. Every industry has a series of norms, rules and regulations in order to achieve safe, reliable, economic and effective power supply, which must be followed and implemented [1]. It is normal to hold different opinions. Rational and effective protection configuration plan should be formulated through analysis and demonstration in order to ensure the safety of power grid and power transmission and transformation equipment, and ensure reliable power supply to electrical railroad. It is very necessary to point out advantages and disadvantages of various plans. In the paper, experience in power supply design to electrical railroad traction station over the years is summarized. Various plans are analyzed and studied from the perspective of technology, and some opinions are provided for reference [2-5].

2. Electrical railroad load characteristics and its line protection characteristics

Electrical railroad power supply is operated by A-B, B-C and C-A phases in different stages after power grid supply voltage (220kv or 110kv) is decompressed to 27.5kv, and larger negative sequence current can be produced. When locomotives are started, larger impact current can be produced, thereby power line protection device can be frequently started. Once the protection device is started, the shock locking loop can be blocked for certain time [6][7]. If faults are discovered under the condition, protection device can not be tripped, therefore electrical railroad power line protection device is provided with a quick resetting loop. Blocking can be terminated during fault, and thereby the protection device can be tripped [8].

3. Main protection and backup protection concepts

Line main protection generally refers to full line high speed protection, such as optical fiber protection, high frequency protection, etc. Backup protection refers to distance (impedance principle) zero sequence protection; Main protection of main transformer refers to differential protection, back-up protection refers to e impedance protection, current protection, turn-to-turn protection, etc.

Stage I protection in main protection and back-up protection (action time is 60 milliseconds to 125 milliseconds, including protection inherent action time of 10 milliseconds to 60 milliseconds, circuit breaker inherent disconnection time of about 40 milliseconds, arc fault time and margin time) belongs to quick-acting protection, and it is protection to ensure power grid stability. Stage II (action time of 360 milliseconds to 625 milliseconds) and stage III (longer action time) protection in backup protection has longer action time, and the instability of power grid can be caused during fault. Therefore, electrical railroad power lines are equipped with double sets of main protection in order to prevent power grid instability due to tripping of backup protection action during fault on line terminal. Main transformer of traction station is equipped with double sets of main protection in order to ensure the safety of main transformer. High voltage side of 220kv traction station main transformer is equipped with fault protection and remote tripping for preventing main transformer burning or power grid instability due to action of main transformer protection and fault of circuit breakers on high voltage side. Power grid side circuit breaker can be quickly tripped by fault protection and remote tripping under the condition, and main transformer fault should be removed.

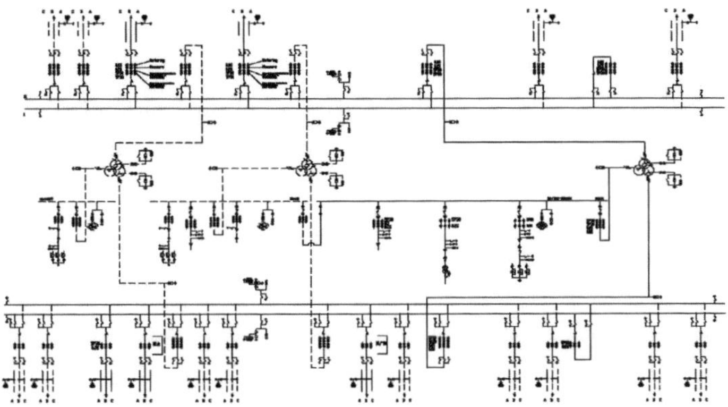

Fig. 1. Main connection

4. Power grid instability factor

Faults leading to power grid instability should be quickly removed. If electrical railroad power supply line terminal has fault, traction station main transformer has fault, and stage II action tripping can not become instable by protection on the side of line power grid according to calculations, line power grid side also can be equipped with double sets of distance zero sequence protection only. However, the power grid is changing, and the current stability can not ensure stability in the future. Therefore, protection capable of removing faults quickly should be configured for stability.

5. Optimization and simplification protection configuration ideas

Characteristics of adopting line-transformer group mode and double power supplies in electrical railroad power supply should be considered. Simplification protection configuration plan can be adopted during fault under the principle of ensuring power grid stability and not expanding power fault scope.

If both ends of the electrical railroad power supply line are equipped with optical fiber differential protection, optical fiber differential protection 'remote tripping' function can be utilized for saving traction station fault protection and remote tripping.

Line power grid side distance zero sequence stage I protection scope is extended into traction station main transformer for tuning. The mode is adopted for saving optical fiber differential protection and traction station side line protection; Fault protection can be further saved, remote tripping can be reserved, and remote tripping even can be saved for power grid safety. The above modes can lead to synchronous tripping of switches on the high voltage side and switch on electrical railroad power supply line power grid side during fault of traction station main transfer. Power fault scope can not be expanded aiming at line-transformer group. In addition, removal of traction station main transformer faults can be accelerated by saving fault protection action time, which is beneficial for safety of main transformers.

The capacity of 110kv electrical railroad traction station main transformer is 40 ~ 50 million volt ampere or so. Its impedance is generally much bigger than lines, and it can be larger for several times under the condition of short circuit especially. Line power grid side protection stage I can be extended into traction station main transformer by 80 % for tuning. Since it is difficult to obtain exact parameters of the main transformer immediately, high accuracy is not required on concrete operation. Main transformer/line impedance ratio is roughly estimated. It can be extended into main transformer as far as possible (it should not be extended to low voltage side of main transformer). Protection stage I has enough sensitivity (such as 3 ~ 5, power grid instability can not be caused due to fault on protection scope terminal). Short-circuit current level is lower aiming at fault on the remaining part of main transformer (it is not extended into line power grid side protection stage I scope). If the high voltage side switch of the main transformer is failed, backup protection on the line power grid side is tripped.

6. Protection configuration plan

6.1. *Basic configuration plan*

220kv line and traction station main transformer supplying power to the traction station are equipped with double sets of main and backup integrated protection. Line protection should adapt to electrical railroad load characteristics. Traction station main transformer high voltage side switch should be equipped with fault protection and remote tripping devices. Traction station main transformer high voltage side switch should be equipped with fault protection and remote tripping devices. Power grid side switch can be tripped when the traction station main transformer protection is acted, and high voltage side switch is failed.

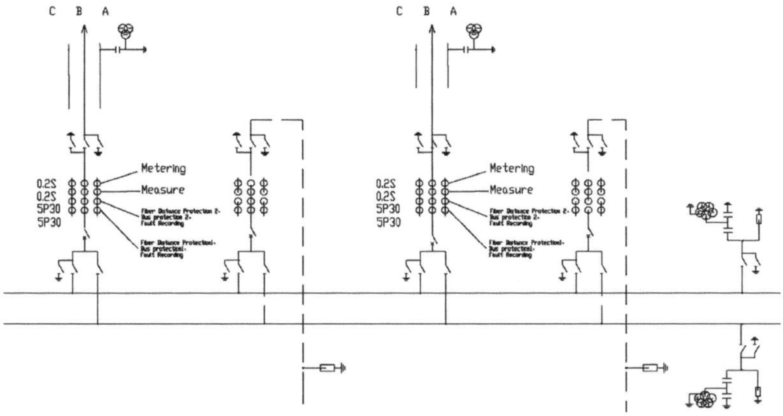

Fig. 2. Protection configuration

The plan conforms to provisions of relay protection and safety automatic device technical regulations and 220kv system protection configuration principle. Fault of main transformer high voltage side switch is solved for solving the safety of main transformer on one hand, quick tripping can be achieved during fault of electrical railroad power supply line or traction station main transformer on the other hand. Even if the power grid is changed, and power grid stability is not calculated again immediately, power grid stability problem can not be produced.

6.2. *Optimization allocation plan I (recommended plan)*

220kv line supplying power to traction station should be equipped with double sets of main and backup integrated protection suitable for electrical railroad load characteristics. Optical fiber differential protection is used in main protection. Distance zero sequence protection is adopted as back-up protection. Traction station main transformer is equipped with double sets of main and backup integrated protection. Traction station main transformer protection is acted, and high voltage side switch is tripped, meanwhile power grid side switch is tripped through optical fiber differential protection 'tripping' function.

Traction station main transformer high voltage side switch fault protection and remote tripping device are saved in the plan. Fault problem can be solved more quickly. However, when traction station main transformer protection acted and tripped, and power grid side switches can not be remotely tripped even if the high voltage side switch is not failed.

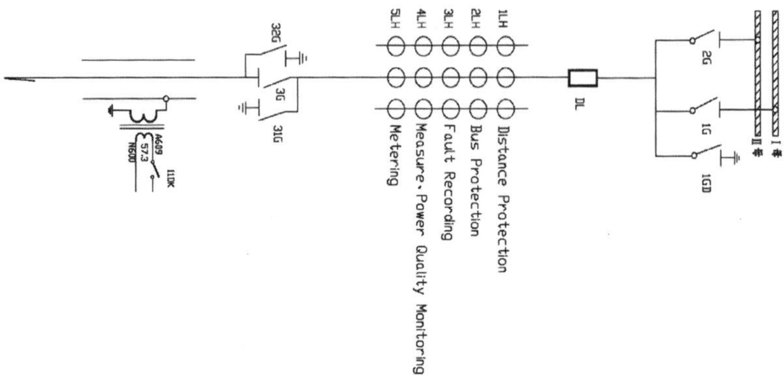

Fig. 3. Optimization allocation plan I

6.3. *Optimization configuration plan II (recommended plan)*

220kv line power grid side supplying power to the traction station is equipped with double sets of distance zero sequence protection suitable for electrical railroad load characteristics. Distance protection stage I and zero sequence protection stage I are tuned according to fault sensitivity on the terminal of line. Traction station main transformer is equipped with double sets of main and backup integrated protection. The traction station is equipped with remote tripping devices. Traction station main transformer protection is acted, and high voltage side switch is tripped, the power grid side switch can be tripped through remote tripping.

In the plan, double sets of line main protection, traction station side line protection and fault protection are saved. Quick-acting problem of any one fault at the full power supply line is solved, and fault problem is solved more quickly synchronously. However, when traction station main transformer protection is acted and tripped, power grid side switches can be remotely tripped even if the high voltage side switch is not failed.

6.4. *Simplification configuration plan I*

220kv line power grid side supplying power to the traction station is equipped with double sets of distance zero sequence protection suitable for electrical railroad load characteristics. Distance protection stage I and zero sequence protection stage I are tuned according to fault sensitivity on the terminal of line. Traction station main transformer is equipped with double sets of main and backup integrated protection. The traction station main transformer high voltage side switch is equipped with fault protection and remote tripping devices. Power grid side switch is tripped when the traction station main transformer protection is acted, and high voltage side switch is failed. In the plan, double sets of line main protection and traction station side line protection are saved. Quick-acting tripping problems of any one fault at the full power supply line also can be solved.

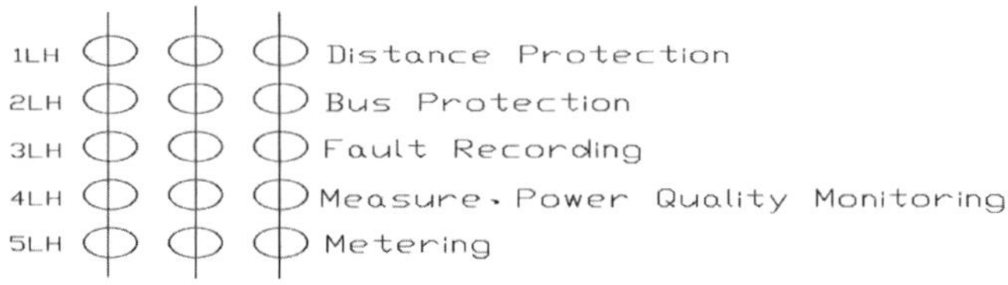

Fig. 4. Simplification configuration plan I

6.5. *Simplification configuration plan II*

Power grid stability is considered, 220kv line power grid side supplying power to the traction station is equipped with double sets of distance zero sequence protection suitable for electrical railroad load characteristics under the precondition of not approving traction station configuration line protection, fault protection and remote tripping devices in the railway industry. Distance protection stage I and zero sequence protection stage I are tuned according to fault sensitivity on the terminal of line. Traction station main transformer is equipped with double sets of main and backup integrated protection.

In the plan, double sets of line main protection, traction station side line protection, fault protection and remote tripping device are saved. It is the simplest and most economical plan. Although quick-acting tripping problem of any one fault at the whole power supply line and traction station main transformer high voltage part fault can be solved by the plan, power grid stability can not be affected, and high voltage side switch fault problem of traction station main transformer can not be solved. The current change may be very small during turn-to-turn fault of main transformer. Faults can not be detected by line power grid side backup protection in principle. Therefore, action and tripping can not be achieved. Risk of main transformer burning can be produced, and the railroad industry select between investment and risk bearing.

7. Conclusion

Some power transmission and transformation equipment may not suffer from faults for several decades and even during the whole service life, some equipment may suffer from faults at the beginning of operation in actual operation. Relay protection and security automatic devices act as detectors and security guards of power grid and power equipment. Related departments should pay much attention to them, and power supply to electrical railroad traction station is not an exception.

References

1. F. C. Schweppe, J. Wildes, D. B. Rom. Power system static-state estimation, part I, II and III, IEEE Transactions on Power Apparatus and Systems, vol. PAS-89, no. 1, January 1970, pp. 120-135.
2. M. Hurtgen and J.C. Maun. Advantages of power system state estimation using Phasor Measurement Units. 16th PSCC, Glasgow, Scotland, July 14-18, 2008.
3. Jialli He, Yongli Li, Bin Li. Relay Protection for UHV Transmission Lines Part Two Disposition of Relay Protection, Automation of Electric Power Systems. 2002, 26(24) , pp. 1-6.
4. Zhang Baohui, Wang Jin, Li Guanghui. Cooperation of relay protection for grid-connected wind power with low-voltage ride-through capability. Electric Power Automation Equipment, 2012, 32(3) , pp. 1-5.
5. Gao Dong-xue, Zhi Quanzhong, Zhu Lijun. Configuration project of protect unit according to levels for smart substation. Power System Protection and Control. 2012, 40(1) , pp. 67-70.
6. Li Feng, Xie Jun, Lan Jinbo. Prospect and discussion of relay system configuration for intelligent substation. Electric Power Automation Equipment, 2012, 32(2) , pp. 84-89.
7. Wang Wei, Dou Hui, Sun Yumin. Finite Element Analysis of Lateral Longitudinal Beam in Power Protection Cabinet. Electro-Mechanical Engineering, 2015, (3) , pp. 102-106.
8. Zhang Xin, Jiang Quanyuan. Capacity configuration of V/V transformer-based railway power conditioner and optimal energy compensation strategy. Electric Power Automation Equipment, 2014, 34(1) , pp. 23-29.
9. F. C. Schweppe, J. Wildes, D. B. Rom. Power system static-state estimation, part I, II and III. IEEE Transactions on Power Apparatus and Systems, vol. PAS-89, no. 1, January 1970, pp. 120-135.
10. M. Hurtgen, J.C. Maun, “Advantages of power system state estimation using Phasor Measurement Units”, 16th PSCC, Glasgow, Scotland, July 14-18 2008.

Muiltidimensional Collaborative Constellation Design and Implement for MIMO Indoor Visible Light Communications

Wang-Feng Liang†, Yi-Jun Zhu and Heng-Bin Cai

National Digital Switching System Engineering & Technological Research Center, Zhengzhou, Henan Province (450000), China

†*E-mail: 787666734@qq.com*

In conventional Multi-Input Multi-Output (MIMO) Visible Light Communication (VLC) techniques such as Repetition Coding (RC), Spatial Multiplexing (SMP), and Spatial Modulation (SM), the energy utilization is not optimal. In this paper, in order to overcome this defect, a novel constellation design for MIMO indoor VLC systems is proposed, in which the symbols from different spaces collaborate so closely that the average optical power is minimized under the constraint that the minimum Euclidean distance is fixed. Therefore, this new constellation is called space-Collaborative Constellation (CC). Computer simulations demonstrated that: 1) CC gets the optimal average optical power when compared to RC, SMP and SM under the same condition; 2) CC also performs better than other schemes under different spectrum efficiencies, except for the case at low one when compared to RC; 3) The performance of CC gets better with dimension increasing.

Keywords: Average optical power; Collaborative Constellation (CC); Light-Emitting Diode (LED); Multi-Input Multi-Output (MIMO); Visible Light Communication (VLC).

1. Introduction

With the increasing use of energy efficient Light-Emitting Diodes (LEDs), Visible Light Communication (VLC) is emerging as a promising technique for high speed indoor wireless communications [1], [2]. Since Intensity Modulation with Direct Detection (IM/DD) has its own important features like low-complexity and low-cost implementation, it is commonly used in indoor short-range VLC systems. In contrast to Radio Frequency (RF) where information is embedded in the carrier, IM/DD systems carry information on the intensity of the optical power [2], [3].

It is known that RC is the simplest MIMO transmission technique, which simultaneously emits the same signal from all transmitters, to achieve good performance in free-space optical communications [5], [6]. In [7], it was shown that RC outperforms orthogonal space-time block codes (OSTBCs). In SM, the conventional signal constellation diagram is extended to the spatial dimension, in which only one transmitter is activated at any symbol duration. SM has a lower decoding complexity compared with other MIMO schemes [8], [9]. Since independent data streams are simultaneously emitted from all transmitters in SMP, it has the highest spectrum efficiency when channel correlation sufficiently low[5]. In RC, SM and SMP, symbols among the transmitting LEDs are generated from constellations independently, it is not optimal when considering about energy utilization.

In order to overcome this defect, we propose a novel constellation design for MIMO indoor VLC systems-collaborative constellation (CC). The simulation results show that the performance of CC is always better than SM and SMP for different channels, and is also better than RC in general except for conditions with high correlation channel at low spectral efficiency [10]. In order to confirm the performance of collaborative constellation more, we design muiltidimensional collaborative constellation and compare it with RC, SMP, and SM in random channel.

2. Channel Model

Let us consider an indoor visible light communication system with M transmitting antennas (LEDs) and N receiving antennas (photo-detectors). Its discrete-time baseband MIMO channel model can be represented as

$$y = Hs + n \tag{1}$$

where n is the sum of ambient shot light noise and thermal noise with zero mean and a variance σ^2 [5]. It is assumed that M data symbols $s_1, s_2, \cdots, s_M$ are chosen randomly, equally-likely and independently to form an input data vector $s = [s_1, s_2, \cdots, s_M]^T \in \mathrm{A}^M$, where A is a given modulation constellation. The received signal vector is denoted by $y = [y_1, y_2, \cdots, y_n]^T$. Notation H denotes an $N \times M$ channel matrix, in which h_{nm} is the on-negative and real coefficient between the mth transmitter antenna and the nth receiver antenna. In this letter, we assume an indoor environment with LOS characteristic, where the channel gain of a VLC link is determined by [4], [6].

$$h_{nm}=\begin{cases}\frac{(k+1)A}{2\pi d_{nm}^2}\cos^k(\phi)\cos(\psi), & 0\le\psi\le\Psi_{1/2}\\ 0, & \psi>\Psi_{1/2}\end{cases} \tag{2}$$

with φ being the angle of emergence with respect to the mth transmitter axis and ψ being the angle of incidence with respect to the nth receiver axis. In addition, d_{nm} is the distance between mth transmitter and nth receiver. $\Psi_{1/2}$ is the field-of-view (FOV) angle of the receiver, A is the detector area of the receiver, and $k=-(\ln 2)/\ln(\cos(\Phi_{1/2}))$, in which $\Phi_{1/2}$ is the half power angle of the transmitter.

Channel correlation is an important factor to the decline of the system Bit Error Rate (BER) performance, and the condition number of the channel matrix directly reflects the correlation of the channel. The greater the condition number of channel matrix, the more sensitive to the noise the system is , namely under the condition of the same SNR, the BER performance gets worse. The expression of channel matrix condition number is

$$cond(\mathbf{H})=\|\mathbf{H}\|_F\bullet\|\mathbf{H}^{-1}\|_F \tag{3}$$

3. Multidimensional Collaborative Constellation Design

In conventional wireless digital communication systems, average electric power P_{ae} is an important figure of merit to measure power performance, which is defined as $P_{ae}=\mathrm{E}\left[\|s\|_2^2\right]/T$, where $\mathrm{E}[\bullet]$ is an expectation operator, and T_s is the symbol period [3], [4]. In the wireless optical systems, another measurement is the average optical power P_{ao}, which is important for VLC system to meet lighting requirement and eye-safety limitations, it is defined as $\mathrm{P}_{ao}=c\mathrm{E}\left[\|s\|_1\right]/\sqrt{\mathrm{T}_s}$, where c represents the electro-optical conversion factor in watts per ampere (W/A) [3]. We let $c/\sqrt{\mathrm{T}_s}=1$ for simplicity here.

In RC, SM and SMP, the symbols $s_m, m=1,2,\cdots,M$, in the signal vector s are generated from the constellation χ independently. Since we let $c/\sqrt{\mathrm{T}_s}=1$, in this section we write notation $P(\chi)$ to denote a normalized version of the average optical power of constellation χ, i.e., $E[\|\mathrm{X}\|_1]$. Throughout this letter, we assume that each signal point from χ is transmitted with equal probability. In this case, $P(\chi)$ can be expressed by $P(\chi)=\sum_{X\in\chi}\|X\|_1/|\chi|$, where notation $|\chi|$ denotes the cardinality of χ. We use the minimal average optical power as an optimization criterion. Our primary goal in this section is to design a collaborative constellation $\Omega, s \in \Omega$, to achieve optimal power-efficient subject to a fixed minimum Euclidean distance. Hence, the corresponding optimization problem can be formally stated as

$$\begin{aligned} &Minmize \quad P(\Omega)=E[\|s\|_1], \\ &\text{Subject to} \quad |\Omega|=L, \\ &\qquad d_{\min}(\Omega)=1. \end{aligned} \tag{4}$$

where $d_{\min}(\Omega)=\min_{s \neq s' \in \Omega}|s-s'|$ and the symbols $s_m, m=1,2,\cdots,M$, in the signal vector **s** are all non-negative integers, L is the constellation size.

When the transmitter-receiver systems is 3×3 (M = 3 and N = 3), its corresponding collaborative constellation is given by optimization goal :

$$\Omega=\bigcup_{j=0}^{\lambda}\Omega_j,\ \Omega_j=\{(x,y,z)^T : x+y+z=j, 0 \le x,y,z \in \mathrm{Z}\} \tag{5}$$

for $0 \le j \le \lambda-1$ and Ω_λ consists of any $L_\lambda = L-\sum_{n=1}^{\lambda} n(n+1)/2$ solutions of $x+y+z=\lambda, 0 \le x,y,z \in Z$. Then, its corresponding L and λ is derived by $0 < L_\lambda \le (\lambda+1)(\lambda+2)/2$. That is to say, when L=4,8,16,32,64,128,256,512, its corresponding λ is 1,2,3,4,6,8,10,13.

When the transmitter-receiver systems is 4×4 (M = 4 and N = 4), its corresponding collaborative constellation is given by optimization goal :

$$\Omega=\bigcup_{j=0}^{\lambda}\Omega_j,\ \Omega_j=\{(x,y,z,k)^T : x+y+z+k=j, 0 \le x,y,z,k \in \mathrm{Z}\} \tag{6}$$

for $0 \le j \le \lambda-1$ and Ω_λ consists of any $L_\lambda = L-\sum_{m=0}^{\lambda-1}\sum_{n=0}^{m}(n+1)(n+2)/2$ solutions of $x+y+z+k=\lambda, 0 \le x,y,z,k \in Z$. Then, its corresponding L and λ is derived by $0 < L_\lambda \le \sum_{n=0}^{\lambda}(n+1)(n+2)/2$. That is to say, when L=4,8,16,32,64,128,256,512, its corresponding λ is 1,2,3,3,4,6,7,9.

4. Numerical Result

In this paper, the random channel model H is based on the Lambert model. When the transmitter-receiver systems is 3×3 (M = 3 and N = 3) which is located within a 4.0×4.0m×3.0m room. $\Phi_{1/2}=60^\circ$, $\Psi_{1/2}=40^\circ$, $A=1cm^2$. We assume that the three receivers are located at (x, y, z) = (2m, 1.6m, 0.75m), (2m, 2m, 0.75m) and (2m, 2.4m, 0.75m). The LEDs are placed at (Xm, X-0.4m, 2.50m), (Xm, Xm, 2.50m) and (Xm, X+0.4m, 2.50m), which X is a variable independently chosen between 1 and 3. Computer simulations show that the value of channel matrix condition $cond(H)$ is range of [5.9450, 303.8442]. That is to say, the correlation of the channel matrix is randomly generated.

Table 1 shows that CC and RC, SM, SMP, their average optical power under the same spectrum efficiency. Data from Table 1 shows that: Under the same spectral efficiency, the average optical power of CC is the smallest.

Table 1. The corresponding of average optical power under different frequency spectrum efficiency of CC, SM, RC and SMP for 3×3 VLC MIMO

Spectral Efficiency	SM	RC	SMP	CC
2 bit/s/Hz	1.25	4.5	1	0.75
3 bit/s/Hz	1.875	10.5	1.5	1.375
4 bit/s/Hz	3.1875	22.5	2.5	2.0625
5 bit/s/Hz	5.8438	46.5	3.5	2.9063
6 bit/s/Hz	11.1719	94.5	4.5	4.0313
7 bit/s/Hz	21.8359	190.5	6.5	5.4219
8 bit/s/Hz	43.168	382.5	8.5	7.207
9 bit/s/Hz	85.834	766.5	10.5	9.4453

Fig. 1 shows that, when the transmitter-receiver systems is 3×3, the spectrum efficiency is 2, 3, 4, 5, 6, 7, 8, 9 bit/s/Hz (L = 4, 8, 16, 32, 64, 128, 256, 128) under the random channel, the BER performance of CC and SM, RC, SMP contrast figure. the figure shows: when the spectrum efficiency is very low, the performance of RC is best; With the increase of frequency spectrum efficiency and the increase of SNR, the performance of CC gradually improve, and is more better than that of RC, SM, SMP.

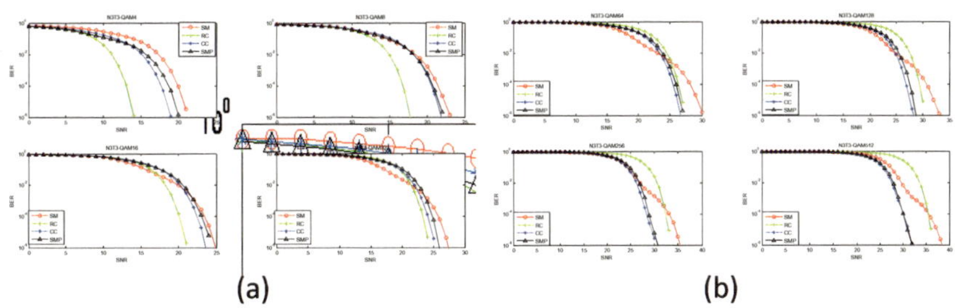

Fig. 1. The performance comparison of RC, SM, SMP and CC for 3×3 VLC MIMO.

When the transmitter-receiver systems is 4×4 (M = 4 and N = 4) which is also located within a 4.0m×4.0m×3.0m room. $\Phi_{1/2}=30^\circ$, $\Psi_{1/2}=60^\circ$, $A=1cm^2$. We assume that the four receivers are located at (x, y, z) = (1m, 1m, 0.75m), (3m, 1m, 0.75m), (1m, 3m, 0.75m) and (3m, 3m, 0.75m). The LEDs are placed at (Xm, Xm, 2.50m), (Ym, Xm, 2.50m), (Xm, Ym, 2.50m) and (Ym, Ym, 2.50m), which X is a variable independently chosen between 0 and 2, and Y is a variable independently chosen between 2 and 4. Computer simulations show that the value of channel matrix condition $cond(H)$ is range of [1.3582, 106.8056].

That is to say, the correlation of the channel matrix is randomly generated. Table 2 shows that CC and RC, SM, SMP, their average optical power under the same spectrum efficiency. Data from Table 2 shows that: Under the same spectral efficiency, the average optical power of CC is also the smallest.

Table 2. The corresponding of average optical power under different frequency spectrum efficiency of CC, SM, RC and SMP for 4×4 VLC MIMO

Spectral Efficiency	SM	RC	SMP	CC
2 bit/s/Hz	1	6	1	0.75
3 bit/s/Hz	1.5	14	1.5	1.25
4 bit/s/Hz	2.5	30	2	1.6875
5 bit/s/Hz	4.5	62	3	2.34375
6 bit/s/Hz	8.5	126	4	3.125
7 bit/s/Hz	16.5	254	5	4.03125
8 bit/s/Hz	32.5	510	6	5.19531125
9 bit/s/Hz	64.5	1022	8	6.486328125

Fig. 2 shows that, when the transmitter-receiver systems is 4×4 , the spectrum efficiency is 2, 3, 4, 5, 6, 7, 8, 9 bit/s/Hz (L = 4, 8, 16, 32, 64, 128, 256, 128) under the random channel, the BER performance of CC and SM, RC, SMP contrast figure. the figure shows: When the spectral efficiency is very low and SNR is larger, the performance of RC is better than that of CC. When the value of spectrum efficiency is gradually increasing, the performance of CC is better than that of RC completely; In any case, CC performance is always better than SM, SMP.

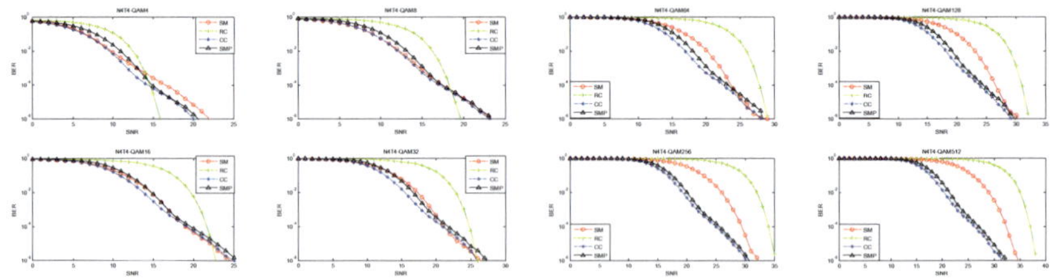

(a) (b)

Fig. 2. The performance comparison of RC, SM, SMP and CC for 4×4 VLC MIMO.

5. Conclusion

In this letter, on the basis of the original [10], we design 3×3 , 4×4 dimensional collaborative constellations. Compared to traditional visible light communication MIMO modulation such as RC, SM, SMP, Computer simulations show that: Under the same conditions, CC achieved the minimum average optical power for a fixed minimum Euclidean distance; Under the different spectrum efficiency, CC performance is mostly better than SM, SMP; In addition to the low frequency spectrum efficiency, the error performance of CC is also better than that of RC. Meanwhile, With the increase of dimension, the BER performance of CC is getting better and better.

Acknowledgment

This work is supported by Grant No. 61271253 and Grant No. 2013AA013603. This work is supported in part by Grant No. 61271253 from China National Science Foundation Council and in part by Grant No. 2013AA013603 from China National "863" Program.

References

1. Thomas Q. Wang, Roger J. Green and Jean Armstrong, MIMO Optical Wireless Communications Using ACO-OFDM and a Prism-Array Receiver, IEEE Journal on Selected Areas in Communications, vol. 33, no. 9, pp. 1959–1971, Sep. 2015.
2. S. K. Routray, The changing trends of optical communication, IEEE Potentials, vol. 33, no. 1, pp. 28–33, Jan. 2014.
3. M. Beko, and R. Dinis, Systematic method for designing constellations for intensity-modulated optical systems, J. OPT. Commun. Netw., vol. 6, no. 5, pp. 449–458, May 2014.
4. T. Komine and M. Nakagawa, Fundamental analysis for visible-light communication system using LED lights, IEEE Trans. Consumer Electronics, vol. 50, pp. 100–107, Feb. 2004.
5. T. Fath, and H. Haas, Performance comparison of MIMO techniques for optical wireless communications in indoor environments, IEEE Trans. Commun., vol. 61, no. 2,pp. 733–742, Feb. 2013.
6. L. Zeng, D. OBrien, H. Minh, G. Faulkner, K. Lee, D. Jung, Y. Oh, and E. T. Won, High data rate multiple input multiple output (MIMO) optical wireless communications using white LED lighting, IEEE J. Sel. Areas Commun., vol. 27, no. 9, pp. 1654–1662, Dec. 2009.
7. M. Safari and M. Uysal, Do we really need OSTBCs for free-space optical communication with direct detection?, IEEE Trans. Wireless Commun., vol. 7, no. 11, pp. 4445–4448, Nov. 2008.
8. R. Mesleh, H. Haas, S. Sinanovic, C. W. Ahn, and S. Yun, Spatial modulation, IEEE Trans. Veh. Technol., vol. 57, no. 4, pp. 2228–2241, July 2008.
9. W. O. Popoola, E. Poves, and H. Haas, Spatial pulse position modulation for optical communications, J. Lightwave Technol., vol. 30, no. 18, pp. 2948–2954, Sep. 2012.
10. Yi-Jun Zhu, Wang-Feng Liang, Jian-Kang Zhang, and Yan-Yu Zhang, Space-Collaborative Constellation Designs for MIMO Indoor Visible Light Communications, IEEE Photonics Technology Letters., vol. 27, no. 15, Aug. 2015.

Motion of Transceivers in MCNS for Underwater Optical Wireless Communications

Hao Xu and Yi-Jun Zhu

Information Engineering University,

Zheng Zhou, HeNan, China

†*E-mail: xuhaobestone@sina.cn*

A method applied in Monte Carlo Numerical Simulation (MCNS) is developed to acquire the impulse response where existing relative constant motion is between the transmitter and receiver. With the simulation results, the motion of transceivers affecting the optical wireless underwater communication is researched.

Keywords: Oceanic optics; Optical communications; Monte Carlo; MCNS; Motion.

1. Introduction

With further exploration into the ocean, the need of underwater wireless communications is growing rapidly. Compared to widespread acoustic communication, the emerging underwater optical wireless communication has the inherent advantage of high bitrate and low latency, which make it an important research area. Most research have applied MC to underwater optical wireless communications

MC is a statistics-based numerical computation method, which is commonly used to acquire the probability distribution function of unknown quantities. With the unique characteristics of highly flexibility, controllability and easy implementation, MCNS is the primary research approach to make estimation, system testing and obtain massive data, especially for the situations where it's hard to conduct field experiment and where it's impossible to make theoretical analysis with mathematical method.

2. MCNS Underwater

The research on underwater optical communication is always restricted with complicated underwater optical channel, and there is no universal effective theoretical model to describe the various environment underwater as far as we know. Since the actual experimental data provided by literature have been decades and lack diversity for different conditions, most research take the results of MCNS for reference and comparison.

The MCNS for underwater optical channel target tracing the properties of a single traveling photon, including position, weight, direction, distance, collision times and everything else we are interested in. Massive independent repeated trials could produce a stable and smooth curve which approximate the impulse response.

This approach have been developed for a long time. At early days people applied it to study simple porous medium, such as fog, cloud and dust, then simulate the light transport in multi-layered tissues, now analysis the optical communication underwater. Though the core thought and methods stay the same, there still exists particular nature of underwater, such as the various scattering phase functions, the affect of scattering coefficient on multiple scattering and the affect of turbidity, which lead to a different way to handle MCNS.

3. Motion underwater

The MCNS have been achieved in many papers.[1] have adopted MC to model the impulse response of underwater optical channel links of different conditions.[2] Applied the double Gamma functions to fit the results of MCNS, which yet didn't work very clearly and was strictly limited to turbid seawater.[3] Come up with the methods and programs of MC modeling of light transport in multi-layered tissues for the first time. And William Charles Cox gives a comprehensive summary of MC modeling of light underwater, including the situations where exists position offset and angular misalignments between the transmitters and receivers[4].

However, to the best of our knowledge, these prior works didn't take into account the relative movement between the transmitters and receivers in real-time communication, and the constant change of channel. Due to the water mobility, devices usually couldn't be easily fixed in one place, but drift with ocean current or swing around an anchor. And it is important to note that ships and underwater vehicles are usually sailing and communicating with each other at the same time. These circumstances put the necessity to study the affect of the motion of transceivers on underwater channel, which would help guide the actual communications, channel quality estimation and get the most efficient ways to navigate a cruise.

4. Implementation of Motion in MC

The typical flow chart of MC shows how to conduct a single simulation experiment of MCNS, like the Figure 4.1 in [4]. To describe the relative motion of transceivers, an extra step is added to the chart as follows:

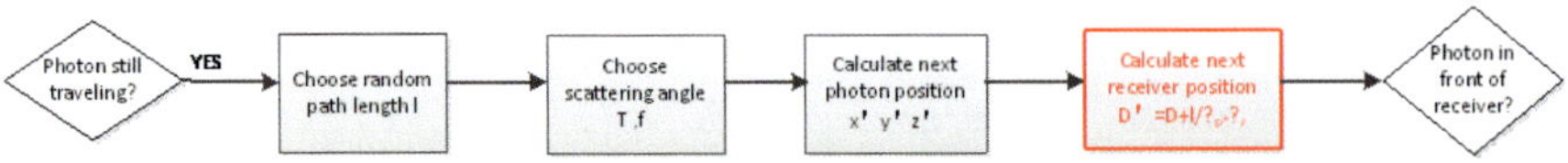

Fig. 1. Modified flow chart of MCNS

Every time a photon travel over a random path length l, the distance between transmitters and receivers D needs to be updated to D' according to the propagation velocity of a photon and the relative velocity of transceivers. The relative velocity of transceivers could be a vector, which can describe complex motions. In this way, the position of receiver plane should be updated constantly, and the trajectories of received photons could be showed as Fig. 2, where dark spots represent photons, broken lines represent trajectories, and the red dotted lines represent a moving receiver. We can intuitively sense that the motion of transceivers may affect the time dispersion of channel, because the motions would magnify or compress the distances of multiple scattered photons just like the doppler effect.

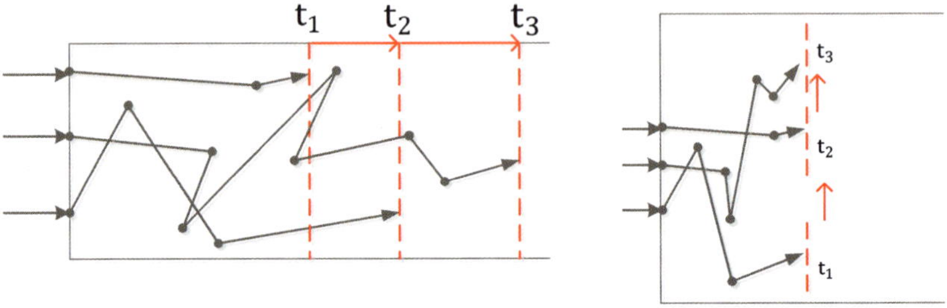

Fig. 2. Trajectory of photons

After massive launching photons and calculation, the receiver shall generate the specific impulse response of its motion patterns.

5. Simulations and results

To acquire accurate and realistic results, the simulations of our MCNS system are introduced in actual measurement parameters, which refer to [5] measured by Petzold. As shown in Fig. 3, we fit the Petzold's measurements of 56 samples to get the seawater spread phase function, which represent the seawater in San Diego Harbor in 1972.

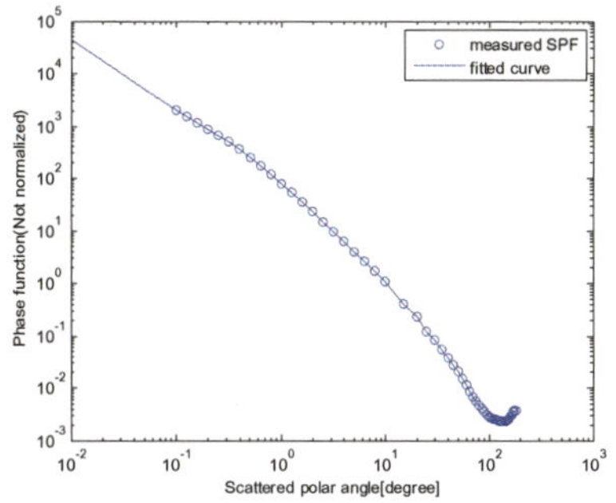

Fig. 3. Seawater phase function from [5] and fitted curve

Considering the light source as gaussian beam and the transmitter is always aligned to receiver. Some other vital parameters are shown in Table 1.

Table 1. Parameters for simulations

Scattering coefficient	Absorption coefficient	Attenuation coefficient	Beam waist
0.407	0.188	0.595	5mm
divergence	aperture	FOV	
10 degree	0.5 m	20 degree	

In the last step, we need to control accuracy and the amount of calculation. Here we set the value of photon weight threshold as and a reasonable time resolution to ensure at least 1000 received photons in each temporal bin of histogram. Without loss of generality, we only consider the various velocities along z-axis which increase link ranges and initial distances of 10m and 20m,then simulating the photon propagation of various velocities in the specific harbor water by MCNS with at least launching photons for each scenario.

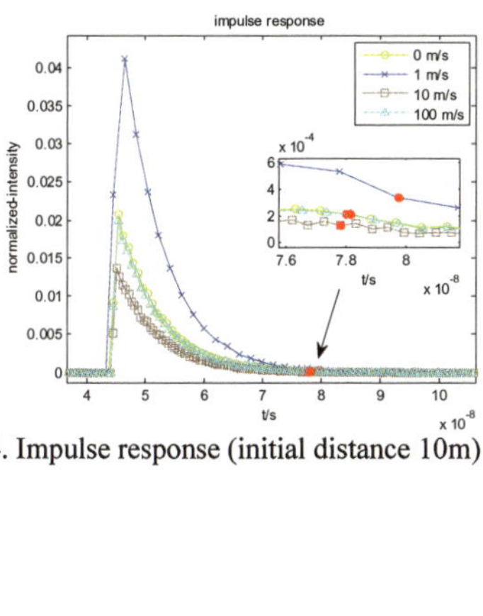

Fig. 4. Impulse response (initial distance 10m)

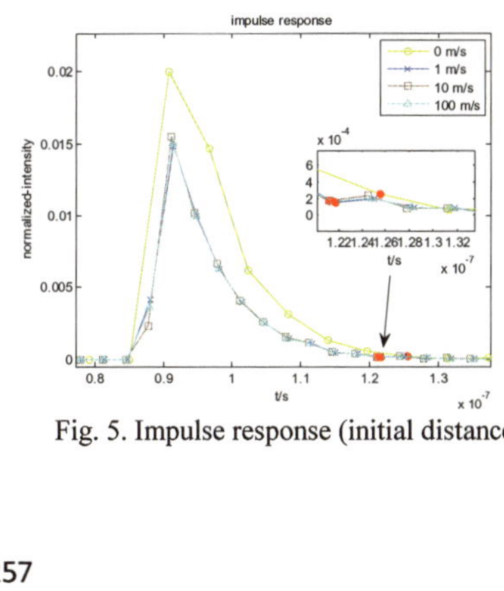

Fig. 5. Impulse response (initial distance 20m)

Based on the above method and parameters, Figs. 4 and 5 and Table 2 show all the simulation results. The received power is normalized by total transmitted power, and the temporal dispersion which is defined as the time interval of impulse response falling 20dB below the peak have been calculated in Table 2, and labeled as red filled dots in Figs. 4 and 5. The peak values of received power are also put in Table 2.

Table 2. Temporal dispersion and peak value of received power for various link ranges and velocities.

Range / Velocity	10 m	20 m
0 m/s	32.71 ns,0.0207	34.75 ns,0.0200
1 m/s	33.36 ns,0.0411	30.23 ns,0.0148
10 m/s	32.83 ns,0.0137	30.02 ns,0.0155
100 m/s	32.91 ns,0.0199	30.17 ns,0.0150

From Figs. 4 and 5, we can directly see that the transmitted optical pulse is dispersed when propagating through water. This process could be concluded into three stages. The first arrival few photons which travel in straight lines take the time just equal to link range divided by light speed in water, then a large number of one or two scattered photons arrive to constitute the impulse. In the last long duration, the slow descending from peak is a consequence of the continuous arrivals of multiple scattered photons.

As the statistics in Table 2 indicated, the impulse responses disperses more heavily as the velocity increases at a short initial distance of 10m, but on the contrary, the dispersion decrease at a long initial distance of 20m when there exists motion between transceivers. The peak value which represent relative total received power reach the maximum at initial distance of 10m with the speed of 1m/s, but at the initial distance of 20m it decrease all along from 0m/s to 100m/s.

We think the reason for that complex behavior is that when initial distance and velocity of transceivers increase, the photons suffer more scattering for a growing transmission distance and the receiver get more time to collect multiple scattered photons, which cause heavy temporal dispersion and high received power.

But when initial distance and velocity of transceivers continue increasing, the multiple scattered photons trouble be scattered enough times that their energy is completely absorbed or they just lose the way into receiver, so the received power and temporal dispersion decrease at the same time.

6. Conclusion

In this paper, we, firstly, consider the importance and possibility to describe the motion of transceivers in MCNS, and provide a simple but effective way to achieve that. Numerical simulation results suggest that the motion of transceivers may seriously affect the temporal dispersion of channel impulse response and received power, which depend on the velocities of motion and initial distances.

There are still a lot of research to do. We want to study how the motion exactly affect the communication performance, and take into account some much more complicated motions like angular motion, accelerated motion and motions of mathematical model.

References

1. Chadi Gabriel, Mohammad-Ali Khalighi, and Salah Bourennane, "Channel Modeling for Underwater Optical Communication", IEEE Workshop on Optical Wireless Communication,2011, pp. 833-837.
2. Shijian Tang, Yuhan Dong, and Xuedan Zhang, "Impulse Response Modeling for Underwater Wireless Optical Communication Links", IEEE Transactions on Communications,Vol.62, No.1, january 2014, pp. 226-234.
3. Lihong Wang, Steven L. Jacques, and Liqiong Zheng, "MCML-Monte Carlo modeling of light transport in multi-layered tissues", Computer Methods and Programs in Biomedicine, Vol.47, 1995, pp. 131-146.
4. William Charles Cox, and Jr, Simulation, Modeling, and Design of Underwater Optical Communication Systems , 2012.
5. T. Petzold, "Volume scattering functions for selected ocean waters," SIO Ref. 72-78, Scripps Institution of Oceanography Visibility Laboratory, San Diego, CA, Oct. 1972.
6. Ghassemlooy, Z, Arnon, S, Uysal, M, Xu, Z, and Cheng, J. "Emerging optical wireless communications – advances and challenges', IEEE J. Sel. Areas Commun., 2015, 33, (9), pp. 1738–1749
7. Cochenour, B., Mullen, L., and Muth, J."Temporal response of the underwater optical channel for high-bandwidth wireless laser communications", IEEE J. Ocean. Eng., 2013, 38, (4), pp. 730–742
8. Reza, A., and Harms, J."Robust grid-based deployment schemes for underwater optical sensor networks". IEEE 34th Conf. on Local Computer Networks, Switzerland, October 2009, pp. 641–648
9. Simpson, J.A., Hughes, B.L., and Muth, J.F.: 'Smart transmitters and receivers for underwater free-space optical communication", IEEEJ. Sel. Areas Commun., 2012, 30, (5), pp. 964–974
10. Gabriel, C., Khalighi, M.A., Bourennane, S., Leon, P., and Rigaud, V."Channel modeling for underwater optical communication". 2011 IEEE Globecom Workshops, Houston, USA, December 2011.

An instantaneous Frequency Extraction Method in the Application of Laser Doppler Velocimeter

Guo-Hua Wang†, Si-Chen Xu and Wei-Shuo Zhang

Instrumentation Science and Opto-electronics Engineering, Beihang University, Beijing, China

E-mail: stevexu@buaa.edu.cn; †wgh@buaa.edu.cn

As the Laser Doppler Velocimeter (LDV) has been applied to flow velocity measurement for many years, it is still difficult to process the Doppler signal in solid velocity measurement, especially with high acceleration. In order to address this issue, an Instantaneous Frequency (IF) extraction method for Doppler signal is demonstrated in this paper. First, the Doppler signal is denoised by wavelet to make the signal smoother. Then, the denoised signal is normalized by empirical Amplitude and Frequency Modulation (AM-FM) decomposition. At last, the improved Direct Quadrature (DQ) method is adopted to extract the IF from the normalized signal. A simulated Doppler signal whose IF is changed from about 25 KHz to 80 KHz in 50 milliseconds is used to validate the effectiveness of this method. The result indicates that this method is able to extract the IF with high accuracy and computation speed.

Keywords: Laser Doppler Velocimeter; Instantaneous Frequency Extraction; Wavelet Denoising; Empirical Amplitude and Frequency Modulation; Direct Quadrature Method.

1. Introduction

Since the laser was first proved can be used in flow velocity measurement, the Laser Doppler Velocimeters (LDVs) have been widely used in medical science, metrology, chemical engineering, and other fields to measure flow velocity [1-2]. However, it has difficulty in measuring solid velocity with high acceleration, as the output signal is nonstationary [3-4].

Several methods have been reported in processing nonstationary signal generated by LDVs. The Short-Time Fourier Transform (STFT) is a traditional method. However, its accuracy is not sufficiently high since the window function cannot be changed [5]. The Continuous Wavelet Transform (CWT) has also been applied to frequency extraction. Nevertheless, its computation speed is not quickly enough [6-7]. Hilbert Huang Transform (HHT) is a newer method in calculating Instantaneous Frequency (IF). Yet, it can only extract IF that changed very slowly due to the rigorous conditions and end effect [8-10].

The Direct Quadrature (DQ) method, proposed by Huang in 2008, overcomes the restrictions of Hilbert transform such as the rigorous theorem limitations and the end effect [11-14]. Thus, it is adopted in this paper for IF extraction. Moreover, the Doppler signal is denoised by wavelet and normalized by empirical Amplitude and Frequency Modulation (AM-FM) decomposition before using DQ method.

The remainder of this paper is organized as follows. Section 2 analyzes the characteristic of Doppler signal. Section 3 describes the procedure of the adopted IF extraction method and an experiment is conducted. The conclusion is drawn in Section 4.

2. Doppler Signal Analysis

For most LDVs, its output signal is an interference signal. The interference signal is composed of a signal light and a reference light, which can be expressed as $E_1 = A_1 \cos(\omega_1 t + \varphi_1)$ and $E_2 = A_2 \cos(\omega_2 t + \varphi_2)$. Where A_i, ω_i and φ_i are the amplitude, angular frequency and initial phase of E_i. According to the optical superposition and the characteristics of the photodetector, the output signal of the photodetector can be expressed as [7]:

$$\begin{aligned} V &= \eta I \\ &= \eta (E_1 + E_2)^2 \\ &= \eta \frac{A_1^2 + A_2^2}{2} + \eta A_1 A_2 \cos\left[(\omega_1 - \omega_2)t + (\varphi_1 - \varphi_2)\right] \\ &= \eta \frac{A_1^2 + A_2^2}{2} + \eta A_1 A_2 \cos\left[\Phi(t)\right] \end{aligned} \tag{1}$$

Where η is the power-conversion coefficient of the photodetector, I is the light intensity of the superposition light signal, $\Phi(t)$ is the time-varying phase function.

On the basis of Eq. (1), the output signal consists of DC component and cosine component. Since only the second item is related to the Doppler shift, we can separate the cosine component from the output signal by high-pass filter. Moreover, the IF of the second item is defined as:

$$f(t) = \frac{1}{2\pi} \frac{d\Phi(t)}{dt} \tag{2}$$

As the Doppler shift can be expressed as $f_D = (\omega_1 - \omega_2)/2\pi$, which is equal to the IF of the second item, we are able to figure out the Doppler shift by analyzing the cosine component of the output signal. It will be quite easy to work out the instantaneous velocity as long as the IF of the cosine component signal can be calculated. The relationship between instantaneous velocity and IF is as follows:

$$u(t) = \frac{\lambda}{2\sin\theta} f(t) \tag{3}$$

Where λ is the wavelength of the laser light, θ is decided by the angle of reflection. The Eq. (3) shows that the instantaneous velocity is proportional to the IF. Thus, the instantaneous velocity can be deduced as long as the IF is known.

3. IF Extraction Method

3.1. *Doppler signal simulation*

According to the analysis in Section 2, a Doppler signal can be simulated in the following form:

$$F(t) = A(t)\cos\left[\Phi(t)\right] + N(t) \tag{4}$$

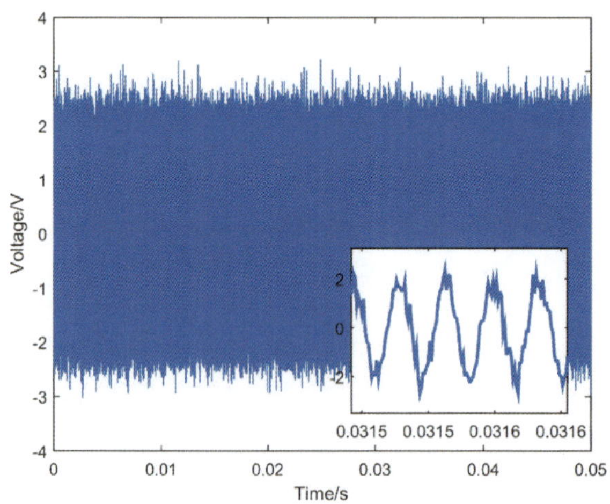

Fig. 1. The original Doppler signal with Gaussian white noise

We set $A(t)$=2 and $\Phi(t) = -20000\cos(8\pi t + \pi/10)$. In order to make the simulated signal closer to the real Doppler signal, Gaussian white noise $N(t)$ is added to the simulated signal. We assume the signal to noise ratio

(SNR) is 10dB, which is not hard to achieve in most LDVs. The simulated signal with Gaussian white noise is shown in Fig. 1.

On the basis of Eq. (2), the theoretical value of the IF is as follows:

$$f(t) = \frac{d\Phi(t)}{dt} = 80000\sin(8\pi t + \pi / 10) \tag{5}$$

The simulation conditions are shown in Table 1. According to Eq. (3), the simulated signal represents an acceleration process. In the remainder of this chapter, we will demonstrate how to extract the IF from the simulated signal.

Table 1. Simulation conditions

Parameters	Value
Initial frequency	≈25KHz
Maximum frequency	80KHz
Simulation time	50ms
Sampling rate	2MHz

3.2. *Wavelet denoising*

Since the IF extraction result is seriously affected by noise signal, wavelet denoising is adopted before subsequent processing. According to previous research, the wavelet denoising algorithm includes three main steps [15]:

(1) Wavelet transform of the noisy signal;
(2) Thresholding the resulting wavelet coefficients;
(3) Inverse of wavelet transform to obtain the denoised signal.

Obviously, the parameters of the wavelet denoising algorithm are determined by the condition of the Doppler signal. The adopted parameters are shown in Table 2.

Table 2. The parameters of wavelet denoising algorithm

Parameters	Value
Decomposition levels	5
Threshold selection type	Soft threshold
Threshold selection rule	Use the principle of Stein's Unbiased Risk
Wavelet name	Symlets or Daubechies
Vanishing moments	15

First, the Doppler signal is decomposed and its scale coefficients is calculated. As the operation time will be increased with higher decomposition levels, the decomposition levels is set to five. Then the signal is processed by thresholding method with soft threshold and the principle of Stein's Unbiased

Risk to smoothen the signal. Finally, we got the denoised signal by the reconstruction of wavelet. After testing, we found that both Symlets wavelet and Daubechies wavelet is effective in Doppler signal denoising. Moreover, the vanishing moments of the wavelet should be more than ten to make the denoised signal smoother. The result of wavelet denoising is presented in Fig. 2.

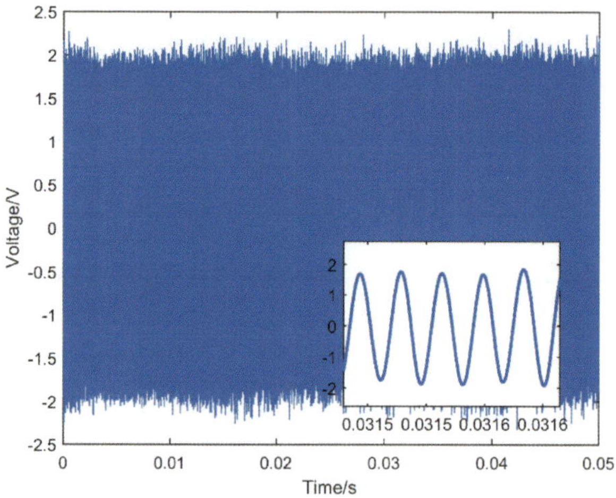

Fig. 2. The Doppler signal after wavelet denoising

3.3. *Signal normalization*

In order to extract the IF by DQ method, the signal should be normalized. The empirical AM-FM decomposition, proposed by Huang for Hilbert transform in 1998, is adopted to extract the envelop and normalize the signal [16]. According to 3.1, the output signal of LDV is a frequency-varying cosine signal. The denoised signal can be expressed as:

$$c(t) = A(t)\cos\left[\Phi(t)\right] \tag{6}$$

The steps of empirical AM-FM decomposition is as follows:

(1) Take the absolute value of $c(t)$ and then find all maximal values of $|c(t)|$. Here a threshold value should be set to diminish the influence of noise.

(2) Extract the envelop signal $e_1(t)$ by using interpolation for all maximal values.

(3) Normalize the signal $c(t)$ by $e_1(t)$, that is $y_1(t) = c(t)/e_1(t)$. If $y_1(t) \le 1$ at any instant, the normalization is completed. The FM component of the denoised signal is equal to $y_1(t)$.

(4) If there is a value greater than 1 existed in signal $y_1(t)$, repeat the previous steps:

$$\begin{cases} y_2(t) = y_1(t) / e_2(t) \\ \quad \vdots \\ y_n(t) = y_{n-1}(t) / e_n(t) \end{cases} \tag{7}$$

The condition $y_n(t) \leq 1$ will be met after finite iterations. Actually, we only need one iteration in most cases. Fig. 3 shows the envelop signal extracted from the denoised signal.

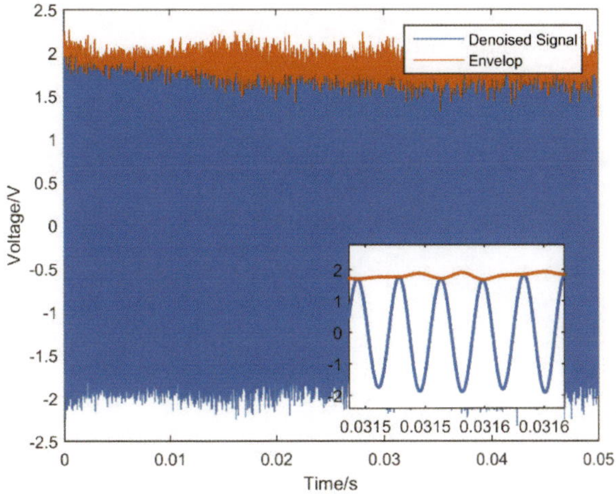

Fig. 3. The denoised Doppler signal and the envelop

3.4. *Direct quadrature method*

As the signal $F(t)$ has already been normalized, its phase function is as follows:

$$\Phi(t) = \arctan \sqrt{1 - F^2(t)} / F(t) \tag{8}$$

Then we are able to extract the IF through Eq. (2). In consideration of the continuity of the function $f(t)$, the procedure of DQ method is improved as follows [17]:

$$f(t)=\begin{cases}\frac{1}{2\pi}\frac{|F'(t)|}{\sqrt{1-F^2(t)}},|F(t)|<0.99\\ \frac{1}{2\pi}\sqrt{|F''(t)|},0.99\le|F(t)|\le 1\end{cases} \tag{9}$$

After this step, the result of IF is shown in Fig. 4. Since the noise signal cannot be filtered completely by wavelet denoising, least squares fitting is adopted to reduce the error. The final result is presented in Fig. 5, indicating that the fitted value is very close to the theoretical value. The accuracy of the fitted result is higher than 0.5%.

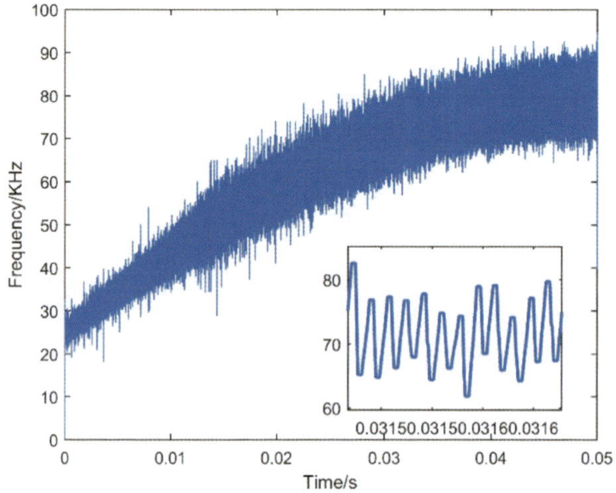

Fig. 4. The IF extracted by improved DQ method

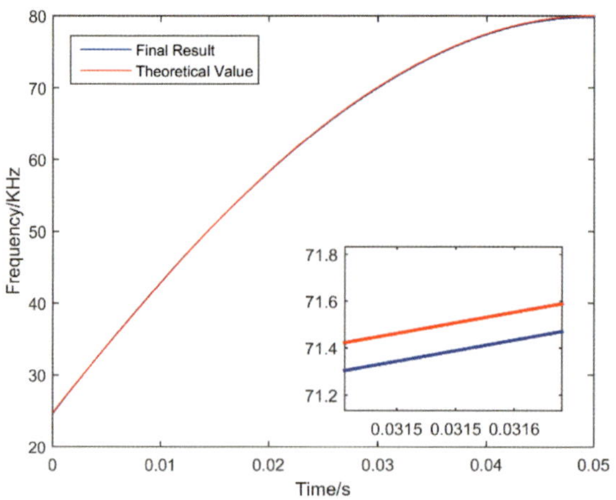

Fig. 5. The final result and the theoretical value of IF

3.5. *Operation time analysis*

In order to test the time complexity of the IF extraction method, the operation times under different data size are recorded in the same computer. The experimental results are shown in Table 3.

Table 3. The operation times under different data size

Data size	Operation times					Average Times
10000	0.5436	0.4999	0.4947	0.4888	0.4873	0.5029
100000	1.2259	1.4965	1.2979	1.2295	1.2768	1.3053
1000000	8.7795	8.7605	8.8202	8.8144	8.9173	8.8184
10000000	87.0267	86.3799	88.0217	88.4695	87.5834	87.4962

For each data size, we repeat the IF extraction method for five times and get an average operation time (in seconds). Then, we analyzed the correlation coefficient of the data size and the average operation time. The result of correlation coefficient is 0.9999935, indicating that the time complexity of this method is linear.

4. Conclusion

In this paper, an improved IF extraction method is adopted to process Doppler signal. The experimental result demonstrates that the adopted method is able to extract the IF from the Doppler signal with high accuracy. In addition, the time complexity of this method is linear, indicating that its computation speed is quite fast.

References

1. Yeh, Y, and H. Z. Cummins, Localized fluid flow measurements with an He–Ne laser spectrometer. *Applied Physics Letters* **4**, 10 (1964) pp. 176-178.
2. Sharma, Utkarsh, et al, Fiber optic confocal laser Doppler velocimeter using an all-fiber laser source for high resolution measurements. *Optics express* **13**, 16 (2005) pp. 6250-6258.
3. Cheng, Chih-Hao, Lyu-Chih Lin, and Fan-Yi Lin. Self-mixing dual-frequency laser doppler velocimeter. *Optics express* **22**, 3 (2014) pp. 3600-3610.

4. Maru, Koichi, and Yusaku Fujii. Wavelength-insensitive laser Doppler velocimeter using beam position shift induced by Mach-Zehnder interferometers. *Optics express* **17**, 20 (2009) pp. 17441-17449.
5. Liu, Shouxian, et al. Analysis of photonic Doppler velocimetry data based on the continuous wavelet transform. *Review of Scientific Instruments* **82**, 2 (2011).
6. Todorovska, Maria I. Estimation of instantaneous frequency of signals using the continuous wavelet transform. University of Southern California, Department of Civil Engineering (2001).
7. Wang Guohua, Jinglin Sun, and Qiang Li. Optical fiber-based system for continuous measurement of in-bore projectile velocity. *Review of Scientific Instruments* **85**, 8 (2014).
8. Xiaoming Nie, Zhou Jian, and Long Xingwu. Application of Hilbert–Huang transform to laser Doppler velocimeter. *Optics & Laser Technology* **44**, 7 (2012) pp. 2197-2201.
9. Kyong, Yongsoo, et al. Mode shape reconstruction of an impulse excited structure using continuous scanning laser Doppler vibrometer and empirical mode decomposition. *Review of Scientific Instruments* **79**, 7 (2008).
10. Roulier, Rémy, et al. Comparison between Hilbert–Huang transform and scalogram methods on non-stationary biomedical signals: application to laser Doppler flowmetry recordings. *Physics in medicine and biology* **50**, 21 (2005).
11. Huang, Norden E., et al. On instantaneous frequency. *Advances in adaptive data analysis* **1**, 2 (2009) pp. 177-229.
12. Wang, Gang, et al. On intrinsic mode function. *Advances in Adaptive Data Analysis* **2**, 3 (2010) pp. 277-293.
13. Bedrosian, Edward. A product theorem for Hilbert transforms. *Proceedings of the IEEE* **51**, 5 (1963) pp. 868-869.
14. Nuttall, A. H. On the quadrature approximation to the Hilbert transform of modulated signals. *Proceedings of the IEEE* **54**, 10 (1966) pp. 1458-1459.
15. Han, Min, et al. Noise smoothing for nonlinear time series using wavelet soft threshold. *Signal Processing Letters, IEEE* **14**, 1 (2007) pp. 62-65.

16. Huang, Norden E., et. al. The empirical mode decomposition and the Hilbert spectrum for nonlinear and non-stationary time series analysis. Proceedings of the Royal Society of London A: Mathematical, Physical and Engineering Sciences. Vol. 454. No. 1971. *The Royal Society* (1998).
17. Dai Haomin, Xu Aiqiang. Comparative research and improvement on the calculation method of instantaneous frequency. Vol. 51. No. 6. *Journal of Sichuan University* (2014)

The Improved Coaxial Bragg Structure

Xue-Yong Ding†, Lian-Sheng Wang and Yuan Wang

Department Of Polytechnic, Sanya university

Sanya, Hainan 572000, China

†E-mail: ding2008ding@163.com

Based on the existing coaxial Bragg structure with sinusoidal ripples, the ripples depth remains unchanged and again, smaller sinusoidal superposition cycle, the improved coaxial Bragg structure is proposed. Based on the mode-coupling method, comparative study is carried out for the frequency response. Results demonstrate that the interval between the band-gaps is broadened, and the band-gap overlap of modes can be efficiently separated by use of the improved coaxial Bragg structure. The residual side-lobes of the frequency response also can be effectively suppressed and the Q factor is greater. These peculiarities of the improved coaxial Bragg structure with Sinusoidal ripples provide potential advantage in constructing a coaxial Bragg cavity with high quality factor for single higher-order-mode operation of a high-power free-electron maser and also are favorable to improve the performance as a reflector or a filter in its applications.

Keywords: Coaxial Bragg structure; Sinusoidal ripples; Reflectivity; Improved structure; Residual side-lobes; Q-factor.

1. Introduction

The Bragg structures have wide application in tunable lasers, photonic devices, integrated circuits, and high-power masers [1-4]. Particularly, a metallic Bragg structure is regarded as a suitable for the construction of overmoded cavities for high-power cyclotron auto-resonance maser (CARM) [3] and free-electron laser [2-4] oscillators in millimeter and sub-millimeter wave ranges. Recently, much attention has been paid to the coaxial Bragg structures due to their attractive peculiarities compared to the planar and cylindrical Bragg structures [3,5-10]. In order to achieve some peculiarities what are favorable to the mode selectivity, the ability of dissipating the ohmic heat on the wall, the oscillating startup of the operating mode, and machining manufacture, the new coaxial Bragg structure with improved coaxial Bragg structure is put forward.

In this paper, based on the existing coaxial Bragg structure with sinusoidal ripples and the depth unchanged, a smaller sinusoidal was superposed. The present paper is to demonstrate frequency response of reflectivity in the improved coaxial Bragg structure.

2. Theory model

Figure 1(a) shows the longitudinal-section view of a coaxial Bragg structure with sole sinusoidal ripple[6], different from the coaxial Bragg structure with sole sinusoidal ripple, this paper envisaged to etch cycle smaller auxiliary sine distribution in the structure of wall surface, and form the improved coaxial Bragg structure. Figure 1(b) shows the longitudinal-sectional view of the improved coaxial Bragg structure, where a_0 and b_0 are average radius of the outer-wall and the inner-rod, l_{out} and l_{in} are the corrugation depth of the outer-wall and the inner-rod, p_b and p_{in} are the main corrugation period and the embedded auxiliary corrugation, and L is the structure length, respectively.

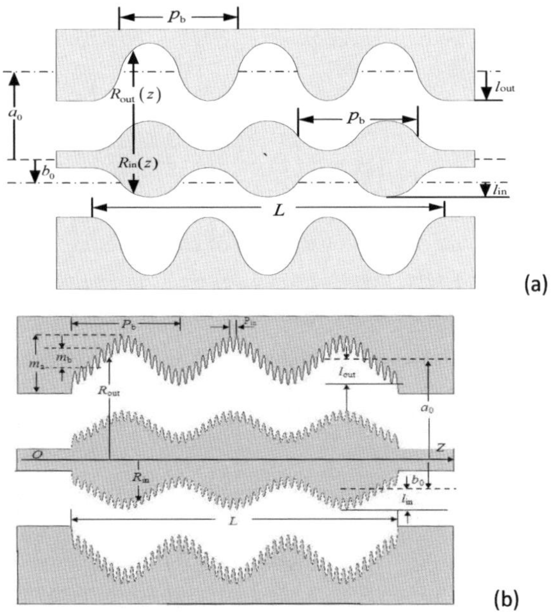

Fig. 1. Longitudinal-section profile of a coaxial Bragg structure ,(a) with sole sinusoidal ripples and (b) the improved coaxial Bragg structure.

For a coaxial Bragg structure with sole sinusoidal ripples, the dependence of the outer-wall radius R_{out} and the inner-rod radius R_{in} of the coaxial Bragg structure on the longitudinal position can be expressed by[6]

$$R_{out}(z) = a_0 - l_{out}\cos(k_b z + \phi_{out}) \tag{1}$$

$$R_{in}(z) = b_0 - l_{in}\cos(k_b z + \phi_{in}) \tag{2}$$

Where $k_b = 2\pi / p_b$.

For the improved coaxial Bragg structure, the dependence of the outer-wall radius R_{out} and the inner-rod radius R_{in} on the longitudinal position z can be expressed by

$$R_{out}(z) = a_0 - l_{out}[m_a\cos(k_B z + \varphi_{out}) + m_b\cos(k_A z + \varphi_{out})] \tag{3}$$

$$R_{in}(z) = b_0 - l_{in}[m_a\cos(k_B z + \varphi_{in}) + m_b\cos(k_A z + \varphi_{in})] \tag{4}$$

where $k_A = 2\pi/p_{in}$, $k_B = 2\pi/p_b$. In order to keep the depth unchanged, we suppose that $m_a + m_b = 1$, and the $K_{am} = \frac{m_b}{m_a}$ is defined as amplitude scale coefficient, in Generally $K_A = NK_B$, and when K_{am}=0, the coaxial Bragg structure with improved coaxial Bragg structure return to the coaxial Bragg structure with sole sinusoidal ripples. The angle θ_{out} between $\hat{r}$ and $\hat{n}_{out}$ and the angle θ_{in} between $\hat{r}$ and $\hat{n}_{in}$ are determined by

$$\tan\theta_{\text{out}} = \frac{dR_{\text{out}}}{dz} = l_{\text{out}}\left[m_a k_B \sin\left(k_B z + \phi_{\text{out}}\right) + m_b k_A \sin(k_A z + \phi_{\text{out}})\right] \tag{5}$$

$$\tan\theta_{\text{in}} = \frac{dR_{\text{in}}}{dz} = l_{\text{in}}\left[m_a k_B \sin\left(k_B z + \phi_{in}\right) + m_b k_A \sin(k_A z + \phi_{in})\right] \tag{6}$$

where $\hat{n}_{out}$ and θ_{in} are the unit vectors normal to the outer-wall and inner-rod surfaces, φ_{out} and φ_{in} are the Initial corrugation phase of the outer wall and Initial corrugation phase of the inner rod respectively.

Based on the coupled mode equations, a Fortran programme code is written to evaluate the reflectivity characteristics of the considered Bragg structure with improved coaxial Bragg structure [6].

3. Characteristics of frequency response in the improved coaxial Bragg structure

For a better comparison, we assume that the incident mode $TE_{6,1}$ is the operating mode with sub-millimeter wave range of 85～115 GHz (the central frequency being 100 GHz), and the other parameters are summarized in Table 1 respectively . As is indicated in Ref. [6], mode coupling should be taken into account for the higher-order mode operation, suppose that, except for the operating mode $TE_{6,1}$, the neighboring modes $TM_{6,2}$ may be involved in the coaxial Bragg structure. Therefore, the Fortran program code is written to evaluate the reflection characteristics of the improved coaxial Bragg structure, instead of the sole ripples shape with sinusoidal employed in previous work [6].

Table 1 Main parameters of the coaxial Bragg structure

Title	Numerical value
Operating frequency	100GHz
Operating mode	TE_{61}
Outer wall radius, a_0	20.0mm
Inner rod radius, b_0	14.0mm
Corrugation amplitude $l_{out} = l_{in}$	0.1mm
Main ripple period p_b	1.52mm
Auxiliary corrugation period p_{in}	0.095mm
Length of cavity	110.2mm
Initial corrugation phase of the outer wall, φ_{out}	0
Initial corrugation phase of the inner rod, φ_{in}	π

Figure 2 shows the frequency response of the reflectivity in the improved coaxial Bragg structure comparing with the sole sinusoidal ripples of the involved modes $TE_{6,1}$ and $TM_{6,2}$. It indicates that there are two main band gaps, which respectively correspond to the coupling between the forward wave and the backward wave of $TE_{6,1}$ ($TE_{6,1} \leftrightarrow TE_{6,1}$ scattering) at lower frequency range, and to the coupling of the forward wave of $TE_{6,1}$ with the backward wave of the competing mode $TM_{6,2}$ ($TE_{6,1} \leftrightarrow TM_{6,2}$ scattering) at higher frequency range. However, when the $K_{am} = 1$, Figure 2 shows that the band-gap of the competing mode $TM_{6,2}$ moves away the band-gap of the operating mode $TE_{6,1}$ when with the improved coaxial Bragg structure, and the reflectivity is almost the same no matter the structure with the improved coaxial Bragg structure or with sole sinusoidal ripples. Therefore, use of the improved coaxial Bragg structure is further efficiently to separate the band-gap overlap, and the residual side-lobes are quite effective to suppress even though without windowing-function technique [7-10].

Because of the BW_{3dB} becomes smaller, Figure 2 shows that the Q factor is get greater and the selectivity of frequency is enhanced when with the improved coaxial Bragg structure than with the sole sinusoidal ripples. Figure 3 shows that the Q-factor versus the frequency for the coaxial Bragg structure with sole sinusoidal ripples and the improved coaxial Bragg structure, and the central frequency of the operating mode $TE_{6,1}$ is 100 GHz and the central frequency of the competing mode $TM_{6,2}$ is 106 GHz.

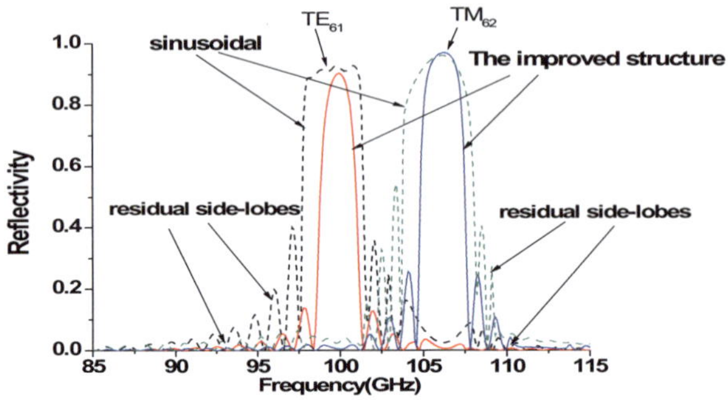

Fig. 2. Frequency response of the reflectivity in the improved coaxial Bragg structure comparing with the sole sinusoidal ripples, where the phase difference between the inner-rod and outer-conductor corrugations $\delta\varphi = |\varphi_{in} - \varphi_{out}| = \pi$, $K_{am} = 1$ and other parameters are the same as in Table 1.

Figure 3 shows that the Q-factor is get greater no matter the mode is the operating mode $TE_{6,1}$ or the competing mode $TM_{6,2}$. It should be pointed out that the sine curve is smooth and continuous curve, therefore, the auxiliary sinusoidal ripples surface is also smooth and continuous sinusoidal slot, it is different from the rough surface machining burrs to make additional loss and reduce the Q value. On the contrary, the auxiliary sine slot properly changed the boundary conditions and improved the field distribution, so the Q value increased.

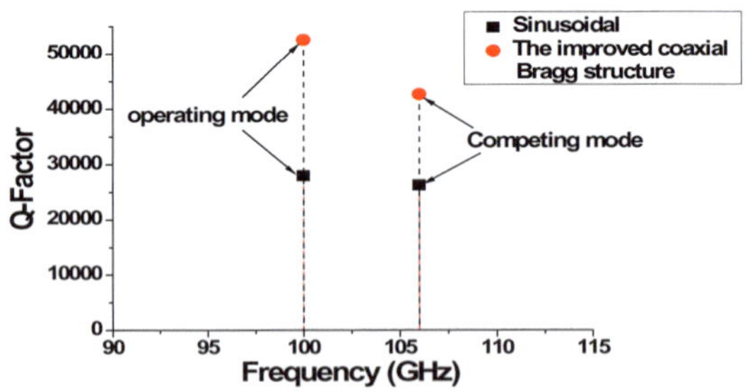

Fig. 3. The Q-factor versus the frequency for the coaxial Bragg structure with sole sinusoidal ripples and the improved coaxial Bragg structure.

Figure 4 plots the reflectivity versus the frequency for different amplitude scale coefficient k_{am} in the improved coaxial Bragg structure. Results show that the reflectivity can be diminishingly and the bandwidth can be narrow by increasing the the K_{am}, but when the $K_{am}>1$, the reflectivity is decreased obviously, so the K_{am} is equal or lesser than 1 is the best selection. Figure 5 shows that the Q-factor can be enlargement by increasing the K_{am}. Combination with the results in figure 4 , we can choose the best range of value of amplitude modulation ratio coefficient $K_{am} \leq 1$.

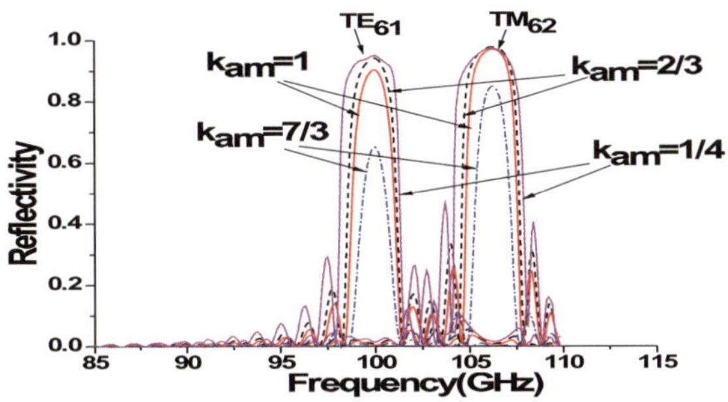

Fig. 4. Reflectivity versus the frequency for different amplitude scale coefficient k_{am} in the improved coaxial Bragg structure, where the phase $\delta\varphi = |\varphi_{in} - \varphi_{out}| = \pi$, and other parameters are the same as in Table 1.

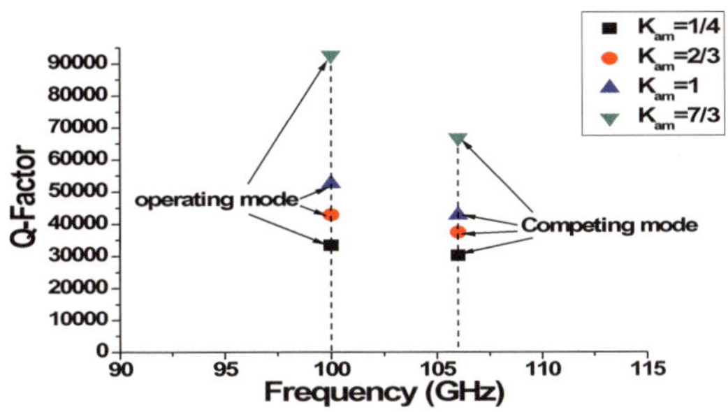

Fig. 5. The *Q*-factor versus the frequency for different amplitude scale coefficient k_{am} in the improved coaxial Bragg structure.

4. Conclusions

In this paper, the main points can be drawn from the simulation results: The interval between the band-gap of the competing mode and the band-gap of the operating mode is expanded by employing the improved coaxial Bragg structure; The reflectivity can be diminishingly and the bandwidth can be narrow by increasing the amplitude scale coefficient K_{am}, and the K_{am} is best to equal or lesser than 1; band-gap overlap in an overmoded coaxial Bragg structure can be efficiently separated, the residual side-lobes of the frequency response of the operating working mode and the competing mode also can be effectively suppressed, and the Q factor is get greater and the selectivity of frequency is enhanced by use of a coaxial Bragg structure with improved coaxial Bragg structure.

These peculiarities of coaxial Bragg structure with Improved coaxial Bragg structure provide potential advantage in constructing a coaxial Bragg cavity with larger dimensional size, narrower bandwidth, larger eigenvalue intervals and high quality factor for single higher-order-mode operation of a high-power free-electron maser and also are favorable to the mode selectivity, the ability of dissipating the ohmic heat on the wall, the oscillating startup of the operating mode, and machining manufacture.

Acknowledgment

This work is supported by the Scientific Research Fund of the provincial Natural Science Foundation of Hainan (No.614252) and the provincial educational reform project funds of Hainan (Hnjg2015-61).

References

[1] A. Yariv and M. Nakamura. Periodic Structures for Integrated Optics, IEEE J. Quantum Electronics, QE-13, 233-253 (1977).

[2] H. Kogelnik and C. V. Shank. Coupled-Wave Theory of Distributed Feedback Laser, J. Appl. Phys., vol. 43, pp. 2327-2335 (1972).

[3] R. B. McCowan, A. W. Fliflet, S. H. Gold, V. L. Granatstein, and M. C. Wang, Design of a waveguide resonator with rippled wall reflectors for a 100 GHz CARM oscillator experiment, Int. J. Electronics, vol. 65, pp. 463-475 (1988).

[4] I. V. Konoplev, P. McGrane, W. He, A.W. Cross, A.D.R. Phelps, C.G. Whyte, K. Ronald, and C.W. Robertson, Experimental study of coaxial free-electron maser based on two-dimensional distributed feedback, Phys. Rev. Lett., vol. 96, pp. 035002 (2006).

[5] J. J. Barroso and J. P. Leite Neto, Design of coaxial Bragg reflectors, IEEE Trans. on Plasma Sci., vol. 34, pp. 666-672 (2006).

[6] Y.-X. Lai and S.-C. Zhang, Multiwave interaction formulation of a coaxial Bragg structure and its experimental verification, Phys. Plasmas, vol. 14, pp. 113301 (2007).

[7] X.-H. Chen and S. C. Zhang, Suppression of residual side-lobes in a coaxial Bragg reflector, Int. J. Infrared Millimeter Waves. 29, 552–557 (2008).

[8] X.-Y. Ding, S.-C. Zhang, and Y.-C. Xie, Characteristics of frequency response in a coaxial Bragg structure with tapered ripples, High power laser and Particle Beams, 20(12), 2051-2054 (2008).

[9] X.-Y. Ding, Hong-Fan Li and Zhen-Su Lv, Effect of Ripple Taper on Band-gap Overlap in a Coaxial Bragg Structure Operating at Terahertz Frequency, Phys. Plasmas, vol. 19, pp.092105 (2012).
[10] X. Y. Ding and S.C. Zhang, Effect of ripple shape and taper on frequency response of reflectivity and transmission in a coaxial Bragg structure[J].J. Appl. Phys,42:085104 (2009).

The Influence of Return Stroke Current Waveforms to Approximation of Spatial Lightning Electromagnetic Field at Near Ranges

Li-peng Wang†, Ya-zhou Chen, Hao-jiang Wan and Xiaojia Wang

Institute of Electrostatic and Electromagnetic Protection, Ordnance Engineering College, Shijiazhuang, 050003, China

†*E-mail: wlp1508710615@163.com*

To further explore the waveform characteristics of the near-field free space lightning electromagnetic pulse field, the mutual relationships between the lightning electromagnetic field at different heights of the near-field region and return stroke channel base current were theoretically researched. Based on TL model, the relationship between the near-field region space LEMP and the return stroke current has been deducted theoretically. By changing the return stroke current waveforms and contrasting precise LEMP waveforms with approximate expressions, it was found that the scope of the approximate expression of the near-field free space horizontal electric field was increasing as the rising time of the return stroke current waveforms were shortened, while the scope of magnetic field was greatly reduced.

Keywords: Lightning electromagnetic pulse field; Return stroke channel base current; Near-field; Free space; Approximate waveform.

1. Introduction

Lightning is one of the most common hazards of electromagnetic pulse in nature[1 - 3]. Therefore, many scholars at home and abroad have done a great deal of research on LEMP[4][5]. About the mutual relationship between the lightning electromagnetic pulse field and return stroke current, Uman et. al., based on the vertical channels, have proved the approximate between the channel base current and the electromagnetic field of the far-field on the surface[6][7]; Leteinturier et. al., has found the approximate nature between the near-field lightning return stroke electromagnetic field and the channel base current waveforms with artificially triggered lightning experiment[8][9].

However, the studies above are all expanded for the LEMP on the surface. To further understand the impact of the return stroke current waveforms on the approximate of near-field free space lightning electromagnetic field, this paper will study the relationship between the return stroke current waveforms and near-field lightning electromagnetic waveforms based on the Vertical channel model.

2. LEMP expression in near-field space

In the condition of vertical return stroke channel, the exact expression of electromagnetic field of any position of the observation point as [10]:

After preliminary studies, the waveforms of vertical electric field component and its approximate waveforms present significant difference when $z \neq 0$, almost no approximate relationship. So this paper only study the horizontal electric field component and the magnetic field component.

The first-level approximate expression as follows:

$$E_r' = [rF(t-\frac{r}{c}) - ri(0,t-\frac{r}{c})(\frac{z}{v}-\frac{r}{c})](\frac{1}{R^3(h_+)}-\frac{1}{R^3(h_-)}) - \frac{ri(0,t-r/c)}{v}[\frac{(h_+)-z}{R^3(h_+)}-\frac{(h_-)-z}{R^3(h_-)}+ \frac{(h_-)-z}{r^2R(h_-)}-\frac{(h_+)-z}{r^2R(h_+)}] \quad (1)$$

$$B_\varphi' \approx \frac{u_0}{4\pi} ri(0,t-\frac{r}{c}) \int_{h-}^{h+} \frac{1}{R^3} \mathrm{d}z' = \frac{u_0}{4\pi} ri(0,t-\frac{r}{c})[\frac{(h_+)-z}{r^2R(h_+)}-\frac{(h_-)-z}{r^2R(h_-)}] \quad (2)$$

With the time increasing, R is tending to $|h+|$ and $|h-|$. Omitting the item $1/R^3$ in the first-level approximate expression of formula (1) and (2), the second-level approximate expression as follows:

$$E_r'' \approx -\frac{1}{2\pi\varepsilon_0 vr} i(0,t-\frac{r}{c}) \quad (3)$$

$$B_\varphi'' \approx \frac{u_0}{2\pi r} i(0,t-\frac{r}{c}) \quad (4)$$

3. The impact of the different return stroke current waveforms on the approximation of near-field spatial lightning electromagnetic field

To compare to the impact of the different return stroke current waveforms on the approximation of near-field spatial lightning electromagnetic field, we adopt two different return stroke currents — first stroke and subsequent stroke — to calculate the lightning electromagnetic field. To facilitate analytic calculation of the return stroke electromagnetic field, the return stroke channel base current is expressed by pulsed current function.

The corresponding return stoke channel base current and parameter settings are shown in Table 1.

Table 1. Return stoke channel base current parameter settings

Parameter	τ_1	τ_2	I_0/kA	η	n	v/(m·s^{-1})
First stroke (10/350μs)	3.371 7×10^{-6}	4.730 1×10^{-4}	100	0.93	10	1×10^{8}
Subsequent stroke (0.25/100μs)	7.813 6×10^{-8}	1.432 4×10^{-4}	25	0.993	10	1.3×10^{8}

Figs. 1 to 6 and Figs. 7 to 12 are given the results of spatial lighting return stroke electromagnetic fields under different distances and different heights, which happen in the first stroke and subsequent stroke respectively (including exact and approximate solution).

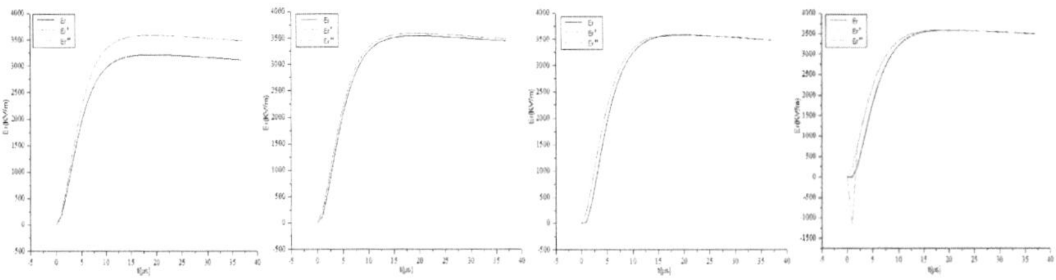

(a) z=10m (b) z=30m (c) z=70m (d) z=100m

Fig. 1 Comparison between the horizontal electric field and two level approximate waveforms of *r*=5m at different heights under the first stroke

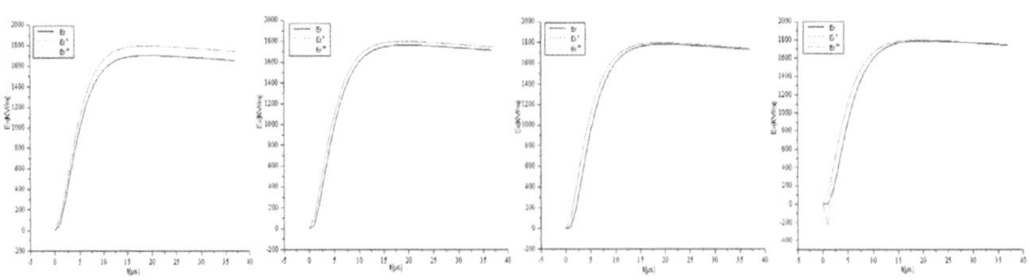

(a) z=30m (b) z=50m (c) z=80m (d) z=100m

Fig. 2 Comparison between the horizontal electric field and two level approximate waveforms of r=10m at different heights under the first stroke

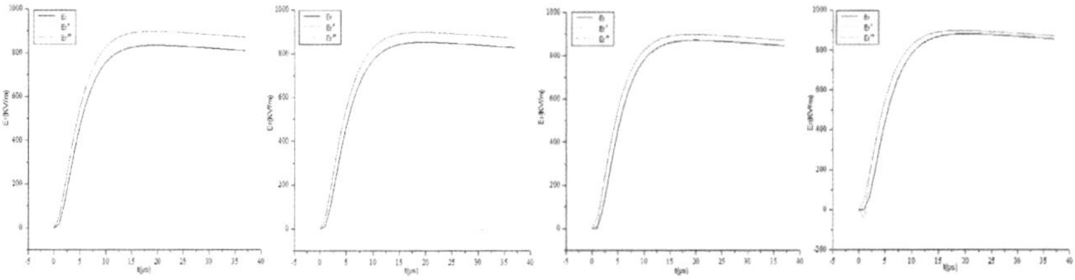

(a) z=50m (b) z=60m (c) z=80m (d) z=100m

Fig. 3 Comparison between the horizontal electric field and two level approximate waveforms of *r*=20m at different heights under the first stroke

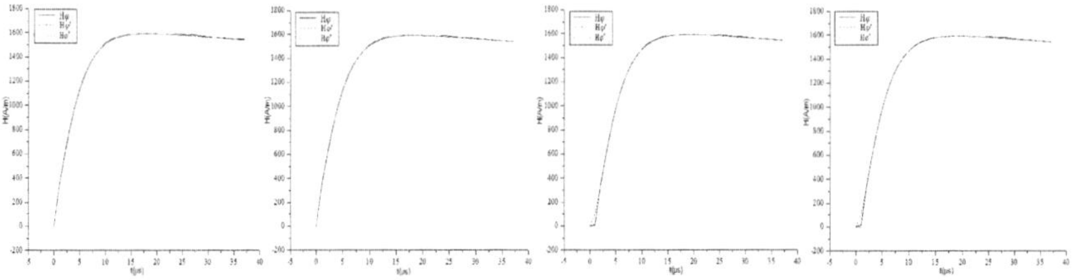

(a) z=10m (b) z=100m (c) z=120m (d) z=150m

Fig. 4 Comparison between the magnetic field and two level approximate waveforms of *r*=10m at different heights under the first stroke

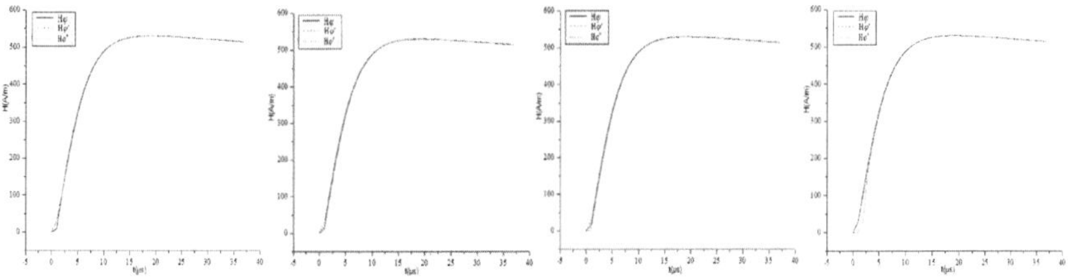

(a) z=10m (b) z=100m (c) z=150m (d) z=200m

Fig. 5 Comparison between the magnetic field and two level approximate waveforms of *r*=30m at different heights under the first stroke

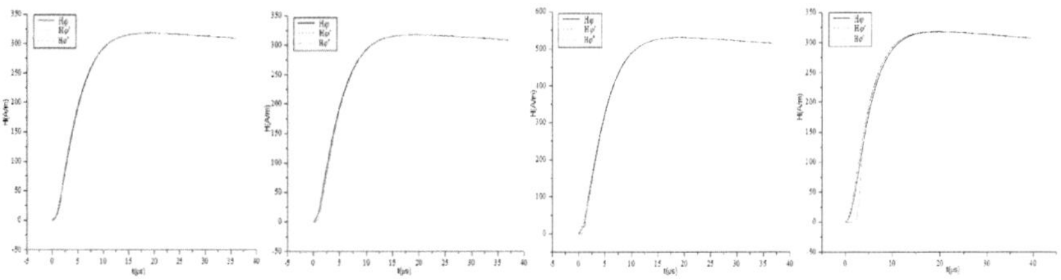

(a) z=10m (b) z=100m (c) z=150m (d) z=200m

Fig. 6 Comparison between the magnetic field and two level approximate waveforms of *r*=50m at different heights under the first stroke

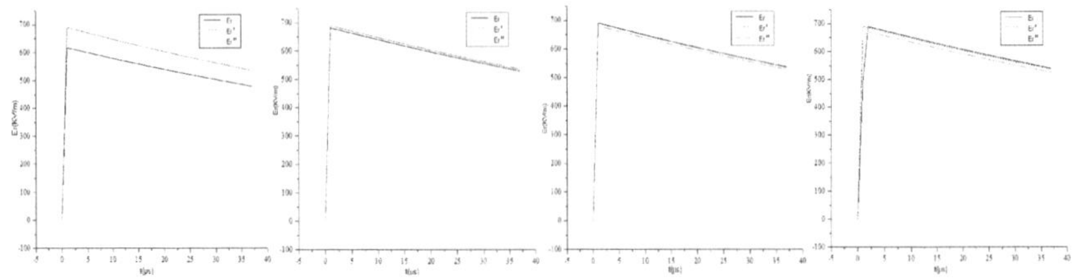

（a）z=10m （b）z=30m （c）z=400m （d）z=500m

Fig. 7 Comparison between the horizontal electric field and two level approximate waveforms of r=5m at different heights under the subsequent stroke

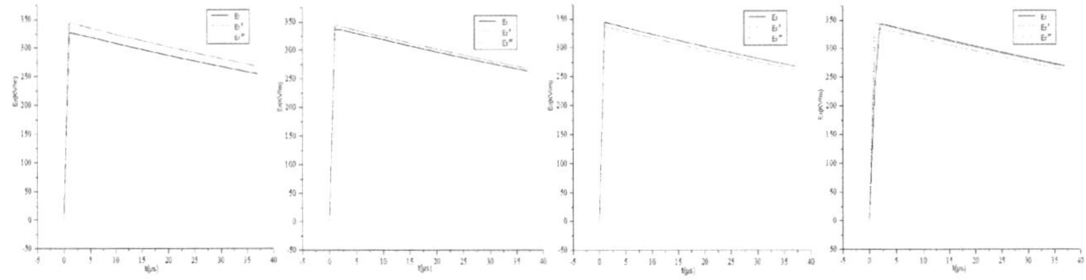

（a）z=30m （b）z=50m （c）z=400m （d）z=500m

Fig. 8 Comparison between the horizontal electric field and two level approximate waveforms of r=10m at different heights under the subsequent stroke

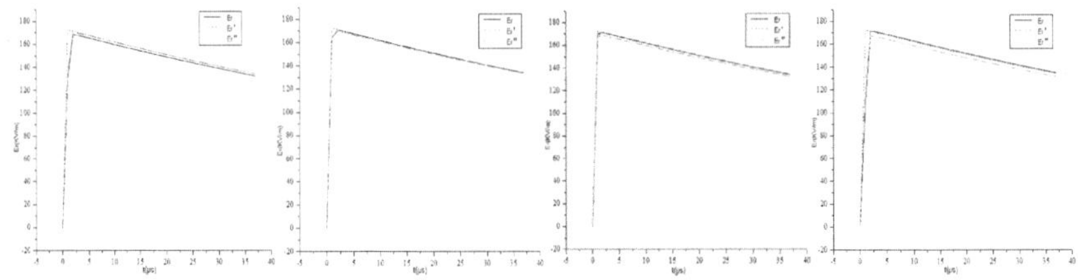

（a）z=100m （b）z=200m （c）z=300m （d）z=500m

Fig. 9 Comparison between the horizontal electric field and two level approximate waveforms of r=20m at different heights under the subsequent stroke

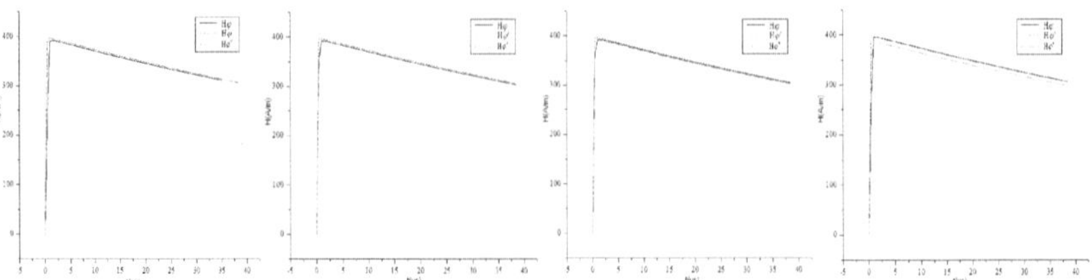

（a）z=10m （b）z=50m （c）z=200m （d）z=500m

Fig. 10 Comparison between the magnetic field and two level approximate waveforms of r=10m at different heights under the subsequent stroke

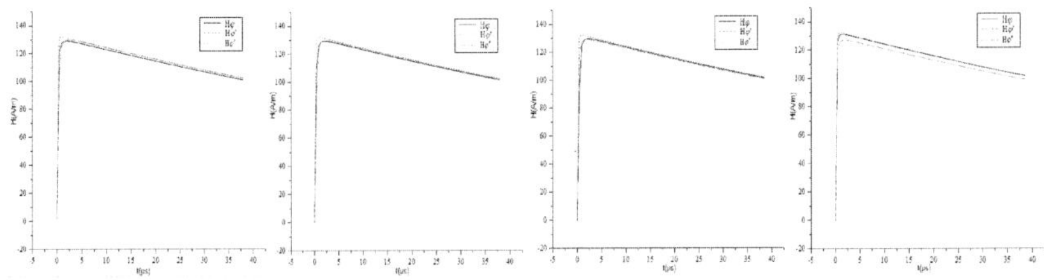

(a) z=10m (b) z=100m (c) z=200m (d) z=500m

Fig. 11 Comparison between the magnetic field and two level approximate waveforms of *r*=30m at different heights under the subsequent stroke

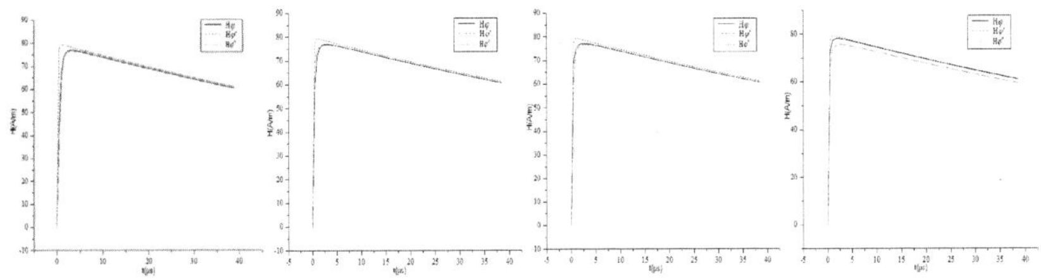

(a) z=10m (b) z=100m (c) z=200m (d) z=500m

Fig. 12 Comparison between the magnetic field and two level approximate waveforms of *r*=50m at different heights under the subsequent stroke

From Figs. 1 to 12, we can find that, under different waveforms — rather the rising time of the return stroke current waveforms are different, the scope of the lightning near-field return stroke electromagnetic approximate analytic expressions (7) to (10) are different. Table 2 shows the scope of the approximate analytic expressions of near-field electromagnetic field under the first stroke and subsequent stroke. It is found that, with the reducing of the rising time of the return stroke current waveforms, the scope of the near-field spatial horizontal electric field approximate expression is increasing, while that of the magnetic field is greatly decreasing.

Table 2 The scope of the approximate analytic expressions of near-field electromagnetic field under the first return stroke and subsequent stroke

	Horizontal electric field		Magnetic field	
	Distances	Heights	Distances	Heights
First stroke (10/350μs)	5m 10m >20m	30—70m 50—80m nothing	10m 30m 50m	<100m <150m <150m

Subsequent stroke (0.25/100μs)	5m 10m >20m	30—400m 50—400m Nothing	10m >30m	50—200m Nothing

4.Conclusion

In this paper, based on the spatial expression of the vertical channel lightning electromagnetic pulse field, we analytically derived the first-level and second-level approximation formula of the near-field spatial LEMP. By changing the return stroke current waveforms and contrasting the precise electromagnetic waveform with its two approximate expressions, we confirmed the scope of the approximate expression of first-level and second-level electromagnetic fields under different return stroke current waveforms. The results showed that: With rising time of the return stroke current waveform is shortened, the scope of the approximate expression of the near-field spatial horizontal electric field is increased, while that of the magnetic field is greatly reduced.

Acknowledgment

This work is supported by National Natural Science Foundation of China (No. 51377171).

References

1. YU Hao. The basis of modern lighting protection technology. (Beijing: Tsinghua University Press, 2005. 1-15.)
2. Rakov V A. Rachidi F. Overview of recent progress in lightning of recent progress in lightning research and lightning protection. (IEEE Transactions on Electromagnetic Compatibility, 2009)
3. Hussein A M. Correlating the characteristics of the CN tower lightning return-stroke current with those of its generated electromagnetic pulse. (IEEE Transactions on Electromagnetic Compatibility, 2008)
4. Willett J C. Submicrosecond intercomparison of radiation fields and currents in triggered lightning return strokes based on the TL model. (J Geophy Res, 1989.)
5. Cooray V. Effects of propagation on the return stroke radiation fields. (Radio Science, 1987.)
6. Uman M A, Mclain D K. Kride E P. The electromagnetic radiation from a finite antenna. (Amer J Phys, 1975.)

7. Uman M A. Time derivative of the electric field 10, 14, 30m from triggered lightning strokes. (J Geophy Res, 2000.)
8. Wang Xiaojia, Chen Yazhou, et. al. Approximate Characteristics of Lightning Electromagnetic Field in Nearby Area Generated by Oblique Channel. (High Voltage Engineering. 2013)
9. Leteinturier, C. Current and electric field derivatives in triggered lightning return strokes. (J. Geophys. Res., 95, 1990)
10. Chen Yazhou, Xiao Xuerong. Characteristic of Approximation Between LEMP in Near Area and Channel-base Current. (High Voltage Engineering. 2007)

Sorting radar signal based on the resemblance coefficient of bispectrum two dimensions characteristic

Chun-Lai Yu, Jun Han and Fang Wan

Air Force Early Warning Academy,

Wuhan, China

†*E-mail: duj81@163.com*

Commom method have low sorting rates and is rensitive to the Signal Noise Radio (SNR), thus, resemblance coefficient of bispectrum two dimensions characteristic is applied to sort unknown complicated radar signal to obtain high sorting rate. The bispectrum of received signal is extracted, and it is predigested to two dimensions characteristic. A rectangle pulse sequence and triangle pulse sequence are constructed. Then, the resemblance coefficient of two dimensions characteristic with rectangle pulse sequence and triangle pulse sequence are gained, and they are used as the sorting parameters. The bispectrum of different signal is distinguishing and it is not sensitive to SNR and the resemblance coefficient is divisible and steady. The advantage of this new method is validated by simulation results and the sorting rate is not less than 86% at SNR of 5 dB.

Keywords: Sorting; Bispectrum; Two Dimensions Characteristic; SNR

1. Introduction

Radar emitter signal recognition is one of the key technologies of electronic warfare system. It can sort and identification of pulse with different radiation sources from intercepted dense radar pulse stream. As the new type of radar system is becoming more and more large. The role of signal sorting and identification of unknown radar emitter signal is becoming more and more important in electronic warfare and more and more problems are also needed to solve. The current radar emitter signal recognition algorithm is based on the analysis of the various conventional parameters of the intercepted signal such as time of arrival, angle of arrival, frequency, pulse width etc. Among then, it is a commom method to use the time of arrival time sorting. PRI sorting algorithm has many kinds. Typica if sequence difference histogram, PRI transform and improved PRI transform algorithm etc. But these algorithms have some defects.

Difficult to adapt to the current complex electromagnetic environment. Comprehensive utilization of AOA, frequency, pulse width, pulse amplitude of four conventional parameters can accurately radar radiation. But when the signal frequency, variable pulse width quick change the recognition accuracy of the method will be greatly reduced.

Intra-pulse characteristics is one of the most characteristic parameters of radar emitter signal. Although the conventional parameters signal are varied, but the Intra-pulse characteristic parameters has a certain stability. At present, there are many scholars entropy, wavelet packet feature, fuzzy function Intra-pulse feature parameters should be to sorting and recognition of radar emitter signals and achieved some results. There methods of identification by increasing the Intra-pulse characteristic parameters are sensitive to noise and the applicable signal modulation styles are limited. In view of the above problems, this text use the characteristics of the bispectrum which are not sensitive to noise to propose a new recognition algorithm. First, extracting bispectrum from received signal. Then translate it into a two-dimensional feature. In order to reduce the computational complexity of the recognition algorithm, farther complex features of the two dimensional features which are rectangle pulse sequence and triangle pulse sequence, and they are used as the sorting parameters. Finally, use SVM to new method gives full play to the characteristics of the bispectrum to the noise insensitive. IT can identify radar emitter signals with different modulation styles accurately under the lower SNR.

2. Bispectrum estimation of radar emitter

The radar emitter signal received by the reconnaissance received, after the main include a variety of clutter, receiving system heat noise, etc. studies have shown that clutter caused by a large number of scattered point, such as weather and so on and receiving system heat noise tents to be distributed by Causs. literature has proved that Higher order spectra as a tool for time series analysis can effectively suppress the effect of Gauss noise. So extracting the bispectrum from the received radar emitter signal firstly in order to effectively suppress causing noise. The bispectrum defined by high order cumulant is as follows:

If of the higher-order $c_{kx}\left(\tau_1,\ldots,\tau_{k-1}\right)$ of a random sequence $\left\{x(n),x(n+\tau_1),\cdots,x(n+\tau_{k-1})\right\}$ cumulants:

$$\sum_{\tau_1=-\infty}^{\infty}\cdots\sum_{\tau_{k-1}=-\infty}^{\infty}\left|c_{kx}\left(\tau_1,\ldots,\tau_{k-1}\right)\right|<\infty \tag{1}$$

spectrum is defined as the k order cumulant (k - 1) discrete Fourier transform, be

$$S_{kx}\left(\omega_1,\ldots,\omega_{k-1}\right)=\sum_{\tau_1=-\infty}^{\infty}\cdots\sum_{\tau_{k-1}=-\infty}^{\infty}c_{kx}\left(\tau_1,\ldots,\tau_{k-1}\right)e^{-\mathrm{j}\left(\omega_1\tau_1+\ldots+\omega_{k-1}\tau_{k-1}\right)} \tag{2}$$

$|\omega_i|\le\pi,\ \ i=1,\cdots,k=1,\ \ |\omega_1+\omega_2+\cdots+\omega_{k-1}|\le\pi$. Bispectrum is three-order spectrum, be defined as

$$B_x\left(\omega_1,\omega_2\right)=\sum_{\tau_1=-\infty}^{\infty}\sum_{\tau_2=-\infty}^{\infty}c_{3x}\left(\tau_1,\tau_2\right)e^{-\mathrm{j}\left(\omega_1\tau_1+\omega_2\tau_2\right)} \tag{3}$$

This article uses Direct estimation of non parametric bispectrum estimation. Using bispectrum to radar emitter signal.

1. Divide data $\{x(0),x(1),\ldots x(N-1)\}$ into segments, M samples per segment, N=KM. The overlap between the two adjacent data is allowed here.

2. Count DFT coefficient

$$X^{(k)}\left(\lambda\right)=\frac{1}{M}\sum_{n=0}^{M-1}x^{(k)}\left(n\right)e^{-\mathrm{j}2\pi n\lambda/M} \tag{4}$$

In the formula $\lambda=0,1,\ldots,M/2,\quad k=1,\ldots,K$.

3. On this basis, find the triple correlation of the DFT coefficient.

$$\hat{b}_k\left(\lambda_1,\lambda_2\right)=\frac{1}{\Delta_0^2}\sum_{i_1=-L_1}^{L_1}\sum_{i_2=-L_1}^{L_1}X^{(k)}\left(\lambda_1+i_1\right)X^{(k)}\left(\lambda_2+i_2\right)X^{(k)}\left(-\lambda_1-\lambda_2-i_1-i_2\right) \tag{5}$$

There into $k=1,\ldots,K;\ \ 0\le\lambda_2\le\lambda_1,\ \ \lambda_1+\lambda_2\le f_s/2,\ \ \Delta_0=f_s/N_0,$ N_0 and L_1 should be selected to meet the value of $M=\left(2L_1+1\right)N_0$.

4. Give the average value of the K bispectrum estimation based on the bispectrum estimation of the given data $x(0),x(1),\ldots,x(N-1)$, be

$$\hat{b}_k\left(\lambda_1,\lambda_2\right)=\frac{1}{\Delta_0^2}\sum_{i_1=-L_1}^{L_1}\sum_{i_2=-L_1}^{L_1}X^{(k)}\left(\lambda_1+i_1\right)X^{(k)}\left(\lambda_2+i_2\right)X^{(k)}\left(-\lambda_1-\lambda_2-i_1-i_2\right) \tag{6}$$

In the formula $\omega_1=\frac{2\pi f_s}{N_0}\lambda_1,\quad \omega_2=\frac{2\pi f_s}{N_0}\lambda_2$.

The estimated bispectrum, not only can effectively restrain the Gauss noise, but also can well reflect the characteristics of different radar signal CW, LFM, FSK, BPSK, QPSK, LFM-BPSK, FSK-BPSK, NLFM as an example. The sampling point for each analysis signal is 5120. Divided it into 40 sections. Each length is 128 points. Finally, obtain the bispectrum of 128*128. As shown in figure 1. From Figure 1, there are obvious differences in the bispectrum of different modulation styles.

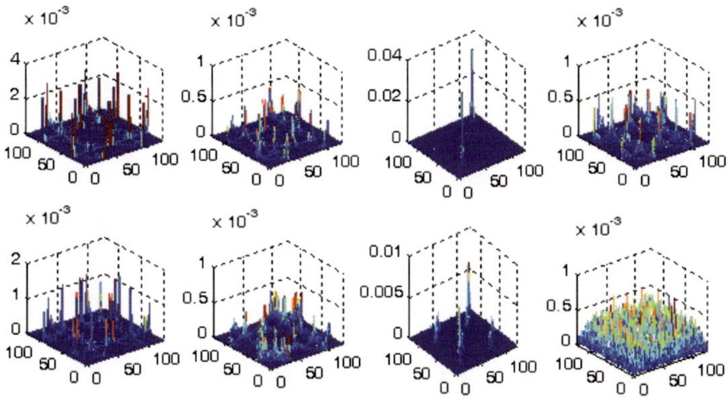

Figure. 1. 8 kinds of radar emitter signals is double spectra

3. Extraction of bispectrum two dimensions characteristic

Bispectrum is 3-dimension characterization. Directly identifying and dealing with have large amount of computation. It is not conducive to real-time processing. Therefore, it is necessary to simplify the two-dimensional feature. Two principles should be grasped in the process of simplification. First, as far as possible to reduce the amount of computation, in order to ensure the spead of the dgorithm. Second, the simplified character can reflect the characteristic of the fuzzy function to the maximum degree. According to these two principles, the equal interval section along the X axis parallel to the YOZ plane. It can get M cross section as the feature vector. Getting a new one dimension feature vector (Two-dimensional feature map).

$$\boldsymbol{R}_x = \max\left(\hat{\boldsymbol{B}}_{\mathrm{YOZ}}\right) \tag{7}$$

When $M = 128$, it can get 128 cross sections, Figure 2 shows the connection diagram of the 128 maximum bispectrum amplitude values of the 8 kinds of radar emitter signals.

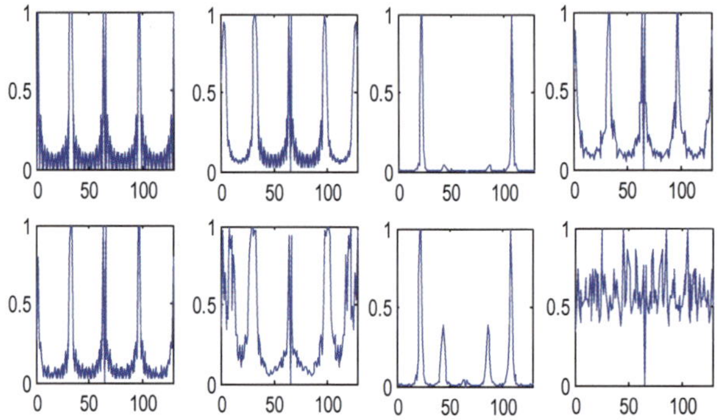

Figure 2. 8 kinds two-dimension characteristics of bispectrum of radar emitter signals

4. Extraction of Complexity

The simplified two-dimension feature can preserve the information of the spectrum, but the dimension is M. When M is larger, both increase the amount of computation, is not conclucive to the subsequent signal recognition. Therefore, it is necessary to further reduce the dimensionally of two-dimension feature. Double spectral feature 8 kinds of radar emitter signals with different geometric, scales and density characteristics that has different complexity characteristics. The box dimension can accurately describe the geometric size of the signal sequence, and has the advantage of simple calculation. Its definition is as follows:

Let (F , D) be a metric spare, ε is a non-negative real number. Make B(f , ε) say a center in f, closed ball with a positive ε, let A be a nonempty set in F. For each positive ε, let M(A , ε) denote the minimum number of closed ball cover A, closed ball of radius ε

$$M(A,\varepsilon) = \{N : A \subset \bigcup_{i=1}^{N} M(f_i,\varepsilon)\} \tag{8}$$

The different parts of $f_1, f_2, \cdots, f_N$ in the formula for F.

Let A be a compact set, and it is a non-negative real number. If there

$$D_f = \lim_{\varepsilon \to 0} \frac{\ln M(A,\varepsilon)}{\ln(1/\varepsilon)} \tag{9}$$

Then caused D_f is the fractal dimension of the set A. That is $D_f = D_f(A)$, A has fractal dimension D_f. This dimension is called box dimension.

The box dimension D_f for the two-dimension feature of the bispectrum is calculated by the following formula.

$$d(\Delta) = \sum_{i=1}^{N} |f(i) - f(i+1)| \tag{10}$$

$$d(2\Delta) = \sum_{i=1}^{N/2} (\max\{f(2i-1), f(2i), f(2i+1)\} - \min\{f(2i-1), f(2i), f(2i+1)\}) \tag{11}$$

$$D_f = 1 + \log_2^{\frac{d(\Delta)}{d(2\Delta)}} \tag{12}$$

Box counting dimension only reflects the bispectrum of 2D feature geometry scale information. If we want to fully reflect the complexity of also need to use information dimension to describe the bispectrum of 2D feature distribute density characteristics.

Let X be a set of R^n, $\{A(i)\}(i = 1, 2, \cdots N)$ is a limited δ - coverage of X. Order Pi to show that the elements of the set X fall in the probability of the set Ai, its value is

$$P_i = \frac{N(X)_i}{N(X \cap A_i)}, i = 1, 2, \cdots, N \tag{13}$$

Where $N(X)_i$ and $N(X \cap A_i)$ respectively the number of elements. Order information entropy

$$S_i = -\sum_{i=1}^{N} P_i \lg P_i \tag{14}$$

Bit entropy as X, if the information entropy satisfies the following relamation

$$S_i(\delta) \sim \lg \delta^{D_i} \tag{15}$$

The information dimension of X is defined as

$$D_i = -\lim_{\delta \to 0} \frac{S_i(\delta)}{\lg \delta} \tag{16}$$

5. Steps of the new method

Comprehensive analysis of the above, for the received radar emitter signal processing, according to the following steps in order to realize the recognition.

Step 1. Extracting bispectrum from received signal;

Step 2. Simplify bispectrum for two dimensional characteristics;

Step 3. Respectively for box dimension and information dimension of bispectrum two-dimension feature;

Step 4. Use SVM to achieve recognition.

6. Simulation testing

6.1. *Simulation conditions*

Simulate of 8 types of radar emitter signal , respectively CW, LFM, FSK, BPSK, QPSK, LFM-BPSK, FSK-BPSK and NLFM signal. The two frequency points of FSK signal are 20MHz and 40MHz respectively. The two frequency points of FSK-BPSK are 25MHz and 35MHz. The rest of the signal carrier 30 MHz, pulse width is 10us, sampling frequency 120 MHz. The bandwidth of the LFM signal is 2MHz, FSK signal encoding rule for [100110]. The phase of the BPSK signal is [11100010010], the phase of the QPSK signal is [01230312211300112012]. The bandwidth of the LFM-BPSK signal is 5MHz, the phase is [11100010010]. The frequency and phase of the FSK—BPSK signal is [11100010010]; NLFM signal is FM. When the SNR is 5dB, 10 dB, 15 dB, 20 dB, each class of signals are generated in 400, 200 of them were used for training, and 200 were for testing.

6.2. *Experiment analysis*

In 5 dB, 10 dB, 15dB, 20dB, first, calculate respectively 8 class shop radar revenue bispectrum of the signal characteristics of each kinds in the corresponding SNR (100 signal), as shown in Figure 3(a), the bispectrum characteristics of two-dimensional complexity of box dimension, in Figure 3(a), Figure 3(b) for the spectral characteristics of two-dimensional complexity information dimension. Figure 3 shows that 8 classes of two-dimensional complexity exist certain differences, with a good degree of separation between classes, for the subsequent identification laid a good foundation. Bispectrum characteristics of two-dimensional complexity are affected little by noise, this is the biggest characteristic parameters, the advantages, and to ensure the identification accuracy affected by the signal to noise ratio is small. In Figure 3, 1-8, respectively, the signals of CW, LFM, FSK, BPSK, QPSK, LFM-BPSK, FSK-BPSK.

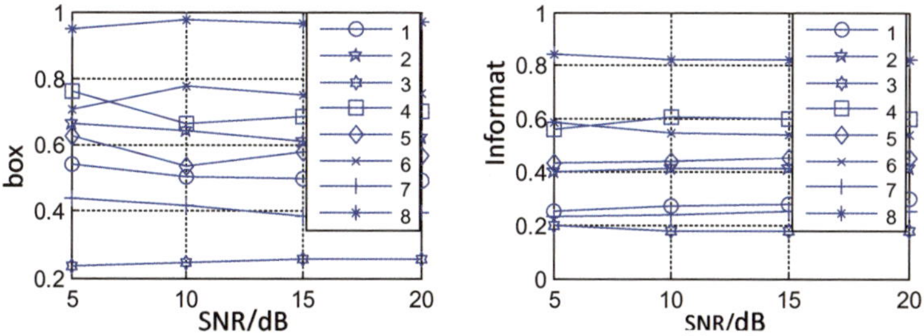

Figure 3(a) Box dimension of   Figure 3(b) Information dimension

Figure 3. Box dimension and information dimension of 8 kinds of radar emitter signals

Choose SVM classifier of 8 kinds of radar emitters signal recognition, using RBF kernel function to identify the characteristic parameters for dual spectral characteristics of two-dimensional box dimension and information dimension. Under the condition of different SNR, 8 classes of radar emitter signal source identification accuracy as shown in table 1. Table 1 shows that when the signal – to-noise radio is 20dB, 8 classes of radar emitter signal source recognition accuracy is 100%. With the reduction of letter shout than accuracy slightly down, when SNR is 15dB, from Figure 3, we find that double's two-dimensional characteristics of LFM and QPSK complexity, there is some overlap, so the second signal identification accuracy is 100%. When the signal-to-noise ratio of 10dB, 8 kinds of signal spectrum and two-dimensional overlapping probability increased complexity, accuracy is generally lower, the lowest was 90%, but the QPSK bispectrum of FSK and 2D still has the characteristics of complexity excellent separating degree, so the accuracy is 100%. When the signal-to-noise ratio down to 5dB, 8 kinds of signal double complexity feature of the overlapping probability to further increase, also further reduce the recognition accuracy, the accuracy is a minimum of 86% but still satisfactory.

Table 1 Recognition accuracy of 8 kinds of radar emitter signal(%)

	CW	LFM	FSK	BPSK	QPSK	LFM—BPSK	FSK—BPSK	NLFM
5dB	91	87	97	86	87	88	89	98
10dB	93	90	100	93	91	94	95	100
15dB	100	94	100	100	96	100	100	100
20dB	100	100	100	100	100	100	100	100

7. Conclusion

Choose intra-pulse characteristics as recognition parameters of radar emitter signals is a very effective way of existing of such algorithm is effective to solve this problem, this paper proposes a new recognition method. According to the characteristics of the bispectrum is not sensitive to noise, receive radar emitter signal to extract the source spectrum and two-dimensional characteristics of complexity, effectively enhance the stability of the algorithm under the condition of low signal-to-noise ratio, through the dimensionality of feature extraction to reduce the computational complexity of the algorithm, the simulation experiment shows that the effectiveness of the algorithm.

References

1. Nelson, D, J. Special purpose correlation functions for Improved signal detection and parameter estimation[J].ICASSP'1993,4:73-76.
2. Kenichi Nishiguchi, Masaaki Kobayashi. Improved Algorithm for Estimating Pulse Repetition Intervals [J]. IEEE Trans on AES, 2000, 36(2):407-421.
3. Guo Jie, Chen Jun-wen. Clustering Approach for Deinterleavin Unknown Radar Signals[J]. Systems Engineering and Electronics, 28(6): 853-856.
4. Chen Chang-xiao, He Ming-hao, Xu Jing, Han Jun. Radar Emitter Signal Recognition Based on the Fuzzy Function Resemblance Coefficient[J]. Chinese Journal of Radio Science, 2014, 29(2): 260-260.
5. Bai Hang, Zhao Xiong-jun. Radar Emitter Recognition Based on Rényi Entropy of Time-frequency Distribution[J]. Journal of Circuits and Systems, 2013, 18(1): 437-442.
6. Shi Yu, Pu Yun-Wei, Zhang Tian Fei. Intelligently searching the slice of ambiguity function main ridge based on superiority inheritance[J]. Infrared Millim. Waves, 2013, 32(1): 80-85.
7. Gexiang Zhang, Haina Rong, Weidong Jin, Laizhao Hu. Radar emitter signal recognition based on resemblance coefficient features, Lecture Notes in Computer Science (LNCS). 2004, 3066: 665-670.
8. Huang Ai-ming. The Basis of Digital Image Processing and Analysis[M]. Beijing: China WaterPower Press, 2005.
9. Zhang Xian-da. Modern Signal Processing[M]. Beijing: Tsinghua University Press, 2002.
10. Lv Tie-jun. Communication Signal Recognition Based on the Complexity Characteristics[J]. Journal on Communications, 2002.

Research on Software Testing of FPGA Embedded System

Hong-Feng Wang, Hui Zhang and Qiang Li

Beijing Space Information Relay Transmission Technology Research Center

Beijing, 100094,China

†*E-mail:econtribution@163.com*

Software testing plays an important role in software quality and reliability assurance, especially the ergodic and effectiveness of FPGA and other embedded software testing need further research. Aiming at tackling the coverage and effectiveness of FPGA embedded software testing, major problems in the process of FPGA software testing are analyzed, testable design and verification program based on digital simulation are brought forward. Top-Down design for testability, combined with system-level simulation and digital quantization is introduced to produce complete input and output test vectors based on system-level model or algorithm. The test vector and hardware description language input and output modules are compared to verify the design correction of FPGA software code module function wholly. A FPGA software testing process is established at all stages of the software development, combined with the actual situation of testing and verification.

Keywords: Embedded system; Software; Software testing; FPGA; Top-down.

1. Introduction

A variety of large, complex systems are utilized to assume real-time state monitoring and significant services safely and reliably. With the rapid development of large scale integrated circuits, embedded programmable devices (FPGA, DSP, etc.) is increasingly widely used, the code size of some of the complex software system equipment is often more than one million rows[1-3].

Sate testing and quality control of hardware equipment can be relatively easy to achieve through rigorous procurement and integration testing. However software testing plays an important role in software quality and reliability, especially the ergodic and effectiveness of FPGA and other embedded software testing need further research[4-7].

This paper is structured as follows; In Section II we will expose the issues existed in FPGA software testing; In Section III we will describe the testability design and verification process based on digital simulations; In section IV we will show how to arrange the workflow of equipment embedded software testing.

2. Main Problems existed in FPGA Software Testing

Currently, there are two major problems in the equipment FPGA software testing as follows:

On one hand, it is difficult to generate an integrate set of test vectors. Instead of using a top-down design[7], FPGA design of some manufacturers in the design stage use the existing IP cores and work like building blocks, thus they lack an external performance model at system level, an lack algorithm models at algorithm level described an advanced language, resulting the difficulty to produce a integrate set of test vectors at system level.

On the other hand, it is difficult to conduct ergodic test with a large number of IP cores. IP cores (such as DSP, RAM, FIFO, encoding, decoding, modulation, demodulation, etc.) are largely utilized in equipment FPGA design[8], and this leads to it difficult conduct ergodic test. Especially for IP cores developed by different manufacturers, there may be exist t irregular code and incomplete test issues. There are potential risks when the developers using these IP cores.

3. Testability Design and Verification Process Based on Digital Simulation

Here, we design testability design and verification process based on digital simulation to solve these problems, combined with common FPGA test methods. The details are as follows.

Functional verification of IP core testing is carried out with conventional method[9]. IP core function and performance verification are performed by compiling test stimulus file or inputting stimulus signals. IP cores are described as a "black box" in advanced language at algorithm level and simulated at system algorithm level and complete functional description[10]. The rest modules are designed with testability approach.

3.1. *FPGA design in testability methods based on digital simulation*

Top-Down design method is adopted. MATLAB, System C, C / C ++ language can be used to build system-level model or algorithm for large scale system,

system-level model simulation and digital quantitation are carried out, verification, optimization and synthesis are also performed, then it will be converted to basic Verilog description with conversion tools, by this way, web chart file can be obtained and be implemented in hardware. Development process is shown in Fig. 1. The transition from the System C and Simulink to HDL hardware language can be done directly with conversion tools, and integrated test environment is provided.

3.2. *FPGA test and validation process based on digital simulation*

Using Top-Down design for testability, combining system-level simulation and digital quantization to produce complete input and output test vectors based on system-level model or algorithm, the test vector and hardware description language input and output modules are compared to verify the design correction of FPGA software code module function wholly. FPGA test and validation process is shown in Fig. 2, and differs from the previous process, test vectors are from the test vectors used in the simulation process with advanced language. Fig. 3 is a proposed optimized FPGA software development and test process.

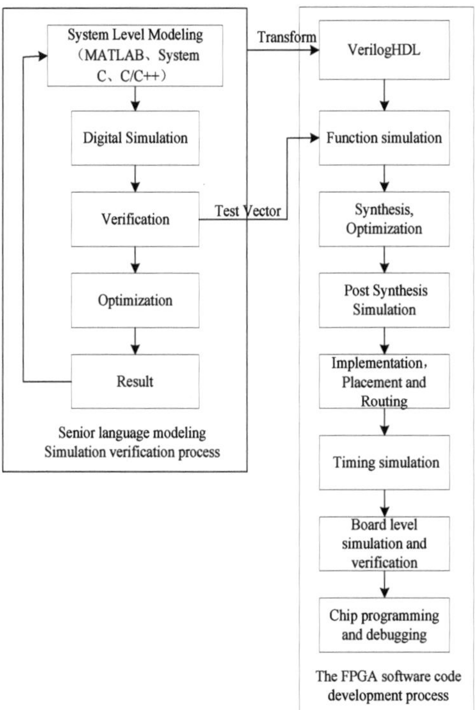

Fig. 1. Testability design process

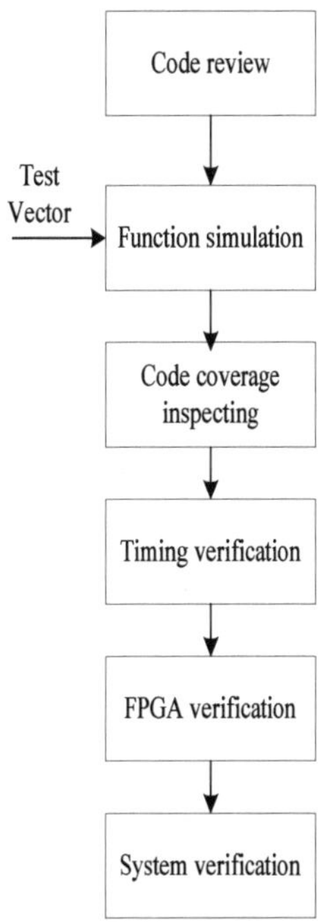

Fig. 2. Test and validation process

3.3. *Validity check for equipment FPGA software test*

Validity of FPGA software test can be checked mainly from the following aspects for developed FPGA software of equipment manufacturers.

(i)Normative test organization;

(ii)Professional test personnel.

(iii)Sets of test documentation.

(iv)Normalization of FPGA software development process

(v) Integrity of code review.

(vi)Integrity of functional verification.

(vii)Integrity of code coverage verification.

(viii)Integrity of timing verification.

(ix) Integrity of FPGA chip verification.

(x)Integrity of system verification.

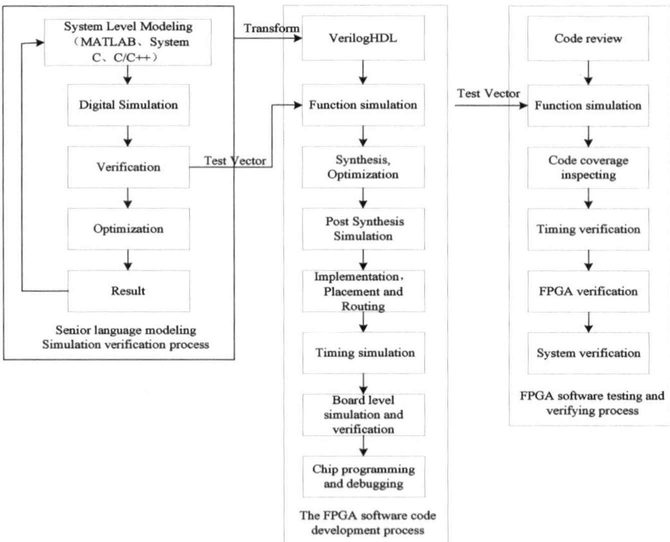

Fig. 3. Optimization process of FPGA software development and test

4. Workflow of Equipment Embedded Software Testing

Combining the actual situation of testing and verification, we divide embedded software testing into the following three stages:

The first stage is the preparation phase before the software development. Testers attend the embedded software design review organized by ground station equipment manufacturers, and review the equipment software demand analysis, the outline design, detail design, etc. In the initial stages of software design, testers actively participate and make full use of their advantages that they are familiar with transport protocols and have testing experience. They make software testing and development work closely together and fully communicate with the manufacturers, so as to reduce the risk of the development and save time and manpower of software development. So it is important to take into account software testing and carry out software development and testing concurrently in the early stage of software design.

Once the embedded software code is finished by manufacturers, the second stage starts. In this stage, software testing is carried out by manufacturers or a third party who has the quality of software testing. After accomplishing device software testing, they submit the device software test reports as the technical basis into the next stage of testing.

The third stage is system-level testing stage of device software organized by the users. After finishing the test, the software test report is submitted to the device manufacturers as the technical basis of software quality.

The following sections describe software testing content and flow of the second and third stage in detail.

4.1. *The second-stage testing*

The testing carried out by manufacturers or a third party is as follows:
(1) The equipment application software testing
(2) FPGA software testing
(3) DSP software testing

The test reports covering above items can be provided by manufacturers or a third party who has the quality of software testing. Users are only responsible for the review.

4.2. *The third-stage testing (system level testing)*

Based on software testing environment equipment, implement functional testing by modifying and adjusting the software parameters of verification system. In addition, design test cases and execute system level testing according to the software interface design or preliminary design provided by device manufacturers.

5. Conclusions

Embedded software quality of some critical subsystems will influence or determine the functional performance and the success or not of the entire system. In this paper, considering the state that the IP cores are used to perform software development like building blocks by a large number of manufacturers, we proposed an improved Top-Down design testability method and thus provide a solution to address the effectiveness and coverage for embedded software testing.

Due to the specialty of FPGA, general testability design methods for ASIC, such as scan insertion, built-in self test, ATPG (automatic test pattern generation), etc are not taken into practical application in FPGA. With the increasing scale of FPGA devices, the above design methods have the potential feasibility for FPGA development.

References

1. Muhammad Rashid, Muhammad Waseem Anwar, Aamir M. Khan . Toward the tools selection in model based system engineering for embedded systems, 2015, 106.
2. Roy Butler, Michael Pennotti. The Evolution of Software and Its Impact on Complex System Design in Robotic Spacecraft Embedded Systems. Procedia Computer Science, 2013, 16.
3. Wang Bo, Bai Xiaoying. Survey on Modeling and Verification Techniques of Composable Embedded Software, Journal of Software, 2014, 25(2).
4. Chen Mingzhu, Zhou Wei. Embedded Software Testing Engineering Practice. Computer Security, 2014, 6.
5. Xu Haiyang, Zhuang Yi. A Formal Modeling Method for Embedded Software Architecture. Acta Electronica Sinica,2014,42(8).
6. Mirko Conrad, Ines Fey, Sadegh Sadeghipour. Systematic Model-Based Testing of Embedded Automotive Software, Electronic Notes in Theoretical Computer Science, 2015, 111(1).
7. Dimitris Gizopoulos. Low-cost, on-line self-testing of processor cores based on embedded software routines, Microelectronics Journal, 2004, 35(5).
8. Sergio Cuenca-Asensi. etc. Soft core based embedded systems in critical aerospace applications. Journal of Systems Architecture, 2011, 57(10).
9. Mahroo Zandrahimi, Hamid R. Zarandi, etc. Two effective methods to detect anomalies in embedded systems. Microelectronics Journal, 2012, 43(1).
10. Tahereh Boroomandnezhad, Mohammad Abdollahi Azgomi. An efficient control-flow checking technique for the detection of soft-errors in embedded software. Computers & Electrical Engineering, 2013, 39(4).

A Judgmental Method of Illegal Electricity Based on BP Neural Network and Errors Theory

Jian-hao Lu[†,1], Sheng-ke Wen[2], Yi-xiong Nie[1] and Yue-hui Yang[2]

[1]*Faculty of Automation,*

Guangdong University of Technology, Guangzhou 510006, China

[2]*Measuring Center, Guangzhou Power Supply Co. Ltd., Guangzhou 510620, China*

[†]*E-mail: 286161231@qq.com*

To identify power theft of sensitive users, this paper proposes a method, which is on the basic principle of "Measurement Error conforms Gaussian Distribution", to search for illegal electricity users. The judgmental method of illegal electricity is to adopt the early stage of the historical load data to create user load forecasting database. In this case, a Load Confidence Interval using BP Neural Network Prediction will be set up to analyze the measured values that come from remote electricity metering. Then, the values, which are identified to be natural, would be concluded as the samples of the database parameters for reappraisal in order to improve the accuracy of the method. The cases listed have proved that the assessment method has a decent practicability in finding illegal electricity usage.

Keywords: Abnormal Metering Device; BP Neural Network; Errors Theory; Abnormal Electricity Users; Illegal Electricity Template; Proceedings; World Scientific Publishing.

1. Introduction

Traditional anti-theft method of energy metering device is the scene of the user periodically checks [1]. However, there is a larger blindness that the possibility of damage caused by maintenance to the metering device exists. And those parameters of reconditioned stealing devices have strong hysteresis as well as retroactive electricity would be implementation difficulties.

At present, many new methods for preventing electricity-stolen appear domestically and internationally. These methods can achieve the basic purpose of the search, but shortcomings exists like its user's data requires a higher degree of match, abnormal judgment threshold means unreasonable, line equipment added would increase the overhead and other feedback problems would come to appearance [2-6].

Under the protection of the economy and reliability of the grid, this paper presents a method for stealing recognition based on BP neural network and Errors Theory. Combined with the method, it takes predictive values produced from BP load forecasting algorithm and measured values for comparison to identify abnormal electrical energy metering device.

2. Implement on power stealing identification method

Electric energy metering device's load line information data has huge quantity which can be viewed as a large sample statistics. Based on the data analysis of large engineering point of view, it may be assumed that the measurement error statistics is in line with the Gaussian distribution [7].

Users vertical and horizontal load values have potentially relevance, which has a subtle effect on load forecasting. In this paper, it adopts BP Neural Network algorithm to match the inner fuzzy correlation.

Historical load values have magnitude differences which could present larger prediction error. In this case, use Maximum & Minimum Method to normalize the load data at horizontal [8].

In general, the holidays load variation is different from the conventional working load variation. Load forecasting model will be in differences. In this paper, only the regular weekday load trends for stealing recognition judgment would be taken into verification.

2.1. *The method of implementation steps*

Step 1: Based on the principle of Gaussian distribution of "Big Data Theory", mean values and standard deviation from load data itself could be used to classify the weekday type: Regular weekday or Sensitive days. For sensitive days, exclude holidays based on historical load, then form sequence E to record the suspected date.

Step 2: Take regular workdays recently into normalization.

Step 3: Use BP Neural Network algorithm to obtain the day forecast value. The value would be the center of the interval and takes the standard deviation from step1 to form Confidence Range. Then record the days listed as suspected anomalies days if out-of-limit and update sequence $\boldsymbol{E}$.

Step 4: Repeat step3. For suspected abnormal day, use BP Neural Network algorithms to obtain the 24-hour forecast values and then record out-of-limit timing to form sequence $\boldsymbol{e}_{1\mathrm{X}k}$.

Step 5: Make analysis and conclusions from sequence ***E*** and e_{1Xk}. If the number of sensitive timing (***k***) is bigger than K_{hour} or K_{day} consecutive days exists in sequence ***E***, mark the sensitive day as high probability abnormal day.

Step 6: Secondary Assessment: For suspected anomalies occur days, analyze weekly load level before and after the suspected abnormal day. If the load fluctuates significantly, energy metering stealing behavior occurs for the high probability event which needs to be verified and site maintenance.

Method flow chart of illegal electricity recognition is shown in Fig. 1.

Consider metering automation systems' approach to processing guidelines of site work [12], set K_{day} to be 3. In consideration of clients' anisotropy, K_{hour}'s setting method should be relevant to the type of customers. Default value is 3.

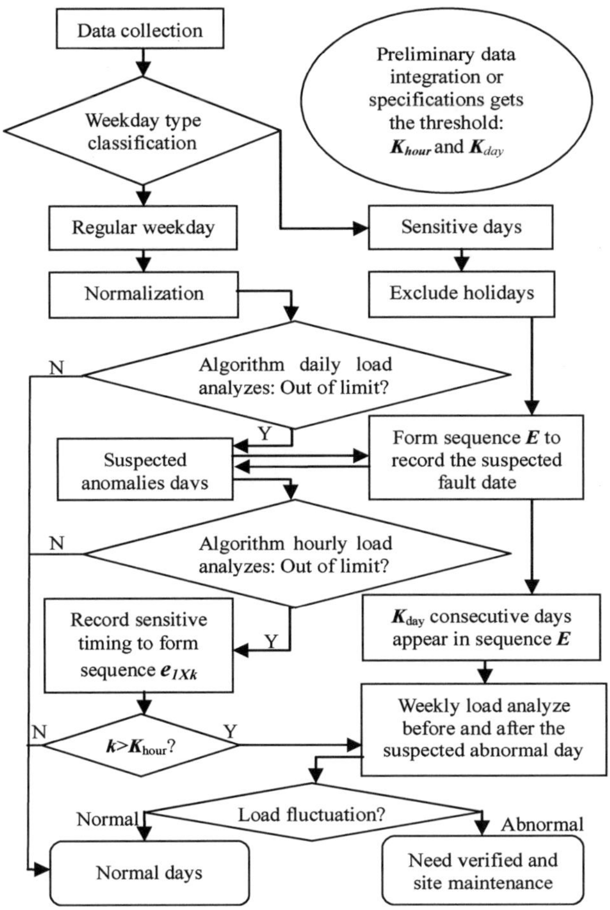

Fig. 1. Method flow chart of illegal electricity recognition

2.2. *Parameter fitting*

At this method, based on the three-dimensional historical data from remote detection, the parameters of BP Neural Network algorithm are obtained from simulation and be determined by the average level prediction accuracy.

2.2.1. *Input Nodes*

Using Maximum & Minimum Normalization for data preprocessing, simulation sets hidden layer nodes as 7 and the number of sample groups as 4. Change training mode to predict the load value on the N_0+1 day with the manner of every-N_0-day. Then simulated average forecast accuracy forms, as shown in Table 1.

Table 1. The comparison of input nodes differences

Input nodes	1	2	3	4	5	6	7	8	9	10
Average accuracy	1.71%	1.71%	-0.02%	1.49%	2.50%	2.20%	2.68%	0.66%	0.38%	-0.67%
Mean error	0.129	0.115	0.085	0.103	0.176	0.176	0.200	0.141	0.149	0.132

Table 1 shows that when the input vector dimension is 3, the prediction error closes to zero and the mean error is less than 0.10. The prediction effect is fit and strong. At this time, the algorithm is easy to converge and the simulation velocity corresponds to the scene need.

2.2.2. *Hidden layer nodes*

Hidden layer nodes are usually obtained by using Trial & Error. According to the relevant empirical formula, set the test number of nodes to be odd.

Simulation adopts every 3-day to forecast the fourth day and set training model of every five-group to get a set of predicted values. An average prediction accuracy of the simulation results are shown in Table 2.

Table 2. The influence of hidden layer nodes for prediction accuracy

Hidden layer nodes	3	5	7	9	11	13
Average accuracy	2.0%	-0.5%	0.4%	1.0%	1.4%	1.5%

By the training results, with the number of nodes increases, the prediction accuracy of trend presents into "Recessed" shape. When the nodes fall in between 5 and 7, they can meet the forecast margin in best precision.

2.2.3. *Number of training sample groups*

Using Maximum & Minimum Normalization for data preprocessing, simulation sets 7 hidden layer nodes to predict the load value on the fourth day with the manner of every 3-day. Changes the number of sample groups to gain an average prediction accuracy of the simulation. Results are seen in Table 3.

Table 3. The influence of the training sample groups for prediction accuracy

Group numbers	2	3	4	5	6	7	8	9
Average accuracy	2.5%	0.0%	0.2%	-0.4%	-0.7%	-1.4%	-3.2%	-3.8%
Mean error	0.119	0.025	-0.004	-0.029	-0.034	-0.113	-0.123	-0.194

Simulation results show that when the number of training sample set between 3 and 5, the mean accuracy error is within 0.5% as well as the average error is under 0.03. When the number of training group is 4, the average error is close to zero, which is in line with the fundamental principle of "Large Data Theory" that data integration errors counteract each in Engineering Field.

Based on the above simulation results, parameters of BP Neural Network algorithm model in the practical application used are: 3 input nodes, 7 implicit nodes and 4 groups of training sample.

3. Instance validation and analysis

Using the given assessment section, take a 42-day load curve of the electronic users with a load capacity of 10 MW as an example (Fig. 2). The analysis of the electrical behavior of enterprises is as follows.

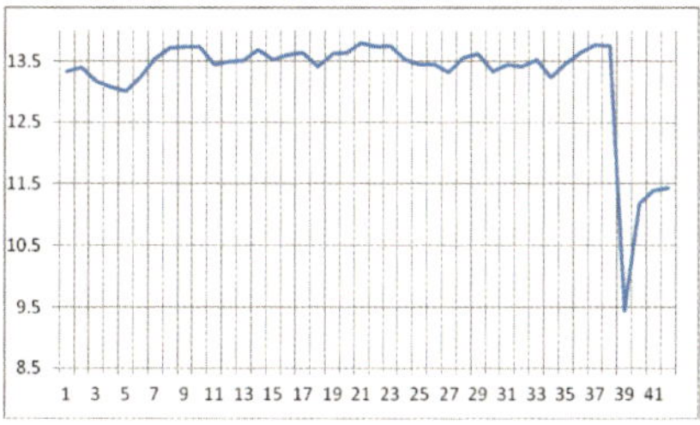

Fig. 2. A electronic enterprise user's load curve

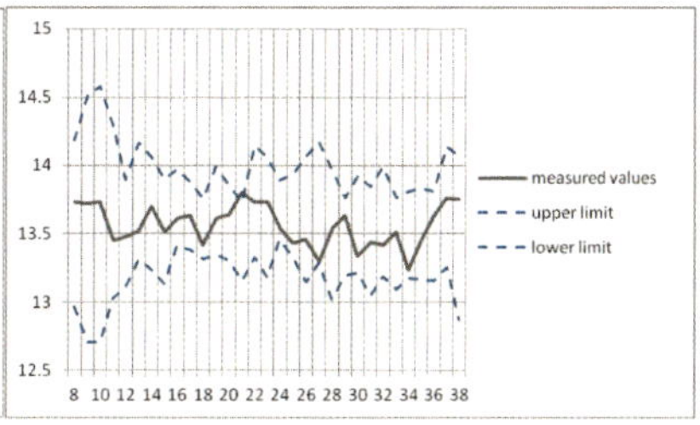

Fig. 3. Load confidence regions

Since the company's production is continuous, there is no holiday factor. After step 1 and 2, weekday property classification found its 39th, 40th, 41st and 42nd were sensitive days. So the date abnormal sequence ***E*** would be {39, 40, 41, 42}.

Excluding the above-mentioned four days, employ the new clusters of data with BP algorithm to predict the next day load values. Then put the value as the center section and take treble standard deviation from adjacent seven days as the center distance. Through step 3, confidence feasible region is formed in Fig. 3. (Prior 7-day is to fit algorithm for internal parameters.)

When the abscissa of Fig. 3 is constant, the midpoint between the upper and lower limit of vertical axis is the predicted value. Longitudinal differential reflects the historical volatility of the daily load. The greater the value, more obvious the fluctuations is.

As the forecast values of Fig. 3 shows, there is a slight out-of-limit on 21st and 27th day. The remaining 30-day measured loads are within the confidence gamut. After step 3, the exception date is updated to the new sequence $\boldsymbol{E} = \{21, 27, 39, 40, 41, 42\}$.

Step 4, use BP Neural Network algorithms to obtain the 24-hour forecast values of 21st, 27th and 39th day. Verify the fitting level of the load forecast and the sensitivity of searching an abnormal time to evaluate the target weekday abnormal or not.

Daily load curve forecasting analysis is shown in Figs. 4, 5 and 6.

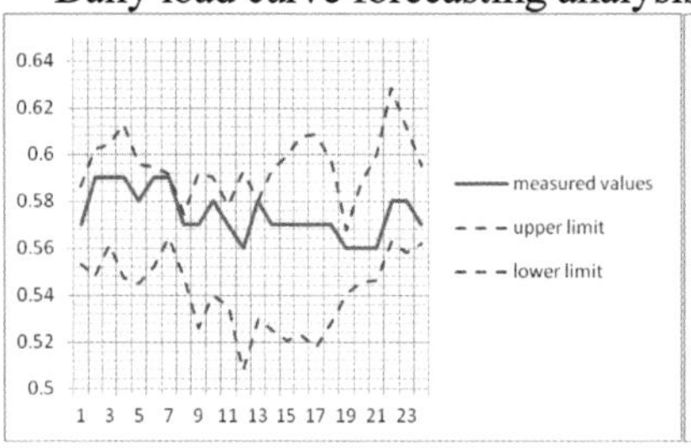

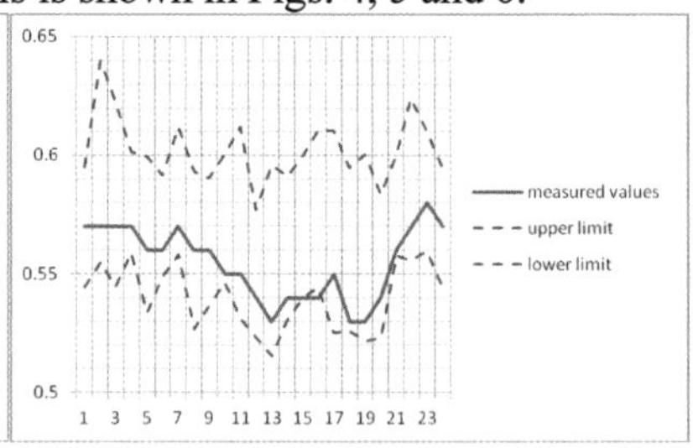

Fig. 4. Hourly load forecasting of the 21st day Fig. 5. Hourly load forecasting of the 27th day

As shown in figs. 4 and 5, when the abscissa of the 24-hour load curve is constant, the midpoint between the upper and lower limit of vertical axis is the predicted hourly value. Longitudinal differential reflects the hourly historical volatility of daily load. The greater the span, more obvious the load fluctuations is at this period of time.

In Fig. 4, the measured curve is within the confidence interval. The load trend is relatively close to the upper limit of the confidence interval, thus the total daily power is slightly out-of-limit. $\boldsymbol{e}$ is an empty set, so it is a normal day.

In Fig. 5, although there is a slightly out of lower limit at 16:00, the remaining sample points are within the confidence interval. The load trend is relatively close to the lower limit of the confidence interval, thus the total daily electricity consumption is similar to lower limit. As $\boldsymbol{e}$ is $\{16\}$, $\boldsymbol{k}$ is $1 < 3$. So it is a normal day either.

In conclusion, the load forecasting algorithm could perform well to fit in hourly load forecast within the allowable range of margin, which could meet the needs of prediction accuracy.

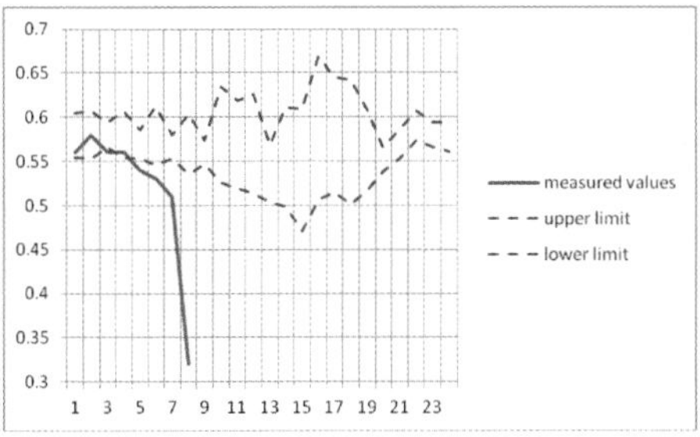

Fig 6. Hourly load forecasting of the 39th day
(Hourly load values were missing on 39th day, only recorded the first eight sample points)

In Fig. 6, load curve from 0:00 to 4:00 is within the confidence interval and then it started to be out of the lower limit. The 8th sampling point (8:00) demonstrates a serious deviation from the lower limit on the degree of 60%. In this case, $\boldsymbol{e} = \{5, 6, 7, 8\}$, $\boldsymbol{k}$ is 4> 3. It could be defined as an abnormal day.

Take step 5, four consecutive days appear in sequence $\boldsymbol{E}$, it can define the 39th day as a high probability of suspected abnormalities day.

On the basis of these facts, it can determine basically that the energy meter appears abnormal.

Take step 6 in measures: The second assessment analysis of weekly load level before and after the suspected abnormal day, which is shown in Table 4.

Table 4. Load analysis before and after the Suspected Day

Time quantum	The previous week	The following week
Mean value	13.54	0.19
Standard deviation	11.33	0.12

Standard deviations are similar that indicates weekly load consumption level before and after the suspected fault day is gentle. The previous week average electricity amplitude is lower than the flowing one at the percentage of 16.3, which shows load fluctuates significantly. So abnormal metering equipment caused by human tampering would be a high probability event. It is necessary for site processing and to be verified.

4. Conclusions

The method is capable of preliminary screening for weekday type. The formation of a confidence domain based on peak, flat and cereal data as well as Errors Theory could realize the function of searching for the abnormal failure time. And the BP algorithm hourly load forecasting could meet the required accuracy within a wide range, which has a certain reference value in the monitoring of human stealing electricity. However, the implementing of the method has higher demand in the integrity of users' data and the raw data should be ensured continuous and accurate. By the case study above, this method for stealing recognition has some site usability.

The case customer mentioned belongs to Class I users. Its base load is so large that daily load is relatively stable. Thus whether this method for Class II and III users has the same simulation effect, it remains related data to verify the common feasibility of the method.

Acknowledgment

Work partially supported by China Southern Power Grid CO., LTD. Science and Technology Projects (K-GZM2014-134).

References

1. JJG596-2012, Verification Regulation of Electrical Meters for Measuring Alternating-current Electrical Energy[S]. State Administration for Quality Supervision and Inspection and Quarantine. 2012.10.08.
2. Mimmi L. M., Ecer S. An econometric study of illegal electricity connections in the urban favelas of Belo Horizonte, Brazil [J]. Energy Policy, 2010, 38(9):5081-5097.
3. B. Bat-Erdene, B. Lee, M.-Y. Kim, T.H. Ahn, D.H. Kim Extended smart meters-based remote detection method for illegal electricity usage[J]. IET Generation, Transmission & Amp; Distribution, 2013, 7(11): 1332-1343
4. Lu C.N., Huang S.C., Lo Y.L. Non-Technical Loss Detection Using State Estimation and Analysis of Variance [J]. IEEE Transactions on Power Systems, 2013, 28(3):2959-2966.
5. Guo Li-cai, Peng Zhi-wei, Fan Qiang. A Survey of Electric Energy Metering and Countermeasures to Electric Power Stealing[J]. High Voltage Apparatus. 2010(46):86-91.
6. Fan Jie, Chen Xiao, Zhou Yu. An Intelligent Analytical Method of Abnormal Metering Device Based on Power Consumption Information Collection System[J].Electrical Measurement & Instrumentation, 2013, (11):4-9.
7. Wang Rong-xin. Mathematical Statistics [M] .Shaanxi: Xi'an Jiaotong University Press, 1986.
8. Zhou Ping. Neural Network Design and Application [M]. Beijing: Tsinghua University Press, 2013.

9. Guo Jun-peng, Li Wen-hua. Principal Component Analysis Based on Error Theory and Its Application[J]. Application of Statistics and Management, 2007, 26(4)
10. Load Management Terminal Specification of China Southern Power Grid Limited Company[S]. China Southern Power Grid Limited Company. 2013. 2.17.

A Simple Frequency Domain Equalization Method for Atmospheric Laser Communication

Hui-Qin Wang[1,†], He Wang[1], Jing-Wei Zhu[1,2], Ke-Jun Jia[1] and Ming-Hua Cao[1]

[1]*College of Computer and Communication, Lanzhou University of Technology, Lanzhou, 730050, China*

[2]*Lanzhou Polytechnic School, Lanzhou, 730050, China*

†*E-mail: whq1222@lut.cn*

To eliminate multipath interference and noise in atmospheric laser communication, a frequency domain equalization method based on Minimum Mean Square Error (MMSE) criterion is proposed. In this method, the channel frequency domain response can be estimated by sending several training sequences. The noise variance can be estimated by setting decision threshold and subsequent received signal separation. Fast Fourier Transform (FFT) technique is utilized to convert the signals to frequency domain, thereafter, the channel compensation can be achieved by multiplying the converted signals with their corresponding equalization coefficients. Simulation results show that the proposed method can effectively improve the system performance under the strong atmospheric scattering channel. Compare to the Zero-Forcing (ZF) equalization method, the proposed method can achieve nearly 4 *dB* Signal-to-Noise Ratio (*SNR*) performance improvement at the Bit-Error-Rate (*BER*) of 10^{-3}.

Keywords: Atmospheric Laser Communication; Atmospheric Scattering; Inter-symbol Interference; Frequency Domain MMSE Equalization; Noise Estimation.

1. Introduction

Atmospheric laser communication has the advantages of large capacity, wide frequency band, low power consumption, good secrecy and flexible installation. However, scatterers which widely exist in the atmosphere will cause multipath effects on the laser beam, resulting in multiple redirects of the scattering lights[1]. This phenomenon will cause time delay and broadening of received optical pulses[1-2], which lead to severe inter-symbol interference (ISI) especially in the sand-dust, fog, rain and snow weather[2-5].

Equalization technique is considered as an effective way to suppress multipath effects caused inter-symbol interference[6-7]. The most existing researches on the channel equalization of the atmospheric laser communication are concentrated in time domain channel equalization. The time domain blind equalization methods for atmospheric laser communication have been researched in reference[8, 9]. Considering the high computational complexity and slow convergence speed of these equalization methods, they may have trouble in tracking and compensating for the fast changing atmospheric channel. Instead, frequency domain equalization (FDE) can greatly reduce the computational complexity and implement difficulty of channel compensation algorithm[11-12]. For this reason, a simple channel compensation scheme is proposed in this paper by using FDE in the atmospheric laser communication channel. *FFT* is firstly applied to convert the received signal into frequency domain. Thereafter, the channel compensation can be achieved by multiplying the converted frequency domain signals with their corresponding equalization coefficients. Simulation results show the proposed method can effectively improve system performance.

2. Frequency domain atmospheric channel equalization

2.1. *Channel equalization scheme*

The proposed atmospheric channel equalization scheme is shown in Fig. 1. Firstly, several training sequences is inserted into the head of data frame for channel estimation. The data is then delivered to the LD for transmission. At the receiver side, the channel state information and noise variance can be estimated by digital signal processing algorithm. Finally, the channel compensation can be achieved by multiplying the corresponding frequency domain signals with their equalization coefficients.

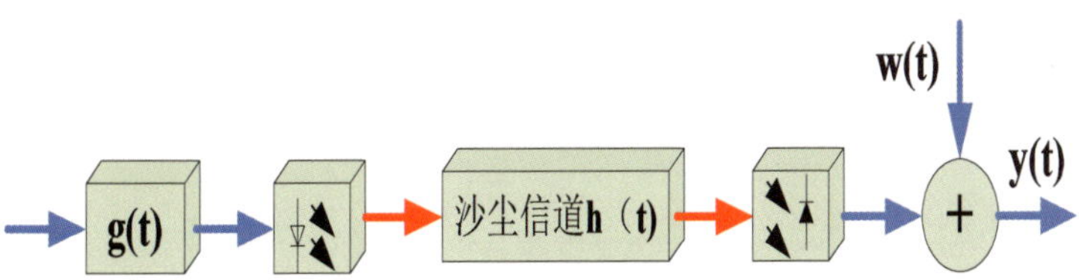

Fig. 1 Model of atmospheric laser communication based frequency domain MMSE equalization.

It should be noted that the aerosol particles such as dust, droplet, ice crystal suspended in the atmosphere, which lead to multipath propagation phenomenon of optical signals[1]. As a result, the multipath atmospheric laser communication channel can be expressed as[10]

$$y(n)=\sum_{l=0}^{L-1} h(l)x(n-l)+w(n) \tag{1}$$

where L is the channel length, $h(l)$ is channel coefficient, $x(n)$ is the transmission sequence, $w(n)$ is the Gauss white noise sequence, which follows the Gauss distribution $w(n) \sim N(\mu,\sigma_w^2)$. Therefore, the corresponding vector of Eq. (1) can be described as

$$\boldsymbol{y}=\boldsymbol{X}\boldsymbol{h}+\boldsymbol{w} \tag{2}$$

where $y=[y(0),y(1),\cdots,y(N-1)]^{\mathrm{T}}$ denotes the received signal vector, $\boldsymbol{h}=[h(0),h(1),\cdots,h(L-1)]^T$ denotes channel impulse response, $\boldsymbol{X}$ denotes the transmitted signals and $\boldsymbol{w}=[w(0),w(1),\cdots,w(N-1)]^T$ denotes noise vector.

2.2. *Channel frequency domain response estimation*

Considering that the atmospheric channel is a fast time-varying channel and the channel covariance matrix is unknown. Therefore, the Least Square (LS) channel estimation method is utilized. As shown in Fig. 1, the data for transmission is composed of the training sequence and the information data. Assume the length of training sequence is N and the length of information data is D. The training sequence is adopted as the optimal training sequence proposed in [13].

The receiver can estimate the current channel impulse response by the first $N+L+1$ bit data. The estimated channel impulse response can be expressed as

$$\hat{\boldsymbol{h}}=(\boldsymbol{X}^{\mathrm{H}}\boldsymbol{X})^{-1}\boldsymbol{X}^{\mathrm{H}}\boldsymbol{y} \tag{3}$$

where $\hat{\boldsymbol{h}}=[\hat{h}(0),\hat{h}(1),\cdots,\hat{h}(L-1)]^T$ is the estimated time domain channel impulse response. The estimated frequency domain channel response can then be obtained by performing M point discrete Fourier transform (DFT) on $\hat{h}(n)$

$$H(k)=\sum_{n=0}^{M-1}\hat{h}(n)e^{-j\frac{2\pi}{M}kn} \tag{4}$$

where M is the length of received signal, and $M=N+D+L\text{-}1$.

2.3. *Noise variance estimation*

Compared to ZF equalization, MMSE equalization has better performance for noise interference considered. However, when determining the equalization coefficients, noise variance must be known. That means, in order to implement the frequency domain MMSE equalization, noise variance must be estimated.

Because the atmospheric channel energy concentration area is difficult to be strictly divided, thus the noise variance estimation method proposed in [12] is difficult for actual implement in atmospheric laser communication system. For this reason, a noise variance estimation method based on threshold decision and sequence separation is proposed. Its implementation steps are as follows:

(1) Threshold decision is done on received signal under decision criteria below

$$\hat{x}(\mathrm{n})_{dec} = \begin{cases} 0 & if: y(\mathrm{n}) < \tau, \mathrm{n} = 0,1,2,\cdots,N+D-1 \\ 1 & if: y(\mathrm{n}) \ge \tau, \mathrm{n} = 0,1,2,\cdots,N+D-1 \end{cases} \quad (5)$$

where $\hat{x}(n)_{dec}$ is the decision output, the decision threshold value τ can be obtained by averaging the data of each received frame

$$\tau = E[y(n)] \quad (6)$$

(2) Using the decision result to separate the "0" and "1" sequence

$$y_0(n) = (1 - \hat{x}_{dec}(n))y(n), (n = 0,1,...,N+D-1) \quad (7)$$

$$y_1(n) = \hat{x}_{dec}(n)y(n), (n = 0,1,...,N+D-1) \quad (8)$$

where $y_0(n)$ is the "0" sequence and $y_1(n)$ is the "1" sequence.

(3) The noise variance can be estimated by the separated "0" sequence

$$\hat{\sigma}_w^2 = E[y_0(n) - E[y_0(n)]]^2 \quad (9)$$

2.4. *Frequency domain MMSE equalization*

Firstly, the received signal is converted to frequency domain by FFT

$$Y(k) = \sum_{n=0}^{M-1} y(n) W_M^{kn}, k = 0,1,\cdots,M-1 \quad (10)$$

where $Y(k)$ is the frequency domain signal corresponding to the received time domain signal of $y(n)$. W_M^{kn} is the rotation factor. For the frequency domain MMSE equalizer, assume the data with unit power, the frequency domain equalization coefficient can be calculated by the following formula[12]

$$C(k) = \frac{H^*(k)}{|H(k)|^2 + \sigma_w^2} \quad (11)$$

where $C(k)$ denotes the equalization coefficient, σ_w^2 denotes noise variance, $H^*(k)$ is the conjugate of $H(k)$. With results of Eq.(4) and Eq.(9), equalization coefficients can be determined. Then channel compensation can be achieved by multiplying $Y(k)$ with its corresponding equalization coefficient

$$X(k) = Y(k) \cdot C(k), k = 0,1,\cdots,M-1 \quad (12)$$

where $X(k)$ is the frequency domain signal after equalization. Finally, the recovered signal $\hat{x}(n)$ can be obtained by convert $X(k)$ back to the time domain by inverse fast Fourier transform

$$\hat{x}(n) = IFFT[X(k)] = \frac{1}{M}\sum_{n=0}^{M-1} X(k) W_M^{-kn} \quad (13)$$

3. Performance analysis

The performance of the proposed atmospheric channel equalization method is simulated. The channel coefficients can be obtained by sampling the channel response with symbol period *T*, the corresponding coefficients matrix is [0.2827,0.7221,0.3560,0.1749,0.0859,0.0487][5]. The laser wavelength set as 3.8*um*, transmission distance is 1.5 *km*, and atmospheric visibility is 1 *km*. The symbol period *T set a*s 5*ns* and modulation type is OOK.

Fig. 2 shows the result of mean square error(*MSE*) of the channel estimation. It indicates that *MSE* of the channel estimation decreases linearly with increase of *SNR*. Given the same *SNR*, the longer training sequence results in smaller *MSE*.

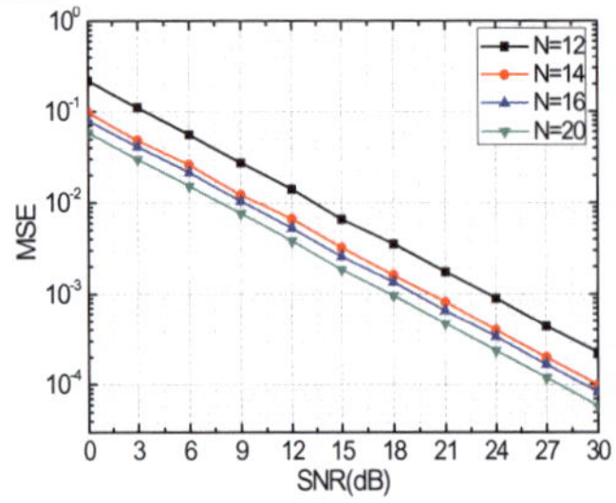

Fig. 2 Mean square error of channel estimation with different length training sequence

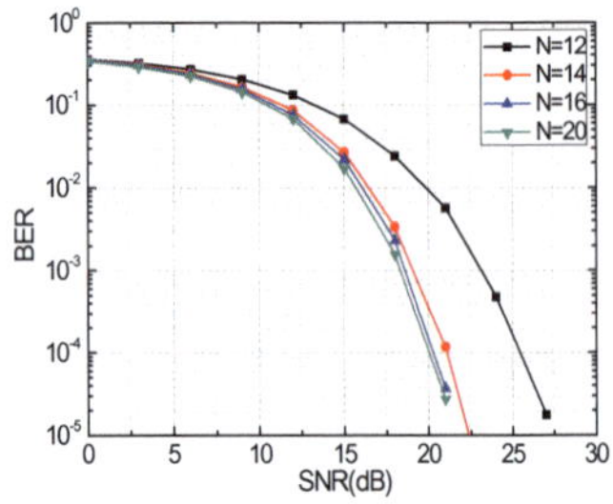

Fig. 3 Bit-error-rate before and after equalization with different length training sequence

Fig. 3 shows the system *BER* of the proposed channel compensation method. Simulation results indicate that the longer training sequence result in lower *BER*. The *BER* curve of *N*=16 is almost the same with that of *N*=20, it means when the training sequence length reach to a certain threshold, addition training sequence leads to limited *BER* performance improvement. Considering the training sequences consumes bandwidth, therefor the recommend training sequence length in the proposed method is 16.

Fig. 4 shows the performance of frequency domain ZF equalization and the proposed MMSE method. It can be seen that the channel compensation can significant improve the system *BER* performance. Moreover, it can be concluded that the proposed channel compensation method can further improve the system BER performance. For example, when *BER* is set as 10^{-3}, the proposed method can improve nearly 4 *dB* system performance compared to the ZF equalization.

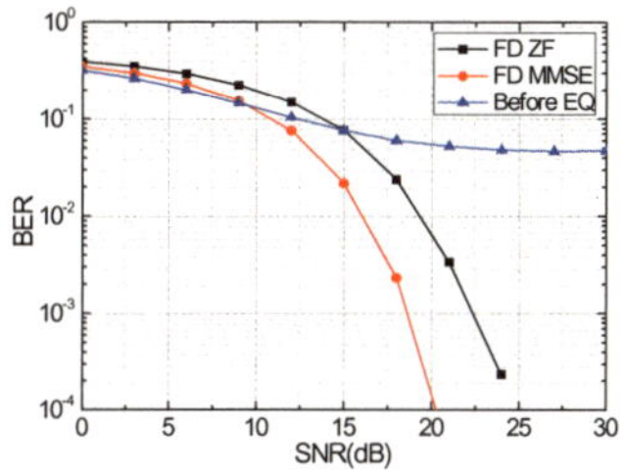

Fig. 4 Performance comparison between frequency domain ZF and MMSE equalization

4. Conclusion

A simple frequency domain atmospheric channel equalization method is proposed. Simulation results show that this method can effectively reduce the system *BER* under strong atmospheric scattering channel. Furthermore, it can achieve better performance compared with ZF equalization.

Acknowledgment

This work is supported by National Natural Science Foundation of China (61265003, 61465007 and 61461026).

References

1. Jiang Huilin, Tong Shoufeng. The technologies and systems of space laser communication[M]. National Defense Industry Press, 2010, pp.1-63.
2. Ke Xizheng, Yin Zhiyun. Coding theory in wireless laser communication system [M], Science Press, 2008, pp.257-367.
3. M. Grabner, V. Kvicera. Multiple scattering in rain and fog on free space optical links[J]. Journal of Lightwave Technology, 2014,32(3):513-520.
4. Xu Zhengwen, Wu Jian, et al. Pulse time broadening characteristics in turbulent medium[J]. Science in China, 2003, 33 (2): 139-147.
5. Wang Huiqin, Wang Yangang, et al. Influence of atmospheric visibility on laser pulse delay and broadening in sand and dust weather[J]. Acta Opt sin, 2015,35 (7): 0701002.
6. Zhang Xianda, Bao Zheng. Communication signal processing[M]. National Defense Industry Press, 2002, pp.226-309.
7. John G. Proakis, Masoud Salehi. Digital communication (fifth edition) [M]. McGraw-Hill, 2008, pp.597-674.
8. P Rvi, JP Leblanc. Blind equalization of PPM signals using third-order moments[J]. IEEE Communications Society,2010,4(9):1298-1302.
9. Chen Dan, Ke Xizheng, Qiao Wei. Blind equalization algorithm of wireless optical communication using sub-carrier modulation based on subspace[J]. Infrared and Laser Engineering, 2015,44 (8):2528-2534.
10. Wang Junbo, Jiao Yuan. Training Sequence Based Channel Estimation for Indoor Visible Light Communication System[J]. Optoelectronics Letters, 2011,7(3):213-216.
11. N Benvenuto,S Tomasin. Iterative Design and Detection of a DFE in the Frequency Domain[J]. IEEE Trans. Commun., 2005,53(11):1867-1875.

12. Zhang Xin, Zhang Xiaoji, et al. Frequency domain response estimation of underwater acoustic channel and noise estimation for single-carrier frequency domain equalization[J]. Acta Phys. Sin.,2014,63(19): 194304.
13. Wang Junbo, Jiao Yuan, Chen Ming. Optimal Training Sequences for indoor Wireless Optical Communications[J]. Journal of Optics,2011, 14(1) :15401-15405(5).

The Prediction of Time Domain Electromagnetic Pulse Waveform Based on the Minimum Phase Method

Qiang Li† and Ming Wei

Ordnance Engineering College

Shijiazhuang, Hebei, China

†*E-mail: 1648754903@qq.com*

With the rapid development of information technology and the wide application of electronic devices, the electromagnetic environment becomes more complex, external electromagnetic pulse may interfere with the electronic device at any time, affecting its normal work[1,2,3]. Therefore, the electromagnetic pulse effect evaluation of electronic equipment is particularly important. In this paper, the complex electronic equipment system is like a "black box". It introduces the minimum phase method, regarding the electronic equipment systems to satisfy the condition of minimum phase signal transmission system[4]. Its system impulse response can be obtained by measuring its input and output signals, and then to predict the electromagnetic pulse waveform of the equipment. In the experiment, the effect of the predicted waveform is verified [5,6].

Keywords: Hilbert Transformation; Waveform Prediction; Minimum Phase; Transfer Function; Time-Domain Response; Model.

1. The introduction

The complex electromagnetic environment is becoming more and more serious, which affects the normal operation of electronic equipment device, so it is very important to evaluate the electromagnetic pulse effect of electronic equipment. The research on Electro Magnetic Pulse (EMP) effect of electronic equipment is usually carried out in three levels: At component level, equipment level (or subsystem) and system level. At present, the test methods and standards for the EMP effect at component level and equipment level have been developed, and the system level EMP effect simulation and evaluation method is relatively lack. The EMP response modeling at system level can be realized through the method of the mechanism analysis and the experimental data statistics. Mechanism analysis is based on the fully understanding the internal structure of the test equipment, by solving the Maxwell's equations or transmission line equation to analyze system capacity coupling channel and predict electromagnetic impulse response. In engineering application, however, we usually don't understand the internal structure of the system. And the method of experimental data statistical do not need to understand the internal structure of the system, we think the complex system as a "black box", by measuring its input and output signal, and use the information provided by the test data to establish the EMP response model. The realization method includes the neural network method, the minimum phase method, the frequency spectrum estimation method, etc.

2. The basis theory

There are many methods to obtain the phase information of the system from the amplitude-frequency characteristic curve of the system, and the minimum phase method is the best one. The minimum phase system is a system that all zeros and poles of the discrete transfer function are in the unit circle[7]. If a linear and time invariant system and its inverse are both causal and stable, so the system is a minimum phase system. However, the actual signal transmission system may not meet the conditions of minimum phase, but any causal and rational system can cascade of a all-pass system and a minimum phase system[8,9]. Because the all-pass system does not affect the amplitude characteristics of the signal, therefore, in most cases, the minimum phase system with corresponding to the original system can be used to approximate the original system[10]. It is assumed that the system meets the minimum phase condition.

Set up the system transfer function $\mathrm{H}(w)$.

The real part $R(w)$ of the minimum phase system transfer function and the imaginary part $I(w)$ can be associated with the Hilbert transform. That is

$$H(w) = R(w) + jI(w) \tag{1}$$
$$R(w) = \hat{I(w)} \tag{2}$$
$$I(w) = -\hat{R(w)} \tag{3}$$

$\hat{I(w)}$ is the Hilbert transform of $I(w)$. $\hat{R(w)}$ is the Hilbert transform of $R(w)$. $H(w)$ written in the form of amplitude and phase is

$$H(w) = |H(w)|e^{j\theta(w)} \tag{4}$$

The natural logarithm of **Error! Reference source not found.** $\mathrm{H}(w)$ is

$$\ln(H(w)) = \ln|H(w)| + j\theta(w) = G(w) + j\theta(w) \tag{5}$$

So the phase Angle $\theta(w)$ of the $H(w)$ and the logarithmic form $\ln|H(w)|$ of the amplitude of the $H(w)$ will be associated by the Hilbert transform

$$\theta(w) = -\hat{G(w)} \tag{6}$$

$\hat{G(w)}$ is the Hilbert transform of $G(w)$.

3. Experimental verification and analysis

In order to validate the method, the following simulation experiment was carried out:

1) First, a set of time domain waveform is set as the input source, we can get a set of real output waveform by the filter, and the frequency response of the filter is known;

2) Based on the filter's amplitude-frequency curve, the method of minimum phase is used to get estimation model of the filter;

3) Set the input waveform through the estimated model, we get the estimated output waveform, and which is compared with the actual output waveform.

The simulation experiment used the double index wave as the input waveform, the expression is

$$y(t)=E_0k(e^{-\alpha t}-e^{-\beta t}) \tag{7}$$

its parameters are: $E_0=5\times10^4V/m$, $k=1.3$, $\alpha=4\times10^7\mathrm{s}^{-1}$, $\beta=6\times10^8\mathrm{s}^{-1}$.

Fig. 1 is the input waveform. Fig. 2 is the actual output waveform. Fig. 3 is the actual output waveform and predicted output waveform.

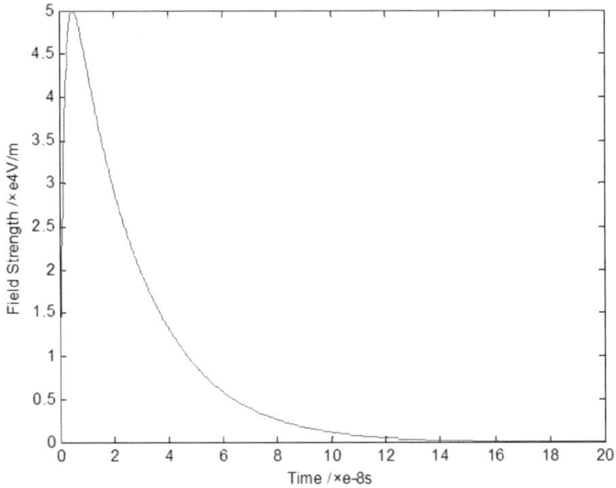

Fig. 1. The input waveform

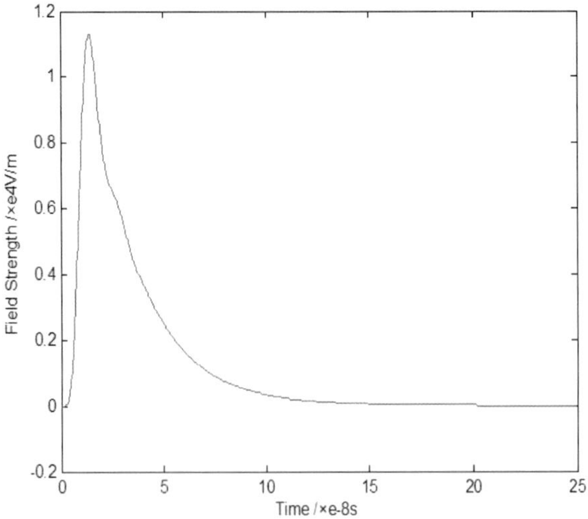

Fig. 2. The actual output waveform

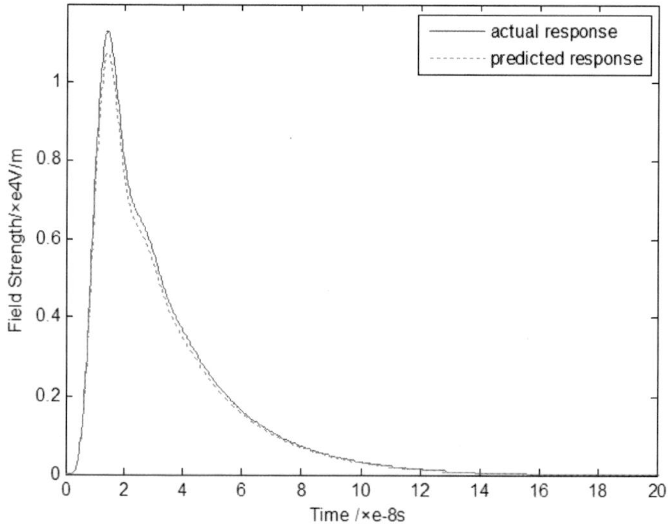

Fig. 3. The actual output waveform and predicted output waveform

Fig. 3 shows that the predicted waveform and the actual output waveform in the rising edge, falling edge and near the peak are basically the same, that is, the predicted waveform and the actual output waveform is basically the same.

4. Conclusion

In this article, through simulation experiment, and we set the filter as a minimum phase system, the simulation of actual output and the estimated results are basically consistent, it proved the modeling of the filter is correct. The experiment verified the feasibility of minimum phase method.

Acknowledgment

This work is supported by National Natural Science Foundation of China(No.51277181).

References

1. Sun Guozhi, Liu Shanghe, Chen Jingping, et. al. Effect of electromagnetic environment in battlefield on the influence of the information warfare. Military Operations Research and System Engineering. 2006, 20(3):43-47.
2. Wei Ming. A survey on simulating arithmetic for electromagnetic pulse effects at equipment level. Journal of Ordnance Engineering College, 2009, 21(6):50-53.
3. Yang Minghui. Research on system identification method based on neural network[D]. Master Dissertation of Beijing University of Chemical Technology, 2000.
4. Chen Xiang. Neural network modeling of the electric equipment EMP energy coupling[D]. Master Dissertation of Ordnance Engineering College, 2009.
5. Shi Lihua, Zhou Bihua, Chen B, et al. Time domain characterization of a system based on the magnitude of its frequency response. Chinese Journal of Radio Science, 2000, 15(4):467-471.
6. Guo Dongyi, Shi Lihua, Guo Yaohua, et. al. Measurement for shielding effectiveness of pulsed magnetic field based on frequency response function[J]. Information and Electronic Engineering, 2010, 2(1): 49-52. (in Chinese)
7. Hayes H M, Lim J S, Oppenheim A V. Signal reconstruction from phase or magnitude[J]. IEEE Trans on Acoustics, Speech, and Signal Processing, 1980, 28(6): 672-680.
8. Xie Yanzhao, Wang Zanji, Wang Shuqun, et. al. Reconstruction of electromagnetic pulse waveform based on the amplitude spectrum data[J]. High Power Laser and Particle Beams,2004,16(3):320-324.
9. Oppenheim A V, Schafer R W. Discrete-time signal processing[M]. Prentice-Hall Inc, 1989.
10. Tesche F M. On the use of Hilbert transform for processing measured CW data[J]. IEEE Trans on EMC,1992,34(3):259-266.

Design and Simulation of Neural Network Unit Based on FPGA

Zhu Fang and Ding Lei,

Faculty of Computer, Computer Technology Guangdong University of Technology

Guangzhou, Guangdong, China

Email: 974313632@qq.com

The artificial neural network has a very good application in pattern recognition and classification, and the characteristics of high-speed parallel computation of FPGA can be used in the hardware realization of neural network. Through the design of hardware implementation of neural network unit, the proposed method was introduced and realized the design of incentive function, neuron in MAC address module and data storage module, as well as hardware. There are also analysis and comparison of the common methods of the realization of the incentive function. This paper proposed to solve the data representation and multiplication accumulation calculation process on the hardware platform. It provides an important foundation and necessary precondition for the construction of hardware neural network.

Keywords: Neural network unit; incentive function; multiply accumulate; data representation; address module.

1. Introduction

Artificial Neural Network (ANN) has been successfully used to solve the problem of pattern classification and recognition, function approximation and prediction. The programmable device (such as FPGA) has the characteristics of low cost, powerful software development tools and parallel execution etc. And their processing power is based on their highly parallel and interconnected architecture. These characteristics and structure make the hardware realization of the neural network is of great value and significance. In the design of FPGA hardware implementation process of the neural network, neuron as the basic unit of neural network, its design is an important part of the hardware implementation of neural networks. The choice of incentive function plays a key role in the neuron. Before people have carried out a variety of ways to try, the paper made the relevant introduction. The representation of the data directly affects the accuracy of the calculation results. In this paper, we use the data representation method to combine the resource consumption and the effective value of the data. Better able to complete the construction of the numerical calculation of neurons.

Based on the analysis of neurons, the artificial neural network is divided into several modules to realize one by one in hardware implementation. The hardware is composed of input neuron control module, memory module, MAC module, excitation function module. Structure as shown in Figure 1.

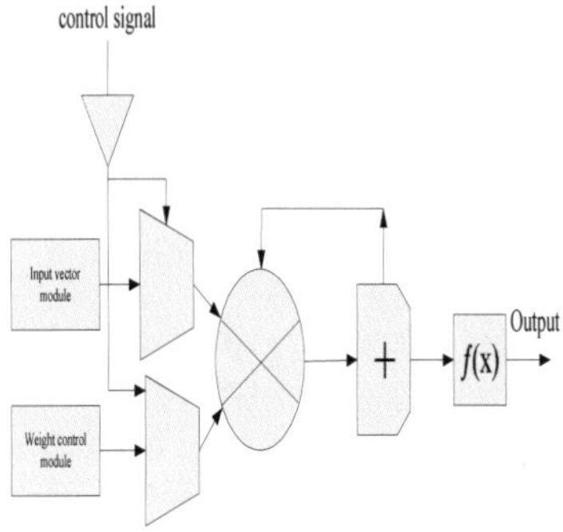

Fig. 1. Neurons each module design structure

During the operation, the corresponding input signal is multiplied by the corresponding weight of the input module and the input signal, and the input signal belongs to the serial calculation. Each time the result of the multiplication is stored in the storage module, when all the input signals and the corresponding weights are finished by multiplying the accumulation module. The input of the module is used as the input of the excitation function module, and then the output of the mathematical operation of the excitation function can be used as the output result of the input signal through a single hardware neuron. The whole calculation process can be described by Figure 2.

In the digital design of neurons, we are equivalent to realize incentive function Sigmoid function alternative. The representation method of input vector data is introduced. At the same time, the implementation of multiply accumulate module and the design of storage module are described.

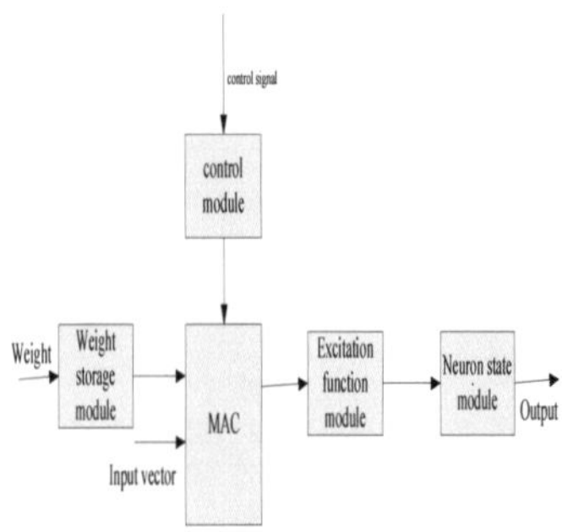

Fig. 2. The flow chart of each module of neuron

2. Selection and implementation of incentive function

The neuron activation function selection according to the processing requirements, the excitation function of each layer in the neural network may be different. In the hidden layer, the function is realized by using the nonlinear function, and the output layer can be satisfied with the simple linear function. For nonlinear functions such as exponential function, FPGA can not be directly in the hardware implementation of this kind of complex transcendental function. So we usually use the approximate equivalent replacement method, instead of realizing the function of exponential function. However, no matter how similar the original function and the original function, the error can not be eliminated completely. Therefore, the design and implementation of the incentive function on the FPGA is the key to the realization of the whole hardware neural network design. Continuous guidance is the necessary condition of the excitation function. In order to satisfy this characteristic, the Sigmoid function is usually chosen as the neural network of the hidden layer, and the derivative of this function can be expressed by itself. The function has many expressions, in this paper, the function expression is:

$$f(x) = \frac{1}{1+e^{-x}}. \qquad (1)$$

Its corresponding derivative function is:

$$f'(x) = \frac{e^{-x}}{(1+e^{-x})^2} = f(x)(1-f(x)). \qquad (2)$$

We use the BP neural network hidden layer activation function for tangent function tansing. By a variety of methods to approximate the function. After the collection of relevant literature, the commonly used approximation methods are as follows:

2.1. *Lookup table method*

Lookup table method is a direct and simple method, which is one of the commonly used methods. It works the principle of the function of the input of the corresponding output calculated in advance, and then through the RAM or ROM will result in the corresponding storage, the input of the excitation function as the storage address. The output of the excitation function is used as the corresponding value of the address, and the approximation of the excitation function is realized by this method. [1]. We can see that the advantage of this method is that it can be used in the most direct expression of the model which can not be expressed by mathematical expression or by mathematical model. Only need to do the corresponding to its input and output, greatly simplify the relevant operation process. It has the advantages of fast running speed and easy implementation. Of course, this approach has its drawbacks, the accuracy of the calculation is based on the size of the storage capacity of the decision. Therefore, to achieve a certain accuracy of the case of the storage of resources consumption will increase. When there are more resources to be occupied, the cost of taking into account the high cost of spending has been limited in practical applications.

2.2. *Piecewise linear method*

The piecewise linear method is to approximate the original nonlinear function by decomposing it into several segments. The general expression of the broken line is a linear function of the first order: $y = kx + b$. The original nonlinear Sigmoid function is replaced by the adjustment of the coefficient. In the hardware implementation, through a multiplier and adder can achieve the function of the broken line. When the function slope coefficient is 2, and with the index of two fold change. In the hardware can by shifting and adder to approximate the sigmoid function, shift instead of the multiplier, this way further save the resource consumption, more simple to be realized. However, this method achieved accuracy is not high, in the practical application of the data only for the high precision of the scene.

2.3. *Other methods*

In sigmoid function approximation, in addition to the above methods, there are some other mathematical methods, such as the Taylor series expansion method, least square method, etc. Some methods are combined with some methods, such as the combination of the lookup table and the linear approximation, and a table driven interpolation method is proposed[2]. There are ways to further optimize the method based on the SBTM (Bipartite Table Method Symmetric)[3] and STAM (Table Lookup Symmetric Addition Method) algorithm [4-5]. They are all based on the extension of the lookup table method. Some people and the two methods of comparative study, the STAM algorithm is applied to the approximation of the Sigmoid function [6]. To achieve the use of less resources to obtain high precision approximation and fast running speed.

Integrated all the methods mentioned above, we can be summed up, no matter what method is used to approximate the sigmoid function, need considering the factors from the accuracy is achieved and the consumption of hardware resources and computing speed. In the field of research or application, we must choose the appropriate method according to the actual demand and the scene to achieve the Sigmoid function approximation.

In this paper, the incentive function is selected by using the method of looking up table to replace the Tansig function. In the literature[7], the sigmoid function of the lookup table and the FPGA implementation of the derivative function are discussed in detail. We can get from Figure 3. No matter how much the input values, the output of the function after Tansig range was between [-1,1]. And through the figure display and mathematical calculation of the value can be intuitive and accurate when the input value is greater than 4, the output value has been very close to the value of 1.By symmetry if the input for negative, absolute value is greater than 4 output results similarly were close to 1, so in the look-up table storage, when the absolute value of the argument in the [0,4] interval is Tansig function value. When the absolute value is greater than 4 stored value directly for 1. The value stored in the lookup table is mainly derived from the corresponding results between the interval [-4,4]. We divide the interval into positive and negative two parts. Due to symmetry we can consider [0,4] half a range of values, it is divided into 40 equal parts. These results are then translated into binary representation of binary 16 bits on the hardware.

The value stored in the corresponding value is stored in the memory, and the corresponding output value is input through the look-up table to get the result.

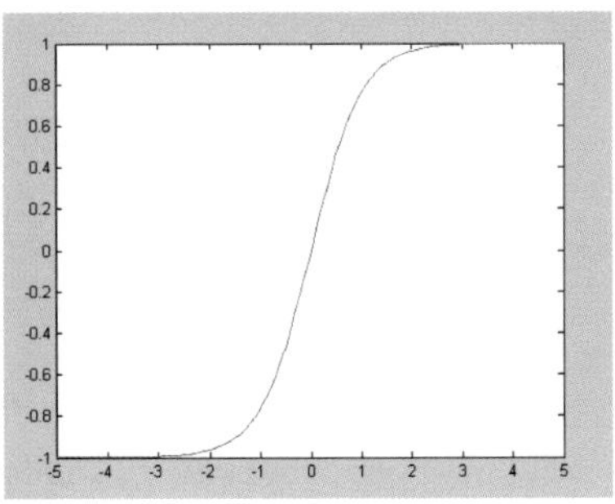

Fig. 3. Graph of function Tansig

3. Data representation

In the operation of the input data, the problem must be considered in advance is the representation of data. We in the hardware of data display and storage must be of positive and negative data symbols and the data bit width (including integer and decimal) of relevant provisions. Since the input weights of each layer are real numbers ranging from -1 to 1, we specify how to represent the neural network data format on the FPGA. Usually we are familiar with the two methods have fixed point algorithm format and floating point algorithm format.

Due to the FPGA is not easy to carry out the relevant operation of floating point numbers, although the choice of this method can get a higher precision, but also will bring the problem of increasing the consumption of hardware logic resources. Comprehensive consideration from the cost, unless there are strict requirements on the accuracy of the general will not choose the floating-point arithmetic format. Studies have showed that in the premise of not weakening the ability of neural network computing processing, the 16 bit fixed point can meet the minimum accuracy requirements of the connection weights [8]. Fixed Point (FXP) algorithm is used to calculate the parameters of the code [9]. There are advantages over floating point operations in computing speed and resource consumption. In this study the choice of the number of bits is set to a 16 bit fixed-point number representation: a symbol of the size of the number of symbols used to represent the 1 symbols, 4 integer digits and 11 bit, the specific format as shown in Figure 4.

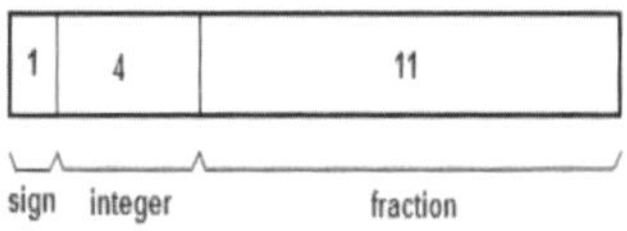

Fig. 4. Fixed-point data format

In the representation of the process, in order to enter the input layer of a weight of 0.1442, for example, into 16 Decimal to 0.0010010011101010. In the hardware, cannot recognize the expression of the decimal, and therefore need to decimal into an integer, prior to the small number of decimal places. So the small number of all the values to the left eleven bits can be regarded as the original number multiplied by 2048(11 of the 2 party) to get the results to take five to four into the results for the transformation of 295, 0.1442*2048=295.3216 in an integer expressed as 0000000100100111.This will be able to express a good hardware.

4. Multiply accumulate module design and implementation

The sum of the product of the input quantity and the weight value is mainly realized by the accumulation module. In the process of neural computing is a core part of the realization of this part, we use the hardware description language Verilog program to achieve multiply accumulate function. This part of the structure for the realization of functions including multiplier, adder, memory and the necessary data processing section. As shown in Figure 5 below:

In order to further verify the accuracy of the module, the design of the test platform, for the convenience of intuitive view results. In the Model simulation diagram, the data are multiplied, and then the results continue to accumulate. When all the input to complete the process of multiplying and adding, and finally get the output is completed by the cumulative results. In Figure 4-3 the validation of the input variables A and B represent the variable weights and the corresponding input signal. To implement product operation within one clock, when finished all the clock run by the cumulative results.

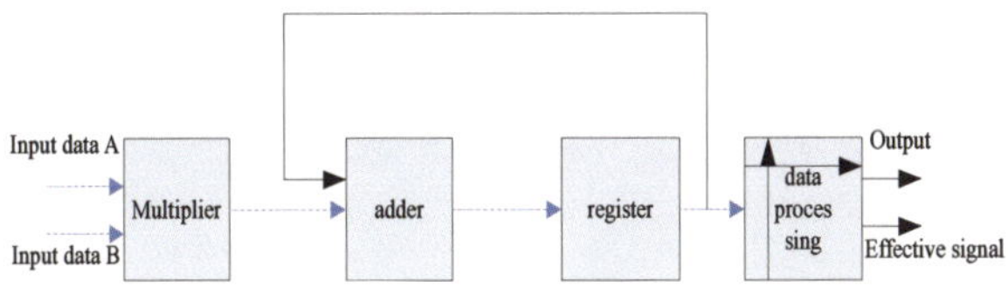

Fig. 5. Multiply accumulate MAC module structure

In order to facilitate the detection, take 12 input, respectively, and the corresponding weights. Figure in the input for the 1 to 12 of the integer, the weights are set to integer 1, the Yellow cursor at the end of the end of the clock. The output value of the position 78 is the result of multiplication and accumulation. Figure 6 has proved by the accumulation of MAC module is accurate.

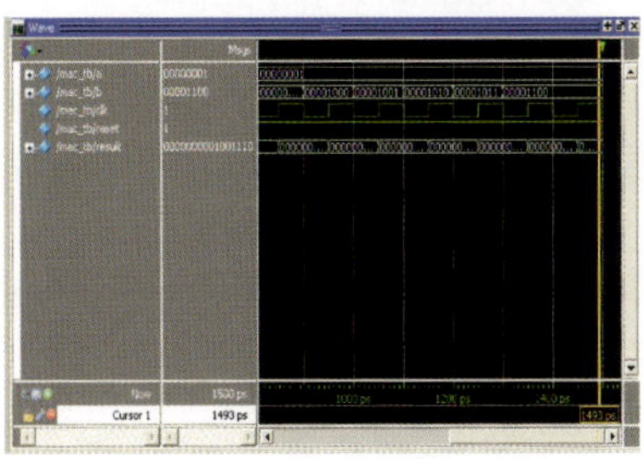

Fig. 6. Test verification results of multiply accumulate MAC module

5. RAM address module design

In the process of neurons, need special module is used to store the weights and threshold. In neural network, each neuron threshold is relatively fixed, it has nothing to do with the level of the hierarchy and the training process. Therefore, we only need a constant value stored in the RAM memory address special, when calculating the binary address acquisition through its 16 bit constant. For the weights, the weights of each neuron are different, and in the course of training, the weights of the hidden layer will be adjusted accordingly. So it is necessary to use the stored RAM value parameters together with multiple stages are stored in the relevant address. For a neuron of the input values and the corresponding weights to be stored at the same time at the same time, otherwise it can not be found in the reverse update phase of the input value. In the weight value, we must initialize the weights in advance, and the initialization values are as close to 0 as possible. Since there is no strict sense of random number generator in FPGA, we use artificial way to generate pseudo-random number in random number. Table 4-2 schematic diagram of the corresponding storage structure.

Table 1. RAM address storage schematic

Address:	Address0	Address1	Address2	...	AddressN
Weight:	Thresholdθ	W_1	w_2	...	w_N

6. Conclusion

In this paper, the design of hardware neurons in the neural network is described in this paper. The data representation of hardware, the selection and design of the excitation function, the implementation of the accumulation module and the design of the RAM storage module are analyzed, and the corresponding elaboration is made. The incentive function of the lookup table is completed, and the test verification of multiply accumulate module is realized. The representation of data and the design of multiply accumulate calculation can meet the requirements of the hardware platform. For the next step neural network hardware implementation provides a necessary precondition, and in the design and implementation of the neural network in the construction of BP neural network to get better application.

References

[1] Luo Zuohuang. Design of BP neural network based on FPGA and its application in water quality prediction [D]. Guangzhou: Guangdong University of Technology, 2009:5-6.

[2] Qian Yuduo. Research on hardware implementation of neural network based on FPGA[D]. Wuhan: Huazhong University of Science and Technology 2012: 30-31.

[3] Michael J. Schulte, James E. Stine. Approximating Elementary Functions with Symmetric Bipartite Tables[J]. IEEE Transactions 0n Computers, V0148, NO.8, Aug,1999.

[4] J.E. Stine and M.J. Schulm. The Symmetric Table Addition Method for Accurate Function Approximation[J]. Journal of VLSI Signal Processing.V01. 21, No.2,167-177, June,1999.

[5] M.J. Schu Ute and James E. Stine. Accurate Function Approximations by Symmetric Table Lookup Addition. Proceedings of the 11th International Conference on Application–Specific Systems, Architectures, and Processor, Zurich, Switzerland,144–153, July,1997.

[6] Yan Ming. Hardware implementation of neural network based on FPGA[D]. Qingdao: Ocean University of China, 2008:31-42.

[7] Zhang Xiao, Huang Xi, Zhong Weihan. The realization of FPGA Sigmoid function and their derivatives. [J]Journal of Fujian Normal University, 2011,27 (2): 62-65.

[8] Medhat Maussa, Shawki Areibi, Kristian Nichols. On the Arithmatic Precision for Implementation Back Propagation Network on FPGA: A Case Study[J]. University of Guelph of Canada,2005.

[9] S. Li, M. Moussa and S. Areibi, Arithmatic Formats for Implementing ANNs on FPGAs [J]. Canadian Journal on Electrical and Computer Engineering, Vol. 31, No.1 pp: 31-40, Winter 2006.ty, 2013 (4): 719-727.

Research on the Harmonic Governance Program for the Enterprise Distribution System with Photo-Voltaic

Yan-Qing Pan[a] and Hui-Bin Sui[b]

School of Electrical Engineering, Shandong University

Jinan, Shandong, China

E-mail: [a]1527042690@qq.com,

[b]suihuibin@sdu.edu.cn

According to the situation of serious harmonic pollution and lower power factor in enterprise's distribution system with photo-voltaic, an analysis and a summary on possible problems of power quality were made. A governance program that is to install active filter on enterprise's distribution system with photo-voltaic has been proposed, and the feasibility of this program has been validated with an instance. Firstly, the authentic date from an enterprise's flied test was analyzed and the present problems of enterprise's distribution system were simulated with matlab. Then, after the active filter is installed and the simulation of the improved systems turned out that the content of each harmonic decreased apparently. At the same time, the power factor improved to the standard on power quality.

Keywords: Photo-voltaic; harmonic; power factor; active filter; matlab.

1. Introduction

Recent years, the problem of over relying on traditional pattern of power generation which expense primary energy has been severity from day to day. Distributed generation [1], micro-grid [2] etc. came into being. The technology of grid connected photo-voltaic systems has developed fairly perfectly. It became the most valuable and available generation patterns by features of widely, independently and reliability [3-5], also, it will be the major source [6] of the future electric power. Many domestic enterprises, considering of their long-term interest, access the photo-voltaic to their distribution systems.

However, photo-voltaic system contains lots of inverters and rectifiers, which bring about numerous harmonics and abnormal reactive power on distribution system side. Inadequate compatibility [7] is appearing to the grid, the most common problems are the power factor in original system export dropped below 0.9. Therefore, the photo-voltaic system put forward a giant challenge to the power quality [8].

Due to its own defects, traditional filter and reactive compensation [9] have difficulty in meeting requirement on power quality. The theory of combining active filter and photo-voltaic have become the focal studied point. Literatures 10, 11 introduced an improved one-cycle control based on DC side active filter, the simulation results show that it can reduce the amount of power switch and optimize the structure of systems. Not only can it improve the compensation effect, but also expand the system stable region. Literature 12 applied the improved genetic algorithm to active filter, and optimized the parameter of active filter in points of the cost, reactive power compensation and filtering effects; active filter performance is increased dramatically. Literature 13 introduces a new-type active filter, based on traditional harmonic current detection algorithm. The current detection method adopts wavelet transform and the results show that it can improve the rapidity and accuracy of current detection.

2. Enterprise power quality

The high power quality requirements continuous and stable sine-waves power supply, the deviation of the voltage and frequency must be within the available range. Due to the equipment and production process, the enterprises have to access nonlinear loads to its distribution systems. Electric arc furnace [14], electronic switch [15] and ferromagnetic saturation [16] is the most three common types of the harmonic source. The principle and the operation performance of harmonic source have various effects on the power quality. Table 1 shows a summary on power quality problems.

Table 1. The summary of enterprise's power quality problems

Types	Property	Characteristics	Reasons	Results
harmonic	steady state	voltage current spectrum	nonlinear load	equipment's overheating, insulatio failure relay protection malfunctic
power factor	steady state	power characteristic curve	lack of reactive power compensation	network loss increase, capacity decreases voltage deviation in consumer side
asymmetry in three phase	steady state	unbalanced factor	asymmetrical load	equipment's overheating, insulatio failure relay protection malfunctio
voltage flash	steady state	fluctuation amplitude modulation frequency	the start of motor and electric arc furnace	servo motor working abnormally.
instantaneous voltage	transient state	instantaneous amplitude duration	the start of motor remote Fault	equipment shutdown part of loads working abnormally.
noise	transient/ steady state	amplitude frequency spectrum	grounding errors solid state switch load	the microprocessor control equipment working abnormally

3. The improved program with active filter

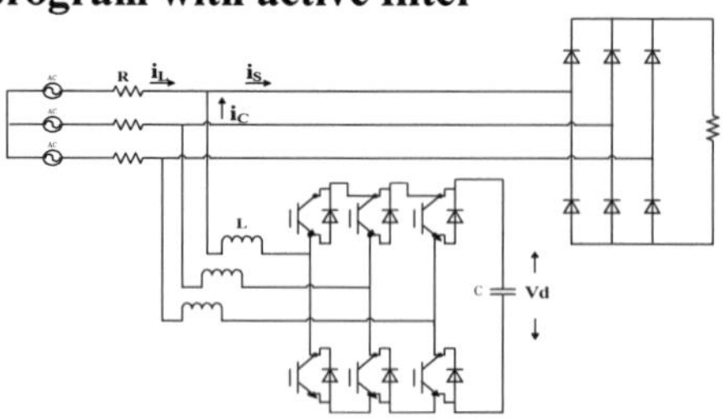

Fig. 1. The principle diagram of active filter

The basic structure of active filter is shown in figure 1. It can be regarded as a current source. The harmonic current i_{Lh} and reactive current i_{Lq} can be separated from load current i_L with detection circuit. Then, they will be reversed and the detection circuit generate modulate wave signal i_C^* of compensation currents i_C, $i_C^*=i_{Lh}+i_{Lq}$. The control circuit output trigger pulse according to the value of i_C^*. The trigger pulse will create compensation current i_C by the power switch of drive circuit. i_C can track i_C^*, $i_C= -i_C^*$, so,

$$i_S=i_L+i_C= i_L+i_C^{*} =i_L-(i_{Lh}+i_{Lq})=i_{Lp} \quad (1)$$

We can draw a conclusion that there is only fundamental load current in i_S. That means the active filter can eliminate harmonic and compensate reactive power well.

4. Case analysis

A new energy electric motor company has installed photo-voltaic on his distribution lines; the access points locates in the workshop's 400V power distribution cabinets. Workshop installed three-phase compensation and distributed LC, the current sampling points is below the 10kV to 400V transformer, as shown in figure 2.

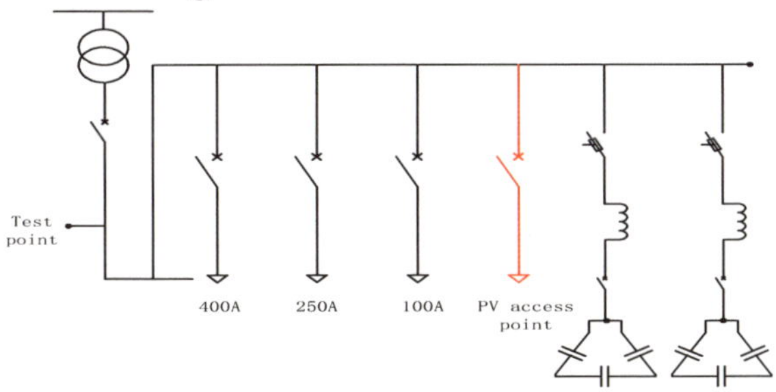

Fig. 2. Field wiring diagram

Before the photo-voltaic is put into operation, the power factor was stable between 0.9 ~ 0.95. When operated, the power factor changed to about 0.2. the current waveform under the condition of normal working in the shop is shown in figure 3, the harmonics contents are shown in figure 4.

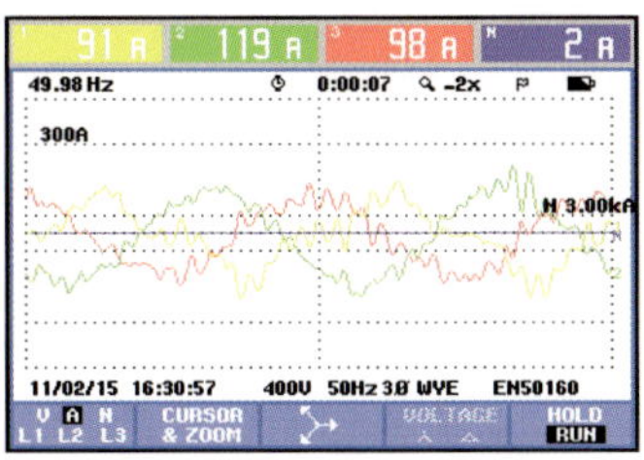

Fig. 3. The current waveform on site

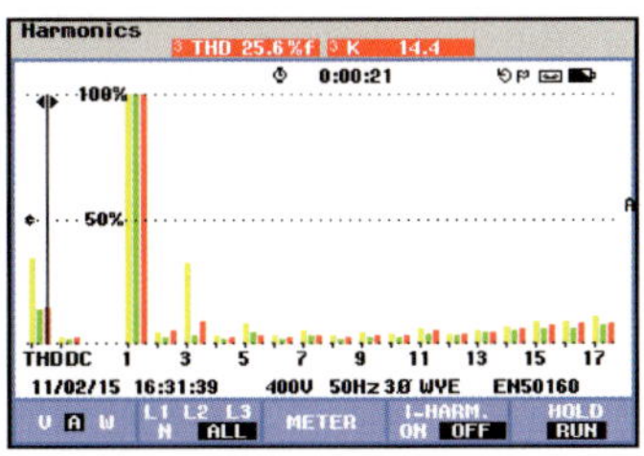

Fig. 4. The harmonic content diagram on site

The connected point locates on the workshop's branch bus-bar. It seems clear that when the load was working on lower power, if photo-voltaic power was equal to or lower than the workshop electricity power, the current is easy to flow back, which is adverse for compensation and acquisition equipment.

The date of electric meters in the upper level bus indicated that when there was no photo-voltaic, the active and reactive power in the line was positive. But once photo-voltaic is accessed to the bus branch, both of them change to negative. Furthermore, the capacitances cut themselves all, the power factor dropped to 0.2, sometimes even turned to negative. Obviously, the active and reactive power detected by LC circuit was incorrect; that resulted in compensation equipment work abnormally. Once photo-voltaic was disconnected and all SVC were removed from systems manually, the active power would be positive. If photo-voltaic was accessed one by one, the active power decreased gradually, the reactive powers kept unchanged basically. When the active power decreased to negative, the reactive power changed to negative as well. These experimental results explained the reasons why the power factor in connected point dropped to 0.2: reactive power compensation controller was not compatible with reverse power, which caused the wrong measurement that misled the capacitor was cut off, and then the power factor dropped.

Considering of restraining harmonic and improving power factor, the active filter should be installed to filter harmonic current produced by nonlinear load.

The problems were simulated on the distribution system with matlab, on this basis, the active filter was installed, and figure 5 shows the simulation diagram.

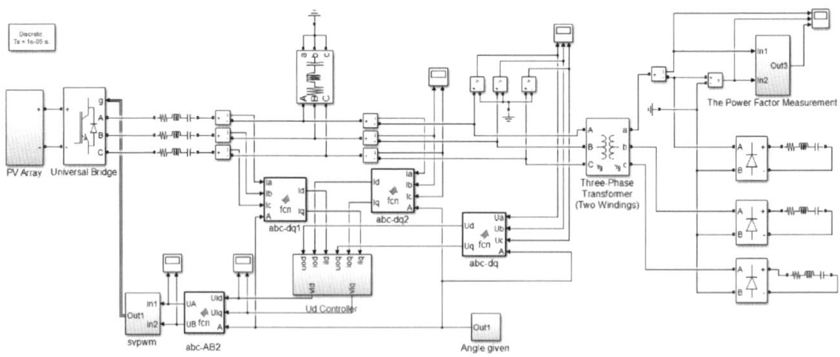

Fig. 5. The simulation diagram of distribution system with active filter

Adjustable rectifier circuit was used to replace the nonlinear load of the enterprise, figure 6 gives the simulation results of current and power factor waveform, figure 7 gives the harmonic analysis diagram. The diagram was consistent with the results of the field test, which was able to simulate the distribution system's harmonic and reactive power problems. After installed the active filter, the current and power factor waveform is shown in figure 8, the harmonic analysis diagram is shown in figure 9.

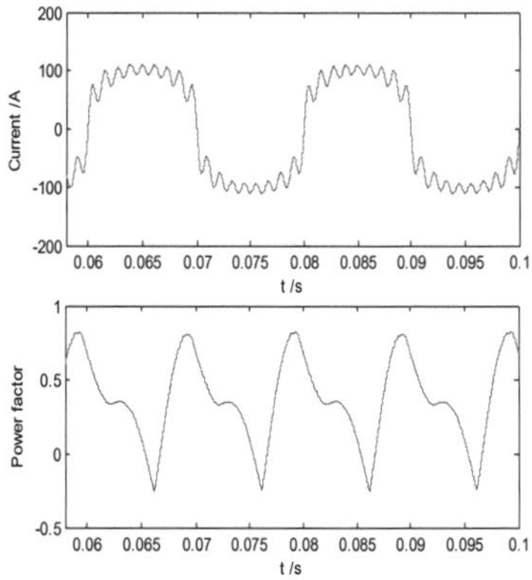

Fig. 6. The waveform of current and power factor before installing active filter

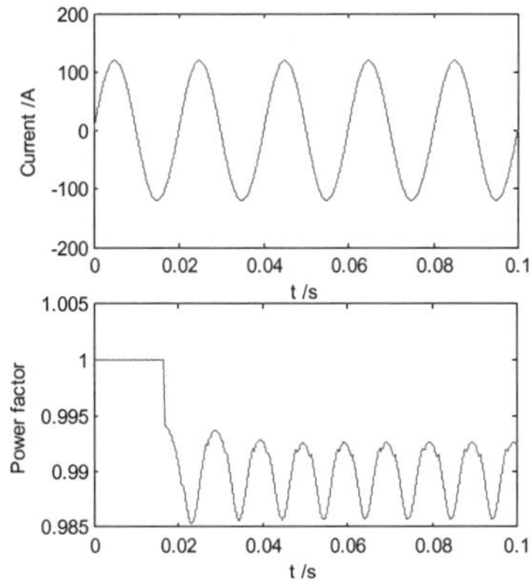

Fig. 7. The waveform of current and power factor after installing active filter

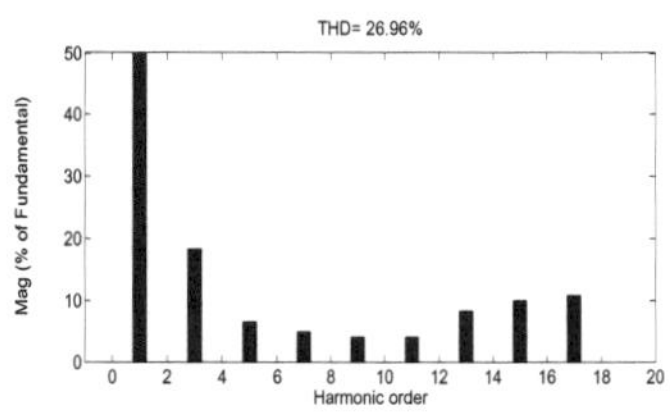

Fig. 8. The harmonic content diagram before installing active filter

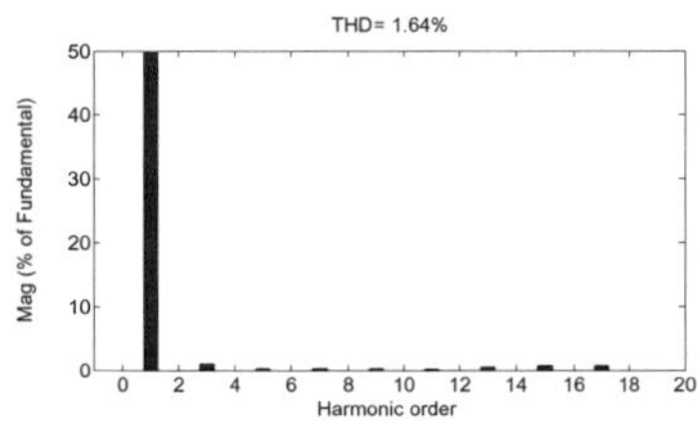

Fig. 9. The harmonic content diagram after installing active filter

The diagrams show that before installing the active filter, the current waveform are distorted apparently, the power factor is lower than 0.9, even negative. Each harmonic was obtained by Fourier analysis, it indicated that the harmonics exceed bid badly, the power factor cannot meet the standard.

Through the comparison and analysis of the simulation results, it can be found that when there were high capacity nonlinear loads in the low-voltage distribution system, for example, welding workshop used a lot of welding machine and coating workshop used frequency converter, etc. if photo-voltaic inverter was put into operation, the reactive power rose a lot. Reactive power compensation cabinets gave a warning, sometimes even cut them all. The reason was that after installing photo-voltaic, the inverter output in unit power factor, and more importantly, there were rare harmonics, so it can be regarded as parts of fundamental harmonic, as thus, the fundamental wave current through the transformer was reduced. However, there was no change in the system harmonic current, so at the test point of filtering and compensation, which was located at transformer low voltage side, the relative contents of harmonics current increased, consequently, the reactive power compensation controller got a bigger THD value. Once it exceeded the warning value, the LC compensator was unable to play the role of reactive power compensation. The processing method on-site was to modify THD protection value of reactive power compensation controller, it ensured that the compensator can work in larger THD value normally, but the risk of damaging capacitor rose.

The simulation results are showed in figure 7 and figure 8, after installing active power, the current waveform was no apparently distortion that can answer for the sine wave. Harmonic analysis results also show that content decreased significantly to a reasonable and acceptable range. The power factor was in accordance with the requirements completely.

4. Conclusion

With the widely application and promotion of the photo-voltaic in enterprises distribution system, power quality issues has increasingly greater impact on the power grid. In this paper, combining active factor with photo-voltaic connected to grid system was confirmed to be an effective solution to enterprise's harmonic pollution and reactive compensation. The experimental data and simulation results provided substantial evidence for this method. In the long term, the application of this method will reduce the enterprise cost greatly; also, it plays an important role in improving the reliability and security of the power distribution system and ensuring high power quality.

References

1. G. Pepermans, J. Driesen, 2005. Distributed generation: definition, benefits and issues, Energy Policy, vol. 33, pp.787–798.
2. T. Nakayama，T. Yagai，2009. Micro Power Grid System with SMES and Superconducting Cable Modules Cooled by Liquid Hydrogen, IEEE Transactions on Applied Superconductivity,19(3):2062-2065
3. LI Shen-sheng, The characteristics of solar energy. Solar energy 2003(5):10-12
4. Zhao Bin, Piao Zai-lin, Liu Na. Prospect application analysis of Small independent distributed power supply system. Agricultural technology and equipment, 2009（4）： 28-29.

5. Kang Jin-song, Yang Shou-jian, The reliability analysis of Solar photo-voltaic power generation system. Mechatronics,2014,20（7）： 14-18.
6. Zhao Yu-wen, The general development situation and future trend of the solar energy utilization. Chain power ,2003,36（9）： 63-69.
7. Zhang Le-kun, Su Nai-bin, The effects and countermeasures of grid connected photo-voltaic on distribution network protection. Management & Technology Xunkan of SMEs, 2015(6): 253-253.
8. Xiao Xiang-ning. XU Yong-hai, The analysis of power quality problems, Power System Technology, 2001,25(3): 66-69.
9. Cao Hai-tang, Research on harmonic suppression and power system reactive power compensation, South China University of Technology, 2005.
10. Zhou Luo-wei, Keyue M Smedly, 2000. Unified constant-frequency integration control of active power filters. IEEE Applied Power Electronics Conference &Exposition, 2000,16(3) :428-436.
11. Du Xiong, Zhou Luo-wei, 2003. An improved one cycle controlled DC side active power filter and research on steady-state and dynamics. Proceedings of the Csee.
12. Hou Shi-ying, Suo Li-Juan, Zheng Han-bo, 2010. Research of refined genetic algorithm in the hybrid active power filter. Power System Protection & control.
13. You Cai-hong, Qiu Xing-liang, 2008. A control method for active power filter based on wavelet transform & gray prediction. Grain Distribution Technology.
14. Wang Yu-fei, Jiang Jian-guo, A new type of AC arc furnace chaotic model used for power quality research. Proceedings of the CSEE, 2008,28(10): 106-110.
15. Chen Cheng, Research on power system nonlinear load harmonic characteristics, Huazhong University of science and technology, 2007.
16. Wang Song, Wang Xiang, Analysis of transformer excitation inrush current mechanism considering the dynamic characteristics of magnetic, Power System Protection and Control.

Design of Imaging System for Infrared Tracker Simulator

Ming-Chang Wang† and Yang-Yu Fan
School of Electronics and Information, Northwestern Polytechnical University,
Xi'an, 710072, China
†E-mail: wang_mc@126.com

IRFPA has become the leading chip in Infrared domain. Infrared trackers use different image size, such as 320×240 and 256×256. In order to simulate these trackers, a method is proposed which uses detector's optional window format. It uses low noise technology to design the peripheral circuit and precision voltage reference to realize the bias voltage. All of the detector's digital signals are insulated by optocouplers. FPGA is used to realize most of the functions, such as interface control, windowing mode, analog to digital convert, video display, RS422 interface and algorithms of non-uniform correction, automatically blind pixel elimination, etc. It can be used in many fields, such as proving the algorithm of recognition and tracking. The results show that the imaging system has the advantages of good image quality, good environmental adaptability, good expansibility and high reliability.

Keywords: infrared; imaging system; FPGA; NUC; blind pixel.

1. Introduction

Infrared Focal Plane Array (FPA) has been applied in many domains, especially military[1][2]. More and more missiles use infrared guidance technology[3] with different window size. In order to simulate these Infrared trackers, special imaging system needs to be designed, which can generate infrared image with different size, such as 320×240 and 256×256. A design of the imaging system for this application is presented. It is made up of optics, circuit, software and structure[4]. This paper mainly introduces the design of circuit and software.

2. Hardware design

Nowadays, detector with 640×512 size has been widely used. In order to meet future requirement, system uses SCORPIO detector of SOFRADIR Corp. It is an infrared opto-electronic detector, sensitive to radiation from 3.7μm to 4.8μm, including a sensitivity FPA comprised of a 640×512 elements connected to a Silicon readout circuit[5]. Composition of the hardware is shown as Fig. 1, including power supply, power filter of cooler, signal preprocessing and conversion, detector interface and image processing.

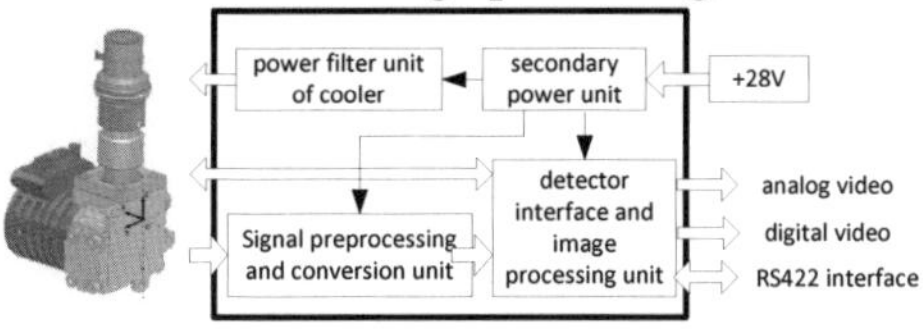

Fig. 1. Block diagram of imaging system

2.1. *Power supply*

This unit changes the input +28V to VDDA, VDDO, VDD, GPOL, VR and +24V for detector, +3.3V, +2.5V and +1.2V for FPGA, +5V for A/D converter, shown as Fig. 2. The system requires the noise to be as low as possible, otherwise image will show a variety of snow spots or oblique stripe[6][7].

Isolation powers, H2824S24BF and H2812S24BF, are used to get +24V and +12V. +24V is used for detector's cooler and +12V is used to generate other powers, such as VDDA and VDDO.

For +3.3V digital power (VDD) and +3.3V analog power (VDDA), system uses LT1763CS8-3.3V chips. For VDDO and VR, system uses LT1763CS8 chip, whose output is adjustable. LT1763 is a micro-power, low noise, low dropout regulator. Its RMS noise from 10Hz to 100KHz is only 20μV. GPOL is a tuned external bias voltage, which must be adjusted to optimize the overall photodiode bias voltage. REF192 is used to generate +2.5V voltage and two resistors divide it to generate GPOL.

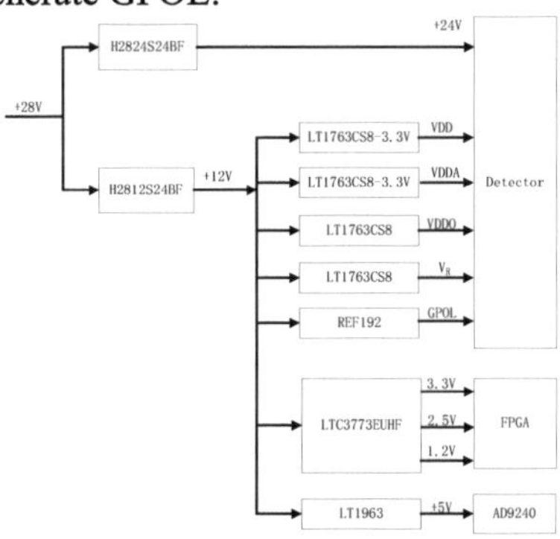

Fig. 2. Block diagram of power supply unit

2.2. *Power filter of cooler*

Filter network is used for the +24V power supply and temperature sensor of the detector to ensure the cooler can work stably. It can also avoid the interference of the cooler to the imaging circuit.

2.3. *Signal preprocessing and conversion*

This unit adjusts detector's analog outputs and convert them to digital signals. System uses AD9240 chip, which is a 10 MSPS, single supply, 14-bit analog to digital converter[8]. It combines a low cost, high speed CMOS process and a fancy architecture to realize the resolution and speed of existing hybrid implementations. The circuit of AD9240 is shown as Fig. 3.

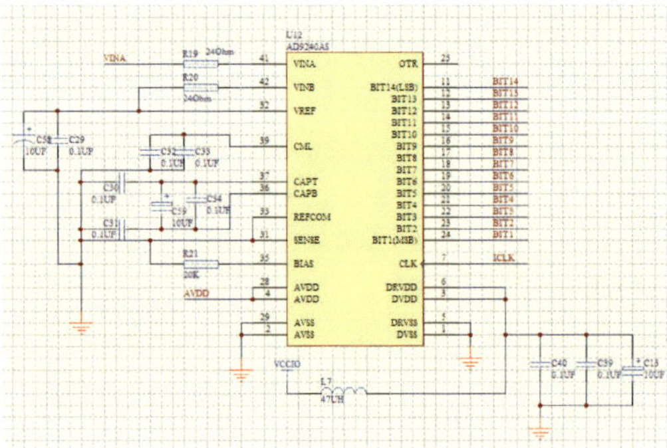

Fig. 3. Circuit of AD9240

2.4. *Detector interface and image processing*

This unit is the core of the hardware circuit, including detector interface, sequential control, display circuit, etc.

2.4.1. *Detector interface*

In order to avoid disturbance, all of detector's digital signals are isolated by way of photo electricity. HCPL-0630 is used, which is suitable for high speed logic interfacing in environments of extremely high ground or induced noise environments.

Detector's digital signals include INT, DATAVALID, MC, etc. INT allows the integration of charges in the pixel input capacitances. DATAVALID, on its high level, indicates the presence of valid data on the analog outputs. Master Clock (MC) is the base for synchronism of the whole circuit. Optional window size mode depends on signals of SIZEA, SIZEB, SCLR, SDATA and ERROR.

2.4.2. *System sequential control*

System sequential control is realized by FPGA[9], using XC6SLX150 chip. External interface is shown as Fig. 4. Flash is nonvolatile memory and stores nun-uniformity correction factors and blind pixel lists. After power on, FPGA reads these information from FLASH and writes into SDRAM, used for NUC.

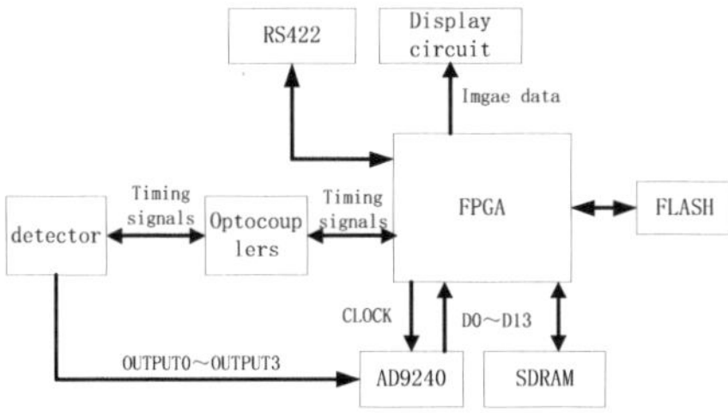

Fig. 4. Block diagram of FPGA external interface

Detector works in accordance with integrate then read mode. The signal of frame N integrated in the input stage is read after the end of the integration time of frame N. FPGA sends MC and INT signals to detector and receives DATAVALID signal from detector. Signal sequential is shown as Fig. 5. TMC is Master Clock period. Integration time equals to high level duration of INT minus 3075.5TMC. Integration time is independent to the readout time.

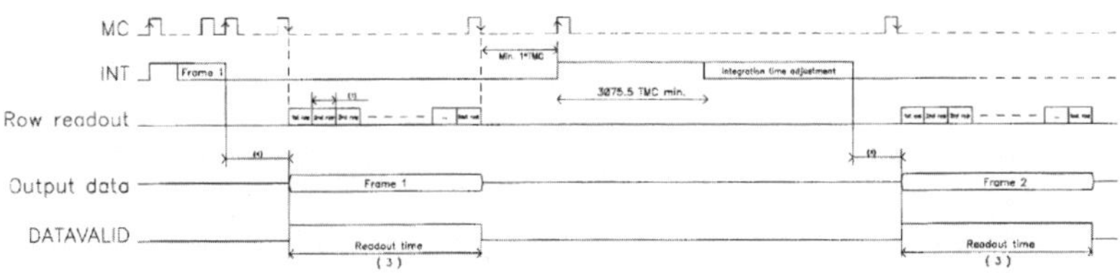

Fig. 5. Basic signal sequential of detector

Three fixed formats and a windowing mode can be selected by means of detector's SIZEA and SIZEB, which allow to choose at each frame between the full window and the partial window defined through the serial interface. For windowing mode, size of selected window is determined by R and C references of the corner points. The coordinates swing from (0;0) for the first pixel and (511;639) for the last one. Minimum size is 1 line×132 columns(within 512 lines ×640 columns). Position of pixel(0:0) relative to the dewar orientation is shown in Fig. 6. (RA;CA) are coordinates of point A and (RB;CB) are coordinates of point B, addressed through values RMIN,CMIN,RMAX,CMAX.

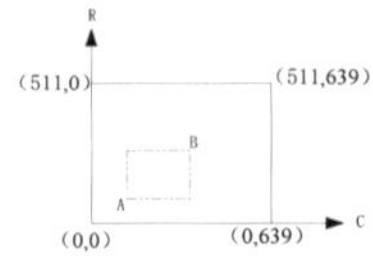

Fig. 6. detector's reference frame

SERDAT receives the serial interface data, entered in particular order: RMIN(9bits), RMAX(9bits), CMIN(8bits), CMAX(8bits), beginning with high weight. The pixels are addressed in the serial interface by blocks of 4 columns (one for each output) among 160. For example, the optional window is 320×256. Coordinates of point A are (160, 128) and coordinates of point B are (479,383), then RMIN=160, CMIN=32, RMAX=479, CMAX=96. Transmission sequence of optional window format is 010100000, 111011111, 00100000, 01100000, shown as Fig. 7.

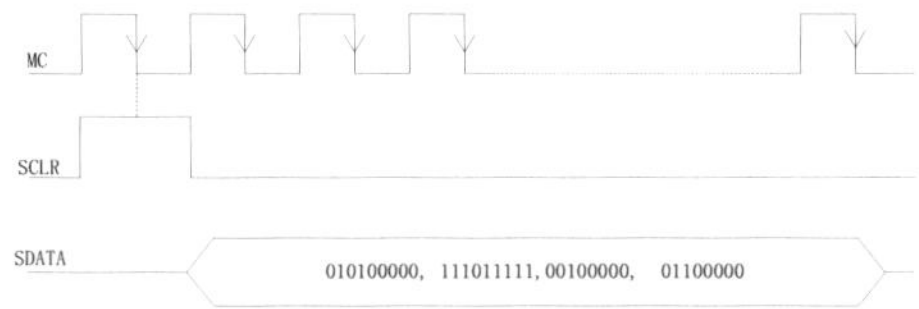

Fig. 7. Sequence for detector's optional window format

2.4.3. *Display circuit*

Block diagram of display circuit is shown as Fig. 8. ADV7122 is used, which is a video speed, digital to analog converter on a single chip. It consists of three, high speed, 10-bit, video D/A converters, a standard TTL input interface and analog output, current sources. A single +5V power supply, an external 1.23 V reference and pixel clock, composite SYNC and BLANK is all that required.

AL422B is used to store image data for display. It is a video frame buffer, consisting of DRAM that works like a FIFO which is long enough to hold up to 819x480 bytes of picture information and fast enough to operate at 50MHz.

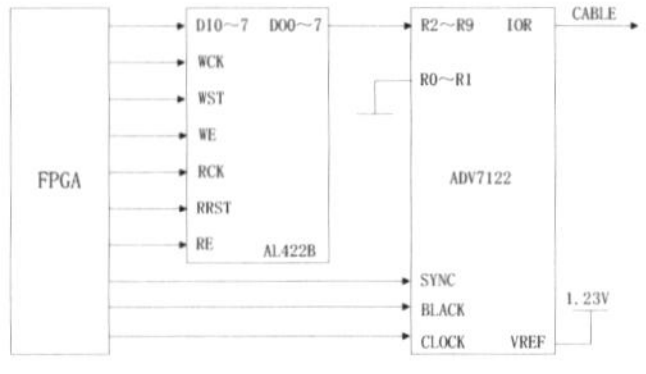

Fig. 8. Block diagram of display circuit

3. Software design

3.1. *NUC algorithm*

Two points correction algorithm is widely used in engineering domains[10]. It only needs to measure two points to calibrate the line and then NUC algorithm can be performed. The formula fow the two points correction is shown as Eq. (1). X_{ij} is output of detector, A_{ij} is gain factor, B_{ij} is offset factor.

$$Y_{ij} = A_{ij}X_{ij} + B_{ij} \tag{1}$$

The correction procedure is:

(a) Computing pixel response

When block-body temperature is T1, we collect M frame data and $X_{ijm}^{T_1}$ is the response of pixel(i,j), the m-th frame. For temperature T_1, average response of pixel(i,j) is shown as Eq. (2). Using the same method, we can get average response $X_{ij}^{T_2}$ for temperate T_2.

$$X_{ij}^{T_1} = \frac{\sum X_{ijm}^{T_1}}{M} \tag{2}$$

(b) Calculate factors

Factors calculation formula is shown as Eq. (3).

$$\begin{cases} A_{ij} = \dfrac{Y_{ij}^{T_2} - Y_{ij}^{T_1}}{X_{ij}^{T_2} - X_{ij}^{T_1}} \\ B_{ij} = Y_{ij}^{T_1} - A_{ij}X_{ij}^{T_1} \end{cases} \tag{3}$$

Commonly, Y_{ij}^{T1} and Y_{ij}^{T2} use average response of frames, $\overline{X^{T1}}, \overline{X^{T2}}$.

However, some factors will drift after a period of time. It makes the correction effect decreased. Basing two points NUC algorithm, single point correction method based on a shutter is adopted. When the image quality drops, host computer sends to the system that uses shutter to cover the field of view. System collects M frames of the real-time shutter image and modifies the offset parameter by Eq. (4). Then system uses B'_{ij} as the offset factor for NUC.

$$B'_{ij} = Y_{ij}^{T_3} - A_{ij}X_{ij}^{T_3} \tag{4}$$

3.2. *Automatic blind pixel elimination algorithm*

There are blind pixels whose response to the radiation is too high, to low or even none. System must calibrate these pixels and replace them by mean value of the neighborhood. However, some time later, new blind pixels still appear. So automatic blind pixel elimination algorithm should be adopted.

It uses the inertia of blind pixels. During period of the motion, image content changes with the movement but blind pixels will not change, algorithm just using this feature. First, algorithm judges whether the scene moves or not, dividing the image into regions of m×m size and calculating the mean value, calculating total number of regions in which the mean difference of the corresponding region in the adjacent two frames is larger than the threshold value. If the statistical result is larger than the threshold value, algorithm considers the scene is in the moving state. If a pixel in the continuous moving images meets the following condition, it will be considered as blind pixel. The condition is that, the pixel is the local extremum for n×n region(such as 16×16), absolute value of difference to other pixels in the adjacent area is larger than the threshold, adjacent regions do not exist similar pixel.

4. Result and analysis

4.1. *Sequential simulation*

After power on, detector works in accordance with 640×512 size. Simulation result of FPGA program is shown in Fig. 9.

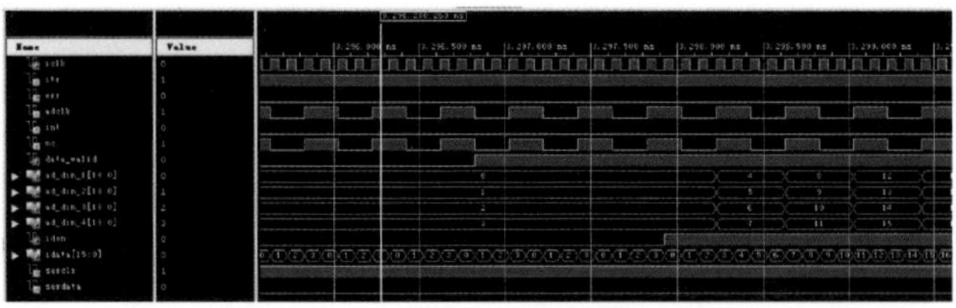

Fig. 9. Simulation result of FPGA program

4.2. *NETD test*

Integration time is 2.5ms. When temperature of block-body is +25°C, image gray value is 5,260 and time domain noise value is 0.55. When temperature of block-body is +30°C, image gray value is 5,680 and time domain noise value is 0.56. NETD value is about 32.7mK.

4.3. *Algorithm test*

Contrasts before and after using automatic blind pixel elimination algorithm are shown in Fig. 10. It shows that the algorithm can effectively eliminate new blind pixels.

Fig. 10. Before and after using of automatic blind pixel elimination algorithm comparison

4.4. *Windowing format test*

In order to test the windowing format function, host computer send corresponding command to the image system. System sets window size according to the command such as 320×256 and 320×240.Result shows that system can work normally.

5. Conclusion

The design of peripheral circuit of the detector has an important influence to the imaging system. Only low noise drive and readout design can make full use of detector performance. System uses optional window mode of the detector and realizes the algorithm of NUC, automatic blind pixel elimination and image enhancement. It can be used to meet requirement of infrared tracking simulator and other equipment. Result shows that system has advantages of good image quality, good environmental adaptability, high reliability.

References

1. Li Xiangdi, Huang Ying, Zhang Peiqing, etc. "Infrared imaging system and applications". Laser & Infrared, 2014, (03):229-234.
2. Zhang Xue, Liang Xiao-Geng. "Development of and Demands for Infrared Detectors". Electronics Optics & Control, 2013,(02):41-45.
3. Gong Hai-mei, Shao Xiu-mei, Li Xiang-yang, etc. "Advanced technology and application of space borne infrared detectors". Infrared and Laser Engineering, 2012,(12):3129-3140.
4. Wang Hua-wei, Cao Jian-zhong, Ma Cai-wen, etc. "Design of infrared imaging system with adaptive correction function". Infrared and Laser Engineering, 2014, (01):61-66.
5. Sofradir. SCORPIO K508 640×512 MW IRCMOS 15um PITCH MULTIMODE Integrated Detector Dewar Cooler Assembly (IDDCA) With Microcooler Type K508 Technical Specification. France: Sofradir, 2008.
6. Zhang Yu, Deng Xi-Ning, Gao Xu-hui, etc. "Design of infrared detector driver circuit for space applications". Laser & Infrared, 2015，45(02):150-152.

7. Zhu Hong-bin, Li Wei, Liu Zi-ji, etc. "Design on a Low-noise Acquisition Circuit of Infrared Focal Plane Device". Infrared Technology, 2010， (03):165-172.
8. ADI. AD9240 Complete 14-Bit, 10 MSPS Monolithic A/D Converter Data Sheet (REV.A). USA:ADI, 1998.
9. Liu Wan-cheng, Tang Shu-wei. "Design of Drive Circuit of Infrared Focal Plane Array Detector Based on Field-programmable Gate Array". Electro-Optic Technology Application,2014， (01)：65-67.
10. Li Xu, Yang Hu. "Application of nonuniformity correction algorithm for IRFPAS based on two points". Infrared and Laser Engineering, 2008, (2): 608-610.

A Simple Implementation of Long Distance Visible Light Communication System

Ziyu Song and Shengliang Peng†

School of Information Science and Engineering, Huaqiao University, Xiamen, 361021, China

†*E-mail: peng.shengliang@gmail.com*

Visible Light Communication (VLC) technology has aroused great attention these days as Light Emitting Diode (LED) is becoming the dominant source for artificial lighting. Previous research on VLC mainly focuses on the high speed transmission, which not only brings about extraordinary implementation complexity but also severely impairs the communication distance. This paper resolves low speed application scenarios, and proposes a simple implementation for VLC with long communication distance. A demo system without extra complex devices is set up as well. Test result demonstrates that the proposed VLC system can realize reliable optical wireless communication between two Personal Computers (PCs) at the distance of several meters, which contributes to the practical development of VLC technology.

Keywords: Wireless communication; VLC; LED; Demo system; FPGA.

1. Introduction

LED is superior to conventional lighting methods in terms of low power consumption, long lifetime, high speed switching performance and rapid response[1]. Besides providing light, LED also can be used as a communication device to transmit data[2]. VLC technology is an emerging approach to use LED for illumination and communication simultaneously.

VLC technology has been widely used in many fields[3], owning to not consuming licensed spectrum bands, causing little interference with traditional wireless communications and good confidentiality[4-6]. The history of VLC dates back to the 1880s when the photophone is invented. More recent work began in 2003 in Keio University, Japan, using LEDs to transmit data by visible light. Since then there have been numerous research activities focused on VLC. In China, the research on VLC starts relatively late. Many universities have made some headway such as PLA Information Engineering University[7], Southeast University[8] and Fudan University[9] in recent years.

However, current researches on VLC mostly focus on high speed data transmission. Complex modulation modes are adopted, resulting in high system cost, short communication range and significant implementation difficulty. This paper deals with low data speed application scenarios, and proposes a simple implementation for VLC with long communications distance.

2. System Description

The overall design of our VLC system is shown in Fig. 1. Both transmitter and receiver include analog parts and digital parts. The analog parts are composed of fundamental components and chips. The digital parts are mainly implemented in the internal Field Programmable Gate Array (FPGA).

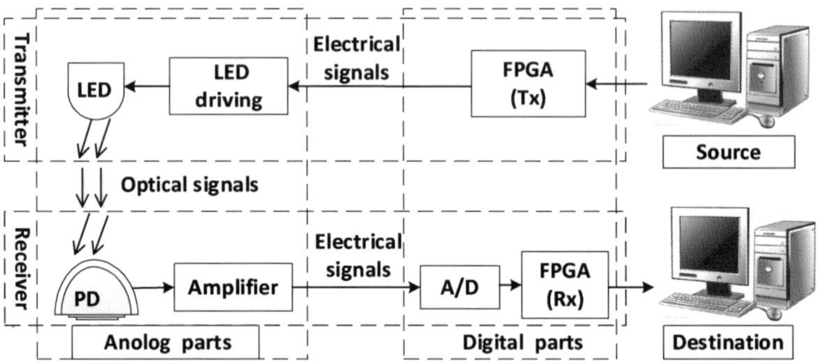

Fig. 1. Architecture of VLC system

The digital data to be transmitted is generated by the source unit, and then sent to the digital part of transmitter, where it is On-Off Keying (OOK) modulated and converted to analog electrical signals. The analog signals make the LED flash through a driving circuit which can switch the electrical signals to optical signals. After propagating in the free space channel, the optical signals will be detected and recovered by analog part of the receiver. Then, an Analog to Digital (A/D) converter is used to change these analog electrical signals to digital form, in that case the receiving data can be demodulated exactly. At last, the digital signals of initial format are sent to the destination, thus a wireless communication process is accomplished.

3. System Implementation

3.1. *Transmitter*

Fig. 2 presents the data processing chain of transmitter. Digital data is processed in the FPGA of ALTERA EP2C5T144C8. Analog part includes LED light source and the driving circuit. The function of each module is as follows.

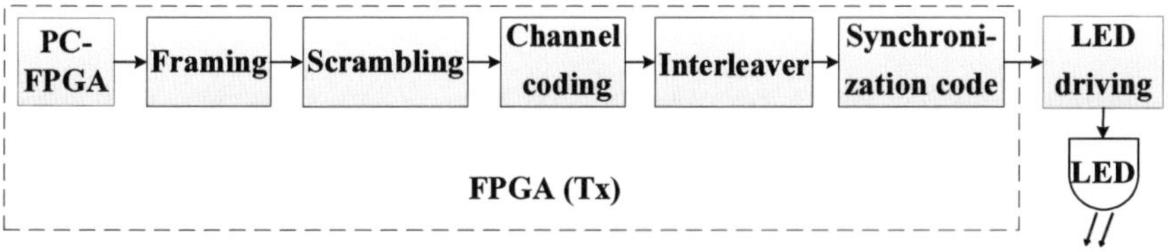

Fig. 2. VLC transmitter

- *PC to FPGA serial communication interface module*

PC is used as the display devices. Generally, Ethernet, USB and UART are the common interfaces for PC to communicate with peripherals. As one of the simplest types requiring no synchronization, UART is adopted.

- *Framing module*

Frame is the basic unit of data transmission. Table 1 shows the frame format used in this paper. Each frame includes 243 bits data. The frame header indicates the start of frame. The next 3 bits is reserved for extension. The frame number and length fields indicate the number of frames already sent and number of valid data bits in the successive Data filed. At last, the whole frame need to be protected by the Cyclic Redundancy Check (CRC).

Table 1. Frame format

Item	Digits
Frame header	16 bits
Reserve	3 bits
Frame number	8 bits
Length	8 bits
Data	192 bits
Cyclic Redundancy Check (CRC)	16 bits

- *Scrambling module*

The polynomialis (1) is used to scramble all bits into a data sequence similar to white noise to avoid too many consecutive '0' or '1' in transmission data sequence, which will cause flickering light clearly.

$$g(x) = 1 + x^4 + x^7 \tag{1}$$

- *Channel coding module*

Communication effect will be impaired due to the presence of background light noise in space optical channel. Channel coding is necessary to reduce the transmission error. This paper uses RS code because it has the characteristics of multiple errors correcting, simplicity in principle and facile implementation.

- *Interleaver module*

Visible light cannot pass through the obstacle. A large section of error will occur once the light is blocked, and the communication is interrupted. Block interleaver is selected to disperse the continuous error.

- *Synchronization code module*

At the receiver, synchronization code is necessary for synchronizing. So 13-bit Barker-code is added in the transmitting sequence for synchronization.

- *Green LED source*

Green LED with narrow bandwidth can maximumly avoid crosstalk. The light source uses single 1W high brightness LED whose luminous efficiency is 80-100 lm/W and wave length is 515-525 nm.

- *LED driving module*

LED is driven by current and its brightness is approximately proportional to forward current. To make the LED light normally, a driving circuit is necessary to maintain a continuous current about 350 mA.

3.2. ***Receiver***

At the receiver, PhotoDetector (PD) is utilized to detect the light intensity and convert the optical signal to electrical signal. Then the electrical signal is amplified by a two-stage amplifier. The analog signal is sampled by the A/D module and processed in FPGA chip of ALTERA EP4CE6F17C8. Fig. 3 shows the block diagram of receiver.

- *Photo-diode*

PD uses the BOSEN G0606M-G detector with green filter which has a light-sensing surface area of 6mm*6mm, the range of spectrum response is 450~650nm. Besides, its peak response is about 550nm, matching the emission wavelength of selected LED.

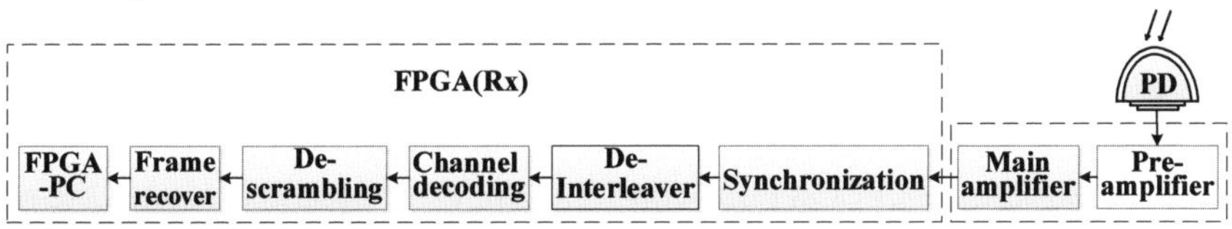

Fig. 3. VLC receiver

- *Amplifier circuit*

The power of received light is generally weak due to the channel attenuation. Signal to Noise Ratio (SNR) at the receiving end is very low as well. Therefore, we use a preamplifier to convert the current signal into a voltage signal and a main amplifier to further amplify the voltage signal.

- *Synchronization module*

According to Nyquist's Theory, the sampling frequency should not be less than twice of the transmission rate. In this paper, four times oversampling is selected, to obtain the best performance. Besides, a phase-locked loop module is called to synchronize clocks between the transmitter and receiver.

The other modules are the reverse of corresponding modules in the sending end, so the details of their implementation will not be listed. The data checked by CRC will be grouped in accordance with the start-stop asynchronous protocol (1 start bit, 8 data bits and 1 stop bit) and sent to the receiving PC for display.

4. Test Results

The VLC demo system is given in Fig. 4. The transmitter includes PC(Tx), FPGA(Tx), LED driving and LED source. Similarly, the receiver includes PD, amplifier, FPGA(Rx) and PC(Rx). The test case for the VLC demo system is to verify whether the receiver output data is identical to the input data transmitted.

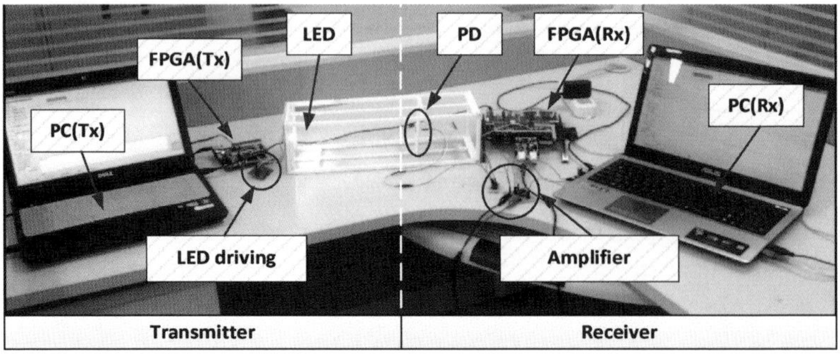

Fig. 4. VLC demonstration system

Serial debugging software has to be installed in the PCs in advance. As shown in Fig. 5, we type "Hello world!" at the sending window and send it. The string is transmitted periodically at the interval of 100 ms.

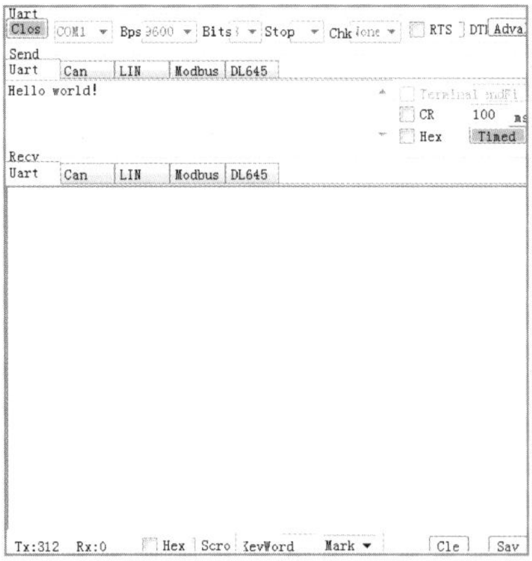

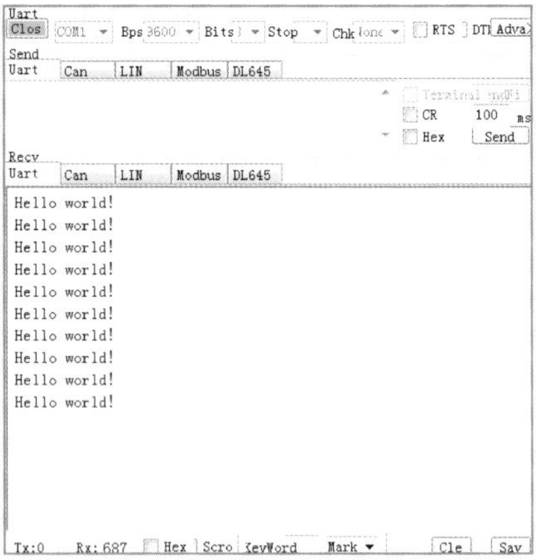

Fig. 5. Sending window

Fig. 6. Receiving window

The test result is shown in Fig. 6. Receiver demodulates and restores the string sent by transmitter. No error occurs and the whole system works well. The effective transmission distance can be up to several meters.

5. Conclusion

In this paper, we propose a simple implementation for VLC with long communication distance. The VLC system structure is simple and easy to implement without complex devices. It is OOK modulated and data rate is 9.6Kb/s. Test result shows that data transmission is done via the visible light between two PCs at distance of up to several meters, which contributes to the practical development of VLC technology.

Acknowledgment

This research is supported by the National Natural Science Foundation of China (Grant No. 61302095, 61362018 and 61401165) and Huaqiao University (Grant No.12BS219 and 13BS101).

References

1. M. Saadi, L. Wattisuttikulkij, Y. Zhao, P. Sangwongngam, Visible Light Communication: Opportunities, Challenges and Channel Models, *International Journal of Electronics and Informatics*, vol. 2, (2013).
2. T. Komine and M. Nakagawa, Fundamental Analysis for Visible-Light Communication System Using LED Lightings, *IEEE Trans. Consumer Electronics*, vol. 50, (2004), pp.100-107.
3. Prasad R, Mihovska A, Compaarative overview of UWB and VLC for data-intensive and security-sensitive applications, *Proc. of IEEE International Ultra Wideband Conference*, (New York, USA, 2012).
4. Klaus Dieter Langer, Jelena Vucic, Christoph Kottke, etal, Advances and prospects in high-speed information broadcast using phoshorescent white-light LEDs, *ICTON*, (2009). Mo. B5.3.
5. Y Tanaka, S Haruyama, M Nakagawa, Wireless optical transmissions with white colored LED for wireless home links, *Indoor and Mobile Radio Communications*, (2000), pp. 1325-1329.
6. Aleksandar Jovicic, Junyi Li, and Tom Richardson, Visible Light Communication: Opportunities, Challenges and the Path to Market, *IEEE Communications Magazine*, (2013), pp. 26-32.
7. Shen Rui, Zhang Jian, Indoor Location Method and the Cramer-Rao Bound Estimation Based on Visible Light Communications, *Laser & Optoelectronics Progress*, (2014), pp. 81-88.
8. Hua Qian, Juan Chen, Saijie Yao, One-Bit Sigma-Delta Modulator for Nonlinear Visible Light Communication Systems, *IEEE Photonics Technology Letters*, vol.27, (2015), pp. 419-422.
9. Chi Nan, Wang Yiguang, Wang Yuanquan, The research on multidimensional reuse of visible light communication, *Optics & Optoelectronic Technology*, (2014), pp. 1-6.
10. Li Shichao, the implementation of Barker code of concentrated insert based on FPGA, *Technology Exchange*, (2010), pp. 33-35.

The Influence of Magnetic Field on the Spatial Resolution Of Pulse-Dilation Framing Tube

Yan-Li Bai

Department of Education Practice, Guilin University of Electronic Technology, Guilin, Guangxi, 541004, China

E-mail: bayaly@guet.edu.cn

Framing tube was used to assess the symmetry of imploding targets in laser driven inertial confinement fusion experiments. A pulse dilation framing tube is introduced, which utilized pulse-dilation technology and the magnetic lens. The spatial resolution of the tube was simulated by the Finite Difference Method (FDM), the Four orders Runge-Kutta (RK4) method and the Modulation Transfer Function (MTF). The influence of magnetic field distribution on spatial resolution of tube is also researched. The uninformed axisymmetric magnetic field is produced by the magnetic lens and computed by the FDM. The trajectories of the electrons were traced by the RK4. The spatial resolution was calibrated by the MTF. The simulation results demonstrated that the obvious influence of magnetic field distribution on spatial resolution and the magnetic field distribution was determined by the Full Width Half Maximum (FWHM) of magnetic field along the axis of a magnetic lens and the numbers of magnetic lenses. The results verified that the pulse-dilation framing tube using magnetic lens is entirely feasible and the tube achieved higher spatial resolution with the smaller magnetic field. The spatial resolution (10–90% rise), about 200 μm for the tube using three or four magnetic lenses was simulated. The pulse dilation framing tube with high temporal and spatial resolution is a possible use for measuring the high-energy electron trans-port rates in fast ignition experiments and analyzing the symmetry of the late-stage implosion of high-energy targets and the burn wave dynamics for igniting targets.

Keywords: Inertial confinement fusion; pulse dilation framing tube; magnetic lens; spatial resolution; the finite difference method; the RK4 method; the modulation transfer function.

1. Introduction

Inertial Confinement Fusion (ICF) is a process in which nuclear fusion reactions are initiated by heating and compressing a fuel capsule containing a mixture of Deuterium and Tritium which can be got from seawater [1-3]. Due to the framing camera has high temporal resolution and two-dimension spatial resolving ability, it has playing an important role as diagnostics of inertial confinement fusion experiments. The conventional x-ray framing cameras based on proximity focused micro-channel plates (MCP) and achieved temporal resolution is limited to 35-100 ps[4-6]. In 2010, the pulse-dilation technology is successfully implemented into a new generation of framing camera with measured temporal resolution of down to 10 ps[7]. Later, the researcher employed the dilation X-ray imager (DIXI) on the NIF and investigated high speed phenomena in laser plasma interactions. They achieved temporal resolution of < 10 ps and spatial resolution of < 300 μm with 500 Gauss magnetic field at cathode [8-9]. Compared with The conventional x-ray framing cameras, the new generation of framing camera has a better temporal resolution and a poorer spatial resolution. Therefore, it has an important significance to study the improvement of the spatial resolution.

The work principle of pulse-dilation framing tube depicted in Fig. 1 and a large uniform magnetic field through the drift area guides the electron signal on the imaging electronics, countered the spatial dispersion as electrons travel through drift area. The work principle of the device is discussed in Ref.[7].The effective temporal resolution of the device is determined by the four parameters, which are the PC bias voltage, the gradient of pulse, the length of the drift area and the MCP gate width. And the effective spatial resolution is determined by the two parameters, which are the PC material and the magnetic field strength [8-9].

The spatial resolution of DIXI is improved through increasing the magnetic field at the cathode [8-9]. However, it is impractical of further increasing magnetic field strength at the cathode because one is high power loss and slowly improved the spatial resolution, the other is enhanced the space charge effect due to larger magnetic field. Therefore, to achieve high spatial resolution with smaller magnetic field, the magnetic lens is introduced in our dilation framing tube. It has been used to improve the focusing ability and imaging quality of optical electronic in streak camera and electron microscope [10-12]. Compared to the DIXI with a larger uniform magnetic field, the magnetic lens has a smaller non-uniform magnetic field. In the article, the spatial resolution of tube is studied, and especially focused on the influence of magnetic field distribution on the spatial resolution. At last, we analyzed the simulation results and discussed the feasibility of the pulse-dilation framing tube using the magnetic lens.

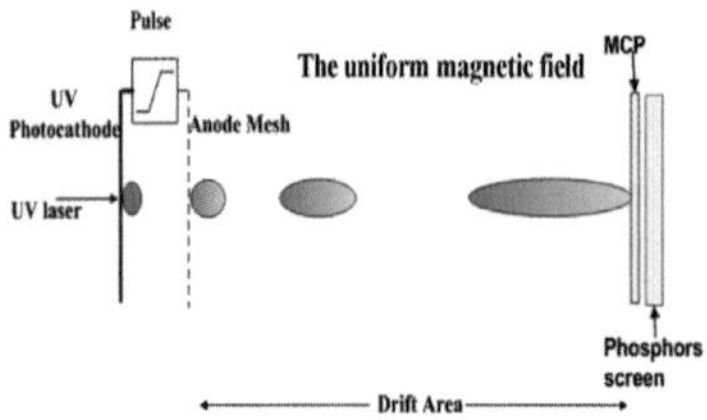

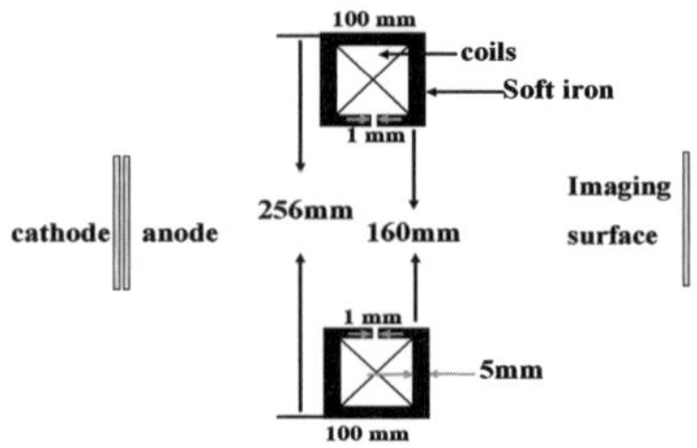

Fig. 1. Working principle Fig. 2. The simulation model

2. The Method of Simulating Spatial Resolution

The simulation model is showed in Fig. 2. The magnetic lens consists of a soft iron frame and copper coil. The inner and outer diameters of the magnetic lens are 160 mm and 256 mm respectively. The axial length of the magnetic lens is 100 mm. There is a 1 mm-wide circular slit in the inner cylinder. The accelerate region between cathode and anode is 1 mm, the drift distance from the anode to the imaging surface is 500 mm. The spatial resolution was simulated by tracing the trajectories of the electrons from the cathode to the imaging surface and Modulation Transfer Function (MTF) [13]. This was done by solving the dynamic equations of motion for the electrons in a non-uniform magnetic field, using the Four orders Runge-Kutta (RK4) method [14]. The magnetic field distribution was calculated by the finite difference method [15-17], assuming that the shield metal has an infinitely high magnetic permeability, and is not saturated. The MTF is a plot of contrast (or modulation), measured in percent vs. the spatial frequency measured in lp/mm (line pairs/mm). This graph is customarily normalized to a value of 1 at zero spatial frequency. We usually calibrated spatial frequency of tube while the contrast is 0.1.The MTF is defined as:

$$MTF = \exp[(\pi \overline{\Delta r} f)^2] \quad \textbf{(1)}$$

In the eq. (1), $\overline{\Delta r}$ is Root Mean Square (RMS) radius (μm) and is defined as:

$$\overline{\Delta r} = \sqrt{\left(\overline{\Delta x}\right)^2 + \left(\overline{\Delta y}\right)^2} \quad \textbf{(2)}$$

$$\overline{\Delta y} = y - y' \quad \overline{\Delta x} = x - x'$$

Where $\boldsymbol{X'}$and $\boldsymbol{Y'}$ is the practical points of the electron trajectory on the imaging surface, respectively. The $\boldsymbol{X}$ and $\boldsymbol{Y}$ are the ideal imaging points on the imaging surface, respectively. The f is spatial frequency (lp / mm).

In the simulation, the initial position of the photoelectrons is Gaussian type, and its initial energy has a β(1, 4) distribution with a 0.6 eV energy spreading. The azimuthally angle of the photoelectrons is evenly distributed within (0~2π). The elevation angle distribution of the emitted electrons is of cosinoidal distribution in (0~π/2) range. The drift energy of electron is ~3 keV. The first imaging point of the magnetic lens is appeared and the imaging magnification is 1:1 when the ampere turn of the magnetic lens is 542*AN*. The peak value of the magnetic field along the axis is ~ 43 Gauss and its' FWHM is 100mm, the trajectories of the electrons from the cathode to the imaging surface, the distribution of electrons on the imaging surface, the graph of MFT (Modulation Transfer Function) are showed Fig. 3(a-d), respectively. From Fig. 3(d), the spatial frequency is 14.71 lp/mm when contrast (or modulation) is 0.1. Therefore the spatial resolution at the centre of cathode is: 1000 μm/ 14.71 ≈ 68 μm.

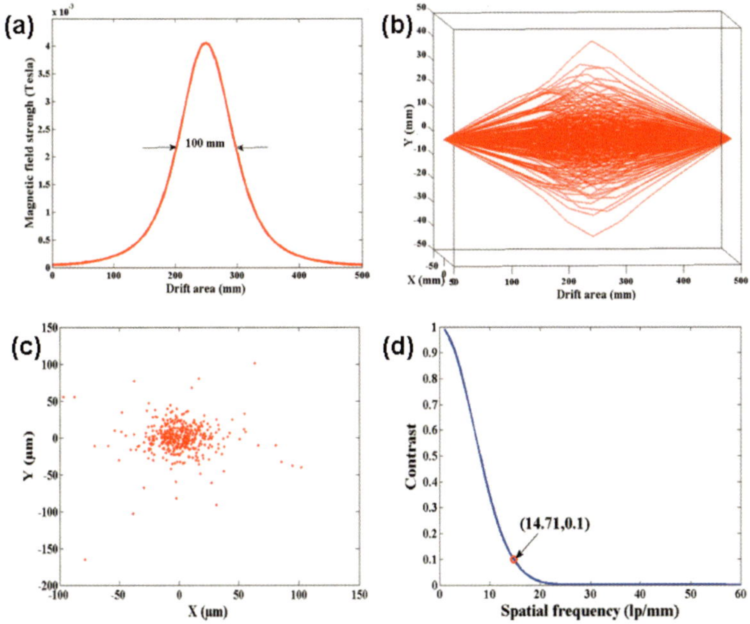

Fig. 3. The simulation of the spatial resolution.(a) the distribution of magnetic lens along the axis in the drift area. (b) The trajectories of the electrons from the cathode to the imaging surface. (c) The distribution of electrons on the imaging surface. (d) The graph of MFT.

3. Simulation Results and Analysis

There are two parameters that determine the spatial resolution of pulse dilation framing tube. The one is the different off-axis point on cathode. As can be seen from Fig. 3(a), the magnetic lens has a non-uniform magnetic field, Therefore, the electrons close to the center are less strongly deflected than those passing the lens far from the axis and the spatial resolution of on-axis point is better than off-axis point. The other is the distribution of magnetic field. As described in Ref.[8-9], there is a large uniform magnetic field in the drift area and the spatial resolution of DIXI depends on the magnetic field strength at the cathode. But, the spatial resolution of pulse dilation framing tube on the non-uniform magnetic strength field is rarely introduced in other literatures. Hence, we mainly studied the spatial resolution in the different magnetic field distribution, off-axis points and imaging magnification of 1:1 in this part.

3.1. *The Single Magnetic Lens*

This part is mainly simulated the spatial resolution of the tube using the single magnetic lens. The inner and outer diameters of the magnetic lens are same as Fig. 2, and the differences are the axial length and the circular slit, only. The parameter of magnetic lens, the magnetic field distribution along the axis and the spatial resolution of different off-axis point (0mm ~ 25 mm) are showed in Fig. 4, respectively. From the Fig. 4(a), the magnetic lenses are marked ①, ② and ③, respectively. The ① lens has the axial length is 100 mm and a 1 mm-wide circular slit in the inner cylinder. The ② lens has the axial length is 200 mm and a 100 mm-wide circular slit. The ③ lens has the axial length is 200 mm and a 180 mm-wide circular slit. The magnetic field distribution along the axis of the ①, ② and ③ are showed in Fig. 4(b) and their FWHM is 100 mm, 150mm and 200 mm, respectively. The spatial resolution of off-axis points through changing the FWHM of the magnetic field are showed in Fig. 4(c). With the off-axis point is changed from 0 mm to 25 mm, the spatial resolution is become poor. The spatial resolution is changed from 68 μm to >2000 μm, especially the spatial resolution is rapid become poor when off-axis distance is > 10 mm from the line①. Compared with the line①, the other lines are gently changed and the spatial resolution is >1000 μm when off-axis point >20 mm, only. In a word, the spatial resolution of tube using a magnetic lens is much difference between on-axis point and off-axis, and the change speed of spatial resolution from 0 mm to 25 mm is determined by the FWHM of magnetic field.

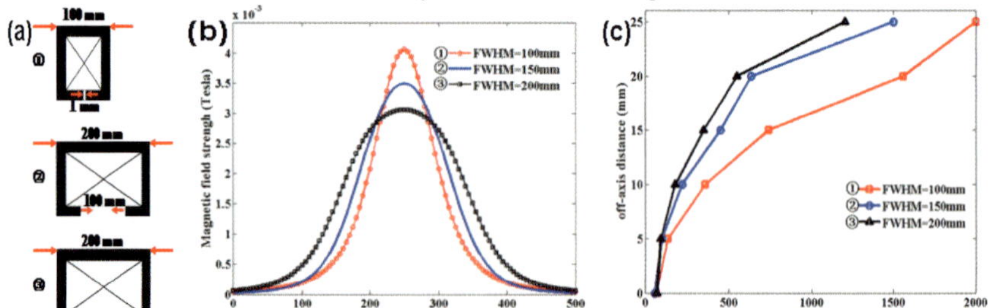

Fig. 4. The spatial resolution of tube is simulated in the different FWHM of magnetic field. (a) The parameters of the three magnetic lenses. (b) The different FWHM of the magnetic field distribution along the axis. (c) The spatial resolution of off-axis points in the different FWHM of magnetic field.

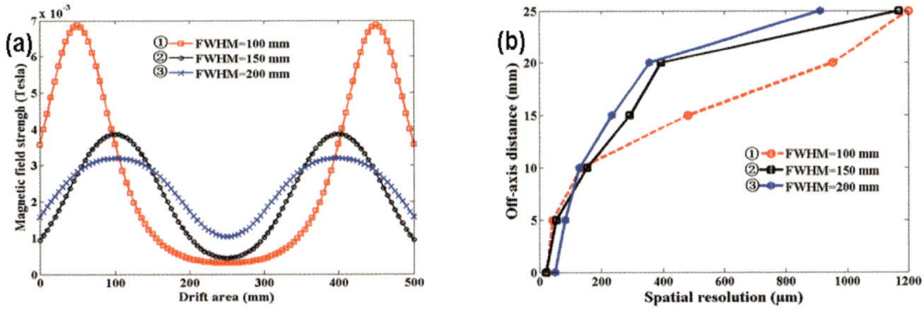

Fig. 5. The spatial resolution of tube is simulated in the different FWHM of magnetic field. (a) The different FWHM of the magnetic field distribution along the axis. (b) The spatial resolution of off-axis points in the different FWHM of magnetic field.

3.2. *The Double Magnetic Lenses*

This part is mainly simulated the spatial resolution of the tube using double magnetic lenses. The parameters of a magnetic lens are showed in Fig. 4(a). The two lenses have the same size and placed near the anode and imaging surface, respectively. The magnetic field distribution along the axis and the spatial resolution of different off-axis point (0mm ~ 25 mm) are showed in Fig. 5. The spatial resolution of different off-axis points in the different FWHM of the magnetic field are showed in Fig. 5(b). Compared with the single magnetic lens, the double magnetic lens has a better imaging and can improve spatial resolution of tube. The spatial resolution is rapid ramp when the off-axis point is > 20 mm, and the spatial resolution is better when off-axis point < 15 mm in the 100 mm FWHM of magnetic field. In a word, the tube using the double magnetic lenses has better spatial resolution and field of view than the single one.

3.3. *The Three and Four Magnetic Lenses*

This part is mainly simulated the spatial resolution at the first imaging point of the three and four magnetic lenses. Considering that the broader FWHM is closed to a uniform magnetic field, the 100 mm FWHM of magnetic lens is used only. The parameters of a magnetic lens are same as the ① in Fig. 4(a). The three magnetic lenses are placed near the anode, imaging surface and in the middle of the drift area, respectively. The four magnetic lenses are averagely placed in the drift area. The magnetic field along the axis and the spatial resolution of different off-axis point (0mm ~ 25 mm) are showed in Fig. 6. The spatial resolution of different off-axis points are showed in Fig. 6(b) , and it has slowly changed from 0 mm to 25 mm. Simulations at smaller along the axial magnetic field give a spatial resolution (10–90% rise) of ~200 μm. Therefore the three and four magnetic lenses have the better spatial resolution than others.

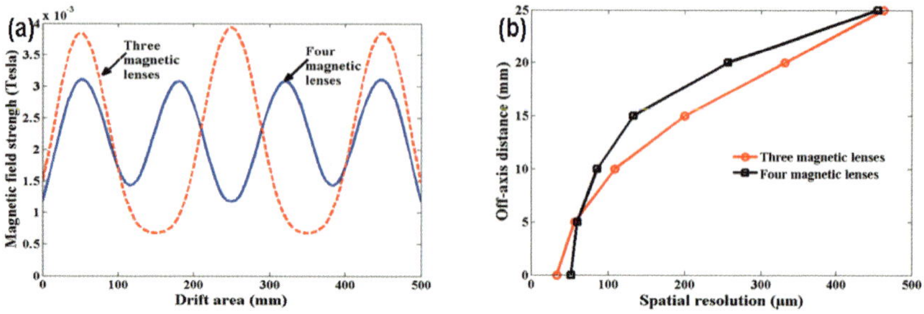

Fig. 6. The spatial resolution of tube is simulated using the three and four magnetic lenses. (a) The magnetic field distribution along the axis. (b) The spatial resolution of off-axis points in the different magnetic field.

As can be seen from simulation results, we have known that the spatial resolution of pulse dilation framing tube is determined by the characteristic of magnetic field along the axis. Considering that a larger field of view, the higher two-dimensional spatial resolution and the simpler structure of framing tube, the three magnetic lenses are suitable for designing the pulse dilation framing tube, which the FWHM of a magnetic lens is 100 mm.

4. Conclusion

Simulation for the spatial resolution of the pulse dilation framing tube using the magnetic lens based on the numerical calculation is described. The simulated results show that the spatial resolution and field of view are changed by the FWHM of magnetic field and the numbers of magnetic lens. The tube with the better spatial resolution and the magnetic field is smaller than the DIXI [8-9], which the three magnetic lenses is used. If want to achieve a better field of view, it can be enlarge diameters of the magnetic lens or reduce size of the image. Additionally, the actual spatial resolution may be larger than the simulated one because of the space charge effect, therefore the future work is to studying the influence of the space charge effect on spatial resolution.

Acknowledgment

This work is supported by Scientific Project of Education Department of Guangxi No.LX2014125.

References

1. S. H. Glenzer, B. J. MacGowan, P. Michel, "Symmetric Inertial Confinement Fusion Implosions at Ultra-High Laser Energies", Science, 327(5970), 1228-1231, 2010.
2. Shaoen Jiang, Longfei Jing, Yunbao Huang, and Yongkun Ding, "Novel free-form hohlraum shape design and optimization for laser-driven inertial confinement fusion," Physics of Plasmas 21, 102710-(1-10), 2014.

3. Andrei N. Simakov, Douglas C. Wilson, Sunghwan A. Yi, John L. Kline, Daniel S. Clark, Jose L. Milovich, Jay D. Salmonson, and Steven H. Batha, "Optimized beryllium target design for indirectly driven inertial confinement fusion experiments on the National Ignition Facility". Physics of Plasmas 21, 022701(1-10), 2014
4. D. K. Bradley, P. M. Bell, O. L. Landen, J. D. Kilkenny, and J. Oertel, "Development and characterization of a pair of 30–40 ps x-ray framing cameras," Review of Scientific Instruments 66 (1), 716-718, 1995
5. J. A. Oertel, T. N. Archuleta, and L. S. Schrank. "The large format x-ray imager," Review of Scientific Instruments 72(1), 701-704, 2001
6. J. P. Holder, K. W. Piston, D. K. Bradley, P. M. Bell, A. K. L. Dymoke-Bradshaw, and J. D. Hares. "Further development of a single line of sight x-ray framing camera," Review of Scientific Instruments 74(3), 2191-2193, 2003
7. T. J. Hilsabeck, J. D. Hares, J. D. Kilkenny, Bell PM, Dymoke-Bradshaw AK, Koch JA, Celliers PM, Bradley DK, McCarville T, Pivovaroff M, Soufli R, Bionta R. "Pulse-dilation enhanced gated optical imager with 5 ps resolution," Review of Scientific Instruments 81(10) 10E317 (1-6), 2010
8. S. R. Nagel, T. J. Hilsabeck, P. M. Bell, D. K. Bradley, M. J. Ayers, Barrios MA, Felker B, Smith RF, Collins GW, Jones OS, Kilkenny JD, Chung T, Piston K, Raman KS, Sammuli B, Hares JD, Dymoke-Bradshaw AK. "Dilation x-ray imager a new/faster gated x-ray imager for the NIF," Review of Scientific Instruments 83(10) 10E116 (1-3), 2012
9. S. R. Nagel, T. J. Hilsabeck, P. M. Bell, D. K. Bradley, M. J. Ayers, K. Piston, B. Felker, J. D. Kilkenny, T. Chung, B. Sammuli, J. D. Hares, and A. K. L. Dymoke-Bradshaw. "Investigating high speed phenomena in laser plasma interactions using dilation x-ray imager," Review of Scientific Instruments 85(11), 11E504 (1-6), 2014
10. Chang, Z., A. Rundquist, J. Zhou, M. M. Murnane, H. C. Kapteyn, X. Liu, B. Shan, J. Liu, L. Niu, M. Gong, and X. Zhang. "Demonstration of a sub-picosecond x-ray streak camera," Appl. Phys. Lett. 69(1), 133-135, 1996
11. Feng, Jun, K. Engelhorn, B. I. Cho, H. J. Lee, M. Greaves, C. P. Weber, R. W. Falcone, H. A. Padmore, and P.A. Heimann. "A grazing incidence x-ray streak camera for ultrafast, single-shot measurements," Appl. Phys. Lett. 96(13), 134102, 2010
12. Konvalina. I, I. Müllerová. "Properties of the cathode lens combined with a focusing magnetic / immersion-magnetic lens," Nuclear Instruments and Methods in Physics Research Section A: Accelerators, Spectrometers, Detectors and Associated Equipment. 645(1), 55-59, 2011
13. Illes P. Csorba. "Modulation transfer function of image tube lenses," Applied Optics, 16(10):2647-2650, 1977
14. Delin T , Zheng C. "On A General Formula of Fourth Order Runge-Kutta Method," Journal of Mathematical Sciences & Mathematics Education Vol. 7 No. 2 1-10, 2012
15. Bohumila L, Michal L. "The sources of errors in first-order FEM for electrostatic lenses". Nuclear Instruments and Methods in Physics Research A 519(1-2) 141–148 (2004)
16. Lencova, B. "Accuracy estimate for magnetic electron lenses computed by first-order FEM". Nuclear Instruments and Methods in Physics Research A 519(1-2) 149–153 (2004)
17. Lencova, B, Zlamal, J. "A New Program for the Design of Electron Microscopes". Proceedings of the Seventh International Conference on Charged Particle Optics (Cpo-7) 1 (1) 315-324 (2008).Trinity Coll, Cambridge, ENGLAND.

Research on Condition Evaluation Technology of Energy Storage Equipment in Photovoltaic System

Hai-Feng Bian†, Rong-Jiang Jin[a] and Chao Zhang[b]

NARI Group Corporation
Beijing, China
E-mail: †bhf2626@163.com, [a]jinrongjiang@sgepri.sgcc.com.cn, [b]zhangchao5@sgepri.sgcc.com.cn

Tian-Chun Xiang[c] and Xu-Dong Wang[d]

State Grid Tianjin Electric Power Company
Tianjin, China
E-mail: [c]tianchun.xiang@tj.sgcc.com.cn, [d]tjwangxudong@sina.com

Based on direct and indirect measurements, this paper proposes condition evaluation methods and grading standards of energy-storage equipment in photovoltaic system, which is relatively rare in current researches. This paper, first, establishes a basic model of condition evaluation; then, calculates the indirect measurements of energy-storage equipment in photovoltaic system. Considering the probable errors resulting from uncertain parameters of battery models when using BP neural network algorithm to evaluate lithium battery's State Of Health (SOH), it adopts Fuzzy-Neural network algorithm optimized by taboo-genetic algorithm. Finally, it grades direct and indirect measurements according to the grading standards. After calculating the weight value of criterion layer by analytic hierarchy process, it uses condition evaluation methods to obtain the equipment's current state of health. The practical calculation proves the effectiveness of this method and these evaluation results provide scientific basis for further risk assessments and maintenance decisions.

Keywords: photovoltaic system; condition evaluation; energy storage equipment; state of health of battery; neural network.

1. Introduction

Condition evaluation of energy storage equipment in photovoltaic system is the process of recording the running condition of equipment and its other performance, classifying the recorded information and conducting corresponding evaluations, with the aim of providing guidance on the risk assessment and condition maintenance of equipment. Although the condition evaluation of electrical equipment has gained much attention during these years[1], unfortunately the condition evaluation of energy storage in photovoltaic system is far from what's needed in practice and relevant theoretical research is rare. Besides, many flaws exist in evaluation technology, in particular, we lack practical theories and methods to guide the condition evaluation work of energy storage equipment in photovoltaic system.

Nowadays, China's condition evaluations of energy storage equipment in photovoltaic system are mainly focused on the analyses and researches of each individual equipment[2] . No photovoltaic condition evaluation system is refined to equipment components and condition parameter and no standard guideline for the condition evaluations of energy storage equipment in photovoltaic system is issued in China. In order to fully and objectively understand the whole running condition of energy storage equipment in photovoltaic system, this paper, based on equipment components and condition parameter, referring to The Guideline for the Equipment Condition Evaluation in Distribution Network, puts forward a condition evaluations system for energy storage equipment in photovoltaic system. In this paper, condition parameters are divided into direct and indirect measurements. Indirect measurements don't refer to those information which can be directly obtained through observation and measurement, but parameters which you have to analyze and calculate based on the direct measurements, such as the SOC (State of Charge) and SOH (State of Health) of energy storage battery.

Many domestic and foreign scholars have conducted researches on the evaluation of SOH and proposed corresponding estimation method[3],[4]. The most popular four are definition method, electrochemical impedance method, rent method and model method. These methods are not recommended for photovoltaic power generation system due to their respective defects[5].

In terms of battery's SOH estimation, Literature 6 estimates battery's SOC through measuring the internal resistance of battery. However, this method requires an internal resistance measurement device which is expensive, huge and hard to popularize. Literature 7 uses Kalman Filter Algorithm to estimate battery's SOC. But it depends too much on the initial value and the accuracy of models. Therefore it's also unsuitable for photovoltaic power generation system.

Neural network, on the contrary, avoids these flaws. It doesn't need perfectly right formula expression between input and output data[8]. It is self-adaptive and less demanding on the accuracy of initial values. It has a variety of learning algorithms and its technology is getting to its maturity. Considering the above merits, the author of this paper believed that neutral network is more suitable for the estimation of SOC and SOH and other parameters in photovoltaic power generation system.

This paper employs Fuzzy- Neural network algorithm (which is based on taboo-genetic algorithm) to estimate battery's SOH and SOC. Most previous researches used different methods in the study of SOC and SOH. It's quite rare to use the same method to estimate both SOC and SOH. However, it is quite necessary to develop a general algorithm for the estimation of SOC and SOH. In terms of algorithm, it is more generally applied and innovative. In terms of engineering application, it's more convenient and efficient for the staff.

2. Establishment of Basic Model of Condition Evaluation

Condition evaluation of energy storage equipment in photovoltaic system includes establishing basic model of condition evaluating, selecting condition evaluation parameters of equipment and determining the weight values of each component of equipment. Selecting condition evaluation parameters of equipment is to seek parameters which can best reflect the condition characteristics of equipment after considering their practical operation experience. Then use these selected parameters to grasp the comprehensive running condition of energy storage equipment in photovoltaic system. After that, choose a proper theoretical method to establish a basic model of condition evaluation of energy storage equipment in photovoltaic system. Since the different components of equipment have different definitions of equipment accidents and their ability levels of reflecting the health condition are varied, this paper applies certain methods to quantify the influence levels and determines the influence weight value of each component of equipment.

Condition evaluation in this paper is mainly focused on the real-time running sate of equipment. Although few condition evaluation theories on energy storage equipment in photovoltaic system can be consulted, condition evaluation methods of equipment in distribution network are numerous. Since equipment's condition is hard to be quantified, most previous researches employed qualitative evaluation method. But it's quite necessary to develop an easy-operated, accurately quantitative method and model of condition evaluation. Under the principle of combining both systematicness and stratification and with the consultancy of many scholars and many relevant research literatures, this paper establishes a basic model of condition evaluation parameters of energy storage equipment in photovoltaic system. As shown in figure 1.

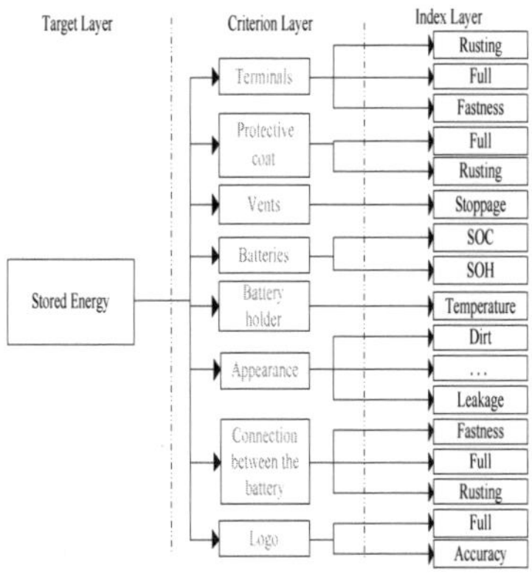

Fig. 1 Model Chart of Energy Storage Equipment's State Parameter in PV System

No excessive explanation of direct measurements (which can be directly measured by inspectors) is made here. On the contrary, indirect measurements cannot be obtained without using some algorithms. Indirect measurements of energy storage equipment can be divided into SOH and SOC. In the estimation of SOH, battery internal resistance, discharge current and ambient temperature will be input to the algorithm. In the estimation of SOC, battery voltage, discharge current and ambient temperature will be input to the algorithm.

2.1. *Fuzzy-Neural Network Algorithm*

Sub-headings should be typeset in boldface italic and capitalize the first letter of the first word only. Section number to be in boldface roman.

2.1.1. *Sub-subheadings*

Typeset sub-subheadings in medium face italic and capitalize the first letter of the first word only. Section numbers to be in roman.

3. Estimation Methods of SOH and SOC of Energy Storage Battery

3.1. *Fuzzy-Neural Network Algorithm*

This paper uses Fuzzy-Neural network algorithm since it's less demanding on samples[9], able to better make use of expert knowledge base, self-learning and highly adaptive. This algorithm is more suitable for working out the results of indirect measurements of energy storage equipment. Adaptive genetic taboo algorithm is also used to instruct rule parameters and weight values of equipment components, in this way it effectively reduces its dependence on samples. The structure is shown in figure 2.

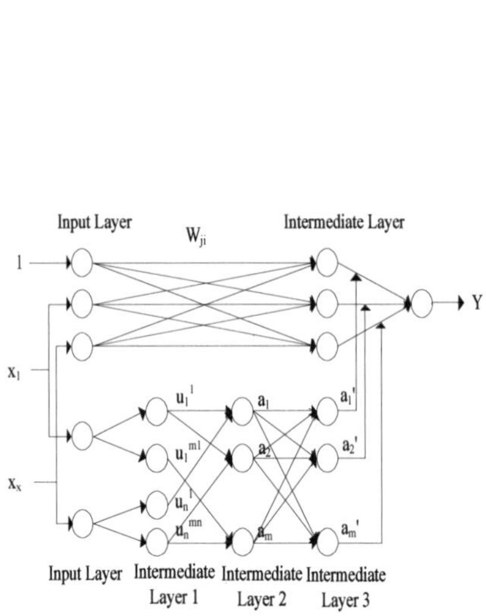

Fig. 2 A fuzzy neural network algorithm structure

Fuzzy-Neural network algorithm structure is composed of two parts: a five-layer premise network and a three-layer consequent network. The fuzzy rule is showed in formula (1) as follows:

$$\begin{aligned} R_j \ : \ \text{IF} \quad & z_1 \ \text{is} \ A_{j,1} \ \text{AND} \ z_2 \ \text{is} \ A_{j,2} \ \text{AND} \ \cdots \ \text{AND} \ z_{nx} \ \text{is} \ A_{j,nx} \\ \text{THEN} \quad & y_j(k) = \omega_{j,0} + \omega_{j,1}x_1 + \cdots + \omega_{j,nx}x_{nx} \end{aligned} \tag{1}$$

The first layer of premise network is input layer and its node number (namely the number of input variables) is nx, as shown in formula (2).

$$O_i^{(1)} = I_i^{(1)} = x_i \quad i = 1, 2, \cdots, nx \tag{2}$$

The second layer is the membership function used to calculate different input components $M_{j,i}(x_i)$. It's fuzzy layer. The fuzzy segmentation of each input component is k_1. The total node number is $nx \cdot k_1$, as shown in formula (3).

$$O_{i,j}^{(2)} = \exp(I_{i,j}^{(2)}) = \exp(M_{j,i}(O_i^{(1)})) = \exp(-\frac{(O_i^{(1)} - c_{i,j})^2}{\sigma_{i,j}^2}) \quad i = 1, \cdots, nx \quad j = 1, \cdots, k_1 \tag{3}$$

The second layer is used to calculate the adaptability. There are $k_2 = k_1^{nx}$ nodes in total, as shown in formula (4).

$$O_m^{(3)} = \beta_m = O_{1,j_1}^{(2)} \Lambda O_{2,j_2}^{(2)} \Lambda \cdots \Lambda O_{nx,j_{nx}}^{(2)} \quad j_1, \cdots, j_{nx} = 1, \cdots, k_1 \quad m = 1, \cdots, k_2 \tag{4}$$

The forth layer is used to conduct unitary processing, as shown in formula (5).

$$O_m^{(4)} = \overline{\beta_m} = \frac{O_m^{(3)}}{\sum_{m=1}^{k_2} O_m^{(3)}} \quad m = 1, \cdots, k_2 \tag{5}$$

The fifth layer of premise network is the third layer of consequent network. It is input layer and the calculation formula is formula 7 as follows.

Similar to the first layer of premise network, the first layer of consequent network is also input layer. Its first input item is supposed to be $x_0 = 1$, which is a constant term. The second layer of consequent network is the calculation of fuzzy rules. The node number is k_2, as shown in formula (6).

$$y_m(k) = \omega_{m,0}x_0 + \omega_{m,1}x_1 + \cdots + \omega_{m,nx}x_{nx} \quad m = 1,2,\cdots,k_2 \tag{6}$$

The third layer of consequent network is the infinite approximation to non-linear system, as shown in formula (7).

$$O^{(5)} = I^{(5)} = \sum_{m=1}^{k_2} \overline{\beta_m} y_m(k). \tag{7}$$

3.2. *Adaptive Genetic Taboo Algorithm Optimizing the Fuzzy-Neural Network*

Both the membership center $c_{i,j}$ and the width $\sigma_{i,j}$ of the second layer in premise network are unknown quantities, so is the weight value $\omega_{m,i}$ in consequent network, $i = 1,2,\cdots,nx,$ $j = 1,2,\cdots,k_1,$ $m = 1,\cdots,k_2$. They have to be calculated by algorithms. This paper uses adaptive genetic taboo algorithm.

Adaptive genetic algorithm is easy to be precocious and its local research ability is a little bit weak[10]. Although taboo algorithm is very dependent on the initial samples, its global research ability is strong which helps to get the global optimal solution. Introducing the memory function of taboo algorithm to the crossover operator and mutation operator of adaptive genetic algorithm can effectively solve these flaws mentioned above.

In adaptive genetic taboo algorithm, chromosome coding is based on the codes of membership center $c_{i,j}$ and the width $\sigma_{i,j}$ of premise network and the weight value $\omega_{m,i}$ in consequent network. Its length is the sum of $c_{i,j}$, $\sigma_{i,j}$ and $\omega_{m,i}$.

The decoding formula is as follows:

$$[c_{1,1},\cdots,c_{nx,k_1},\sigma_{1,1},\cdots,\sigma_{nx,k_1},\omega_{1,1},\cdots,\omega_{k_2,nx}] \tag{8}$$

This paper combines taboo algorithm with the crossover operator and mutation operator of adaptive genetic algorithm. It respectively adds taboo algorithm after crossover and mutation. After crossover, put individual fitness away into a taboo table with the length of L. If the individual adaptive value is higher than the average group adaptive value, it will pass into the next generation and update the taboo table. If the individual adaptive value is lower than the average group adaptive value and belongs to the taboo, optimal individual genetics will be chosen to the next generation. If the individual adaptive value is lower than the average group adaptive value but doesn't belong to the taboo, the progeny will be accepted. Adding taboo algorithm after mutation also uses this method. When the repetition rate of late population mutation in the algorithm is higher than 50%, the second forceful mutation will be conducted, so as to improve the calculation precision of the algorithm.

4. Unit Evaluation Method

Consulting The Guideline for the Equipment Condition Evaluation in Distribution Network and other research papers[12,13], this paper divides the running condition of equipment into four categories: normal condition, attentive condition, abnormal condition and severe condition. Normal condition means that equipment's running data is stable and all the condition parameters meet the required standards. Attentive condition means that several condition parameters of equipment are not up to the required standards but have little impact on the equipment's operation. Abnormal condition means that several condition parameters have obvious abnormity and have affected the equipment's performance indicator or even worsened to severe condition, although they can still work. Severe condition means that the equipment's condition parameters have exceeded the required standards or shown severe abnormality to an extent that the equipment can only operate for a very short time or immediately break down.

Energy storage equipment includes terminal, protective coat, vents, batteries, battery holder, appearance, connection between batteries, logo and etc. The condition parameters of each component is shown in table 1 as follows:

Table 1 The state of the energy storage unit components parameters

Component	Criteria
Terminal	Full, Rusting, Fastness
Protective coat	Full, Rusting
Vents	Stoppage
Batteries	SOC, SOH
Battery holder	Temperature
Appearance	Dirt, Rusting, Leakage, Hollow units, Tympanites
Connection between batteries	Fastness, Rusting, Full
Logo	Accuracy, Full

This paper uses quantitative methods to estimate the condition of energy storage equipment. The initial grades of each component is supposed to be 100. Grades can only be reduced not added[11]. The weight values of each component will be calculated by analytic hierarchy process, as shown in table 2. Attached is the grade standard of condition evaluation using direct and indirect measurement. Due to the classified problem, cannot open.

Table 2 Weights of Different Components in Energy Storage Unit

Component	Weight K*i*
Terminal	0.09
Protective coat	0.09
Vents	0.03
Batteries	0.49
Battery holder	0.09
Appearance	0.09
Connection between batteries	0.09
Logo	0.03

The final grade M_i of component *i* is calculated by formula 9.

$$M_i = m_i \times K_F \times K_T \tag{9}$$

$$m_i = 100 - \text{Maximum deduction in the state of the corresponding parts} \tag{10}$$

$$K_T = (100 - \text{Run the number of years} \times 0.5) / 100 \tag{11}$$

K_F is the ratio of group defects. Equipment components with group defects will get 0.95 and those without group defects will get 1. K_T is the life factor.

If components are in the normal state, the arithmetic mean will be their grade. If one or more component is not in the normal state, the final grade should be the lowest grade of that component.

If all the components of energy storage equipment are in the normal state, then the condition of this energy storage equipment is normal and its final grade $= \sum M_i \times K_i$. Ki is the weight value of component *i*. If the grade of one or more component is lower than the grade in the normal state, the condition of this energy storage equipment will be the condition of the worst component and its final grade $= \min M_i$.

The relationship between evaluation grade and its condition are as follows:

Table 3 Relation between Energy Storage Components and Evaluation Grade and State

Component	85-100	75-85(Contain))	60-75(Contain)	Under 60(Contain))
Terminal	Normal condition	Attentive condition	Abnormal condition	serious condition
Protective coat	Normal condition	Attentive condition	Abnormal condition	serious condition
Vents	Normal condition	Attentive condition	Abnormal condition	serious condition
Batteries	Normal condition	Attentive condition	Abnormal condition	serious condition
Battery holder	Normal condition	Attentive condition	Abnormal condition	serious condition
Appearance	Normal condition	Attentive condition	Abnormal condition	serious condition
Connection between batteries	Normal condition	Attentive condition	Abnormal condition	serious condition
Logo	Normal condition	Attentive condition	Abnormal condition	serious condition

5. Numerical Example

5.1. *SOH Estimation*

This paper conducts charging and discharging experiments on batteries under setting conditions. The experiment objects are a certain type of 50Ah lithium iron power battery. Constant temperature box is used to adjust the ambient temperature of batteries at 20°C,30 °C and 40 °C and programmable charging and discharging device is employed to conduct continuous charging and discharging experiments with a ratio of 0.5 C, 1C and 1.5C. Monitoring software is used to collect the real-time voltage and current of batteries. The battery's internal resistance will be calculated by recursive least squares method. This paper takes the ambient temperature, battery internal resistance and discharge current as the input of fuzzy neutral network algorithm and takes SOH as the output. A total of 300 effective groups of data are collected and Matlab is used to establish algorithm models. Use 250 groups of data to practice the Fuzzy-Neural network algorithm and then use the remaining 50 groups of data to prove the validity of the algorithm.

The selection of fuzzy number in Fuzzy-Neural network algorithm directly influences the time and precision of the algorithm. Through numerous experiments, the fuzzy number of this paper is eventually determined 7.

This paper adopts adaptive genetic taboo algorithm to practice the network. The biggest evolution algebra is 1000. The contrast curves of SOH's actual value and predicted value are as shown in figure 3, from which we can obviously notice that this paper's algorithm is much better than BP neural network algorithm. Figure 4 is the contrast chart of SOH's prediction errors, from which we can see that this paper's algorithm has a prediction errors within 1% while the BP neural network algorithm's SOH prediction error can be up to a maximum of 4.3%.

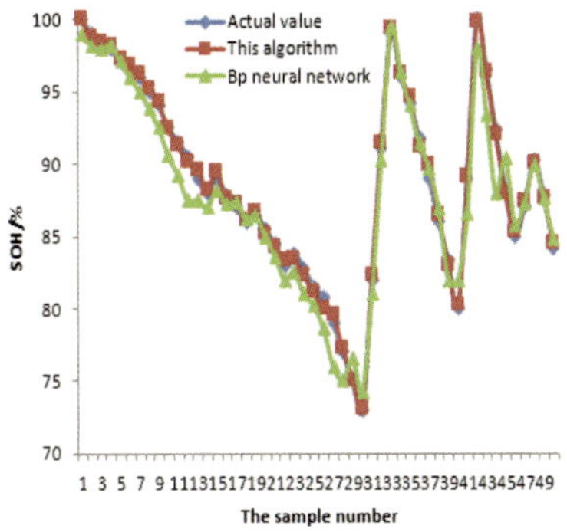

Fig. 3 Contrast Curves of SOH's Actual Value and Predicted Value

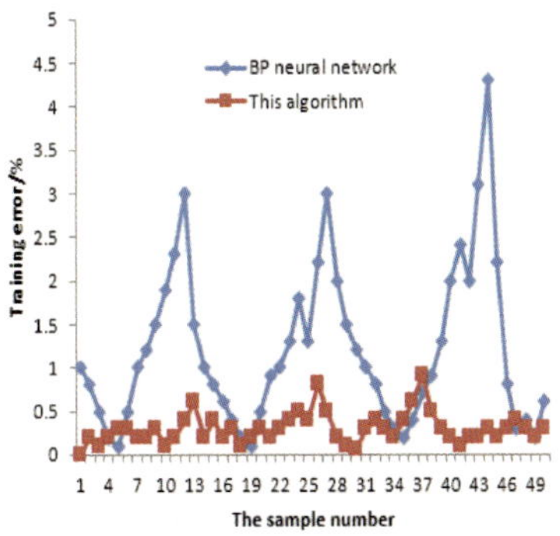

Fig. 4 Contrast Chart of SOH's Prediction Error

Comparing the above two algorithms, this paper's algorithm has obvious advantages. The structure of BP neural network algorithm is unstable and local optimal solution often pops up, as a result its learning effect is unsatisfying. Compared with BP neural network algorithm, this paper's algorithm is more accurate and more stable and able to be used for the estimation of SOC.

5.2. *SOC Estimation*

Use the same method in 4.1 to get 300 groups of effective data and take discharge current, battery voltage and ambient temperature as the input of Fuzzy-Neural network algorithm and take SOC value as the output.

The contrast curve of SOC's actual value and predicted value is shown in figure 5 as follows and SOC's prediction error curve is shown in figure 6.

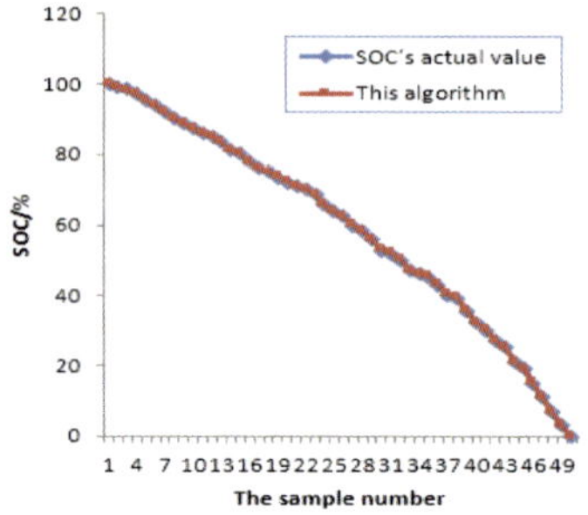

Fig. 5 Contrast Curves of SOC's Actual Value and Predicted Value

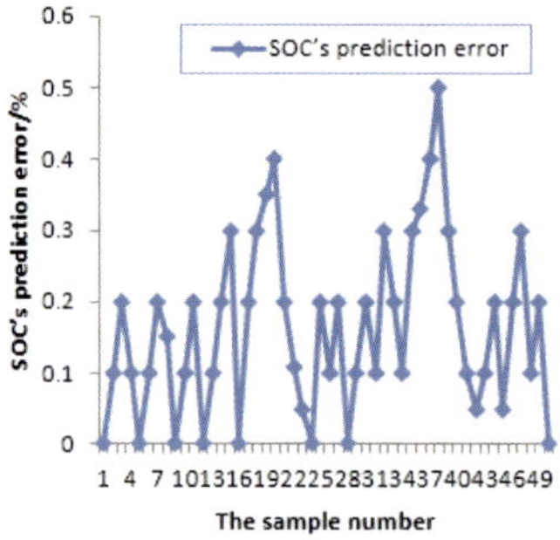

Fig. 6 SOC's Prediction Error Curve

5.3. *Condition Evaluation*

Based on a region inspector's scoring of energy storage equipment in filed photovoltaic system, this paper calculates the final score of different components, calculate equipment's health index and its score of the current condition, as shown in table 4.

Table 4 State Evaluation Results of Energy Storage Units

Component	final score	Equipment health index	Current condition
Terminal	89.55		
Protective coat	89.55		
Vents	99.5		
Batteries	89.55	92.8335	Normal condition
Battery holder	99.5		
Appearance	99.5		
Connection between batteries	99.5		
Logo	99.5		

The equipment's health index and its current condition could provide scientific basis for the next- step risk assessment and maintenance decision-making.

6. Conclusion

Nowadays, the permeability of photovoltaic system is higher and higher in China. But there is no standard estimation theory and method for the condition evaluation of energy storage equipment in photovoltaic system. This paper, based on direct and indirect measurements, provides condition evaluation methods and grading standards for the energy storage equipment in photovoltaic system. This paper, first, establishes a basic model of condition evaluation of energy storage equipment in photovoltaic system. In order to calculate indirect measurement of key equipment in photovoltaic system, this paper uses adaptive genetic taboo algorithm to optimize Fuzzy-Neural network algorithm to estimate batteries' SOH and SOC. Second, the grading standard of direct and indirect measurement is put forward in the appendix. Third, it puts forward the condition evaluation method of energy storage equipment in photovoltaic system and works out equipment's current health index and state, whose results could provide theoretical and data support for the subsequent risk assessment and condition maintenance. Finally, this paper uses practical experiments to verify the versatility and accuracy of Fuzzy-Neural network algorithm optimized by adaptive genetic taboo algorithm in SOH and SOC estimation and a region's application also proves the practicability of the condition estimation method of energy storage equipment in photovoltaic system. Because of my limited ability, condition evaluation methods and grading principles still have much room to improve.

Acknowledgment

The National High Technology Research and Development of China 863 Program (2014AA052003).

References

1. Hao Yuan, Gang Qu. Discussion on the state monitoring content of the two equipment of power grid [J]. Automation of Electric Power Systems. 2014(12).
2. Longxiang Peng. Research on control strategy optimization of grid connected photovoltaic grid connected inverter [D]. North China Electric Power University.2011.
3. WEI Kexin, CHEN Qiaoyan. States Estimation of Li-ion Power Batteries Based on Adaptive Unscented Kalman Filters [J]. Proceedings of the CSEE. 2014(03.)

4. Haifeng D, Xuezhe W, Zechang S. A new SOH prediction concept for the power lithium-ion battery used on HEVs[C]. Vehicle Power and Propulsion Conference, 2009: 1649-1653.
5. Yanqiong Kang. Research on Estimation for SOH of PEV Li-ion Battery Pack[D]. Beijing Jiaotong University.2015.
6. Taesic Kim, Wei Qiao. Online state of charge and electrical impedance evaluation for Multicell lithium-ion batteries(C). IEEE Transportation Electrification Conference and Expo (ITEC),2013:1-6.
7. Novie Ayub Windarko, Jaeho Choi. Li-pb battery SOC evaluation using extended kalman filter improved with var-iation of single dominant. Parameter Journal of Power Elec-tronics, 2012,12(1):40-48.
8. Xiuliang Li. Research on soft sensor modeling method and its application [D]. Zhejiang University,2009:1-12.
9. Li Shi, Tiejun Chen. Intelligent Control Theory and Applications [M]. Tsinghua University Press, 2009.
10. Saiyi Wang, Chengshan Wang. Genetic tabu hybrid algorithm and its application in power network planning [J]. Automation of Electric Power Systems. 2004(20).
11. Xiaobin Yang, Heming Li. Energy efficiency index system of distribution network based on Analytic Hierarchy Process [J]. Automation of Electric Power Systems. 2013(21).
12. Dongliang Lu. Research on quality management of power distribution system [D]. North China Electric Power University.2011.
13. Yan Tang. Research on condition based maintenance decision of distribution network equipment [D]. North China Electric Power University. 2014.

Research on Laser Eavesdropping of Interference Method Based on Linear CCD

Jian An*, Jia-Hao Song and Zhi-Yong Dai

School of Optoelectronic information, University of Electronic Science and Technology of China, Chengdu, 610054, China

**E-mail: ajuestc@163.com*

For improving the technique of laser monitoring and performance of laser monitoring system, a research on a new transceiver laser monitoring system based on Michelson Interferometer is conducted. Firstly, the connection between sound waves and interference pattern is introduced. Then, 650nm DFB laser and 1550nm DFB laser are used in both to realize the system. Two lasers are used for optical path adjustment and real monitoring respectively. Using light splitter to divide one laser to two lasers for differential processing, linear CCDs to collect interference pattern, and sound restore algorithm, the sound waves are finally recovered. The whole research proved the feasibility of the system, which can recover sound waves from vibration of glass, by MATLAB software simulation and real optical structure.

Keywords: Laser Monitoring; Michelson Interferometer; Differential Processing; Linear CCD; Sound Waves.

1. Introduction

In the military filed and in public security, monitoring is a matter of Great significant in obtaining information. With the development of laser technology and optoelectronic measuring technology, laser monitoring came into researchers' sight and went to practical application.

In the late 20th century, researchers did some related study [1-2] and had some achievement but the distance of monitoring is short, and there were not subsequent reports about it. In 2003 in China, Hai-jun Luo published a paper about laser monitoring [3]. He described the basic principle of laser reflecting technique in laser monitoring, concluded the characteristic of this technique and proposed the implementation. In 2008, Chao-fan Zhang published a paper about improvement and implementation of laser monitoring [4], proposed that use "cat's eye effect" and laser interferometry to improve the method of laser monitoring. But this method must need the "cat's eye effect" equipment near the target to realize monitoring, which had big limitation. After that, there were several papers published successively [5-7], which have improvement on optoelectronic devices and signal collection circuit, but they still can't address the underlying issues of laser monitoring.

This paper proposed to use Linearity CCD as optoelectronic measuring device, which differ from previous methods in hardware and algorithms and haven't been reported before. This paper use Linearity CCD collecting the interference pattern and finding the pixel at static work point, then use the changing voltage of this pixel and differential amplifying algorithm recover the vibration of glass, namely the sound waves.

2. Principle of the system

Acoustic waves can cause the vibration of an object, but for windows, vibration amplitude is very small. So the reflective laser monitoring [8], which uses the tiny displacement of the reflection laser caused by the vibration of glass modulating the laser intensity on photoelectric detector, is almost impossible. But the interferometric laser monitoring, whose principle is different from the reflection-type, can demodulate the microvibration of glass. In our system, we use Michelson interferometer to realize monitoring, as shown in Figure 1:

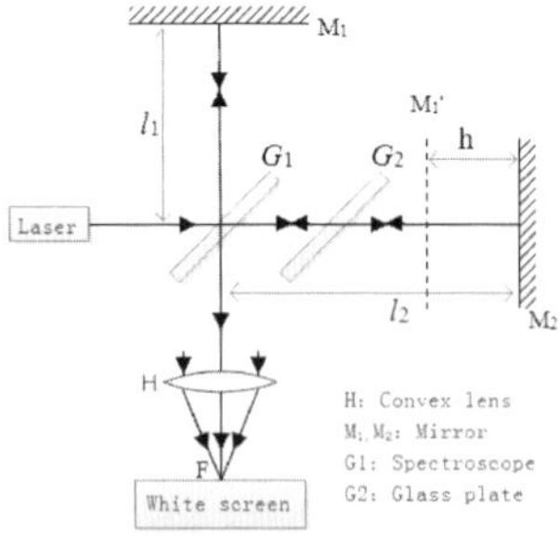

Fig. 1. Michelson interferometer

According the principle of Michelson interferometer, the distribution of intensity on the focal plane F is:

$$I = I_1 + I_2 + 2\sqrt{I_1 + I_2}\cos\frac{2\pi\Delta}{\lambda}. \tag{1}$$

I_1 and I_2 are the intensity of two reflected light, λ is the wavelength of laser. Δ is the optical path difference of two reflected light and $\Delta = 2nh\cos\theta$ .n is refractive index, h is the distance between M_2 and M_1, θ is the laser divergence angle relative to the optical axis. When the value of h is fixed, laser of different divergence angles form interference rings.

When optical path difference $\Delta = m\lambda$, m=0, 1, 2 … namely:

$$2nh\cos\theta + \frac{\lambda}{2} = m\lambda. \tag{2}$$

The interference pattern is bright fringe. When optical path difference $\Delta = (m+1/2)\lambda$, m=0, 1, 2…, the interference pattern are dark fringes. Light fringes and dark fringes form the interference rings. When the M_1 and M_2 are strict vertical, by the formula (2) we can know that if M_2 move forward (h is smaller), circle stripes contract to the center, and disappear one by one in the center. There is one fringe disappear in the center each $\lambda/2$. If the M_2 move backward (h is bigger), the fringes continuously appear in the centre, and the closer to the center of interference pattern , the bigger shift distance of fringes is [9].

3. The design simulation of laser eavesdropping

Due to the system as a whole is complex, this paper begins with theoretical analyzes. We build an interference model on MATLAB to validate the feasibility of the system and test the system's main parameters.

3.1. *Build an interference model*

Set the focal length of convex lens in Figure 1 to f, the distance between the focal point, which is the point that parallel light focuses in focal plane of the lens, and optic axis is R. By image-forming formula of infinite objective of optical system, we can know:

$$R = f \tan\theta. \tag{3}$$

Take the formula (3) into formula (1), we can get:

$$I = I_1 + I_2 - 2\sqrt{I_1 + I_2}\cos\frac{4\pi nh\cos(\arctan(R/f))}{\lambda}. \tag{4}$$

We assume that λ=650*nm*, I1=I2=1, n=1, and set optical path difference *nh*=10m as the distance from the monitoring object. Take above four equations into the formula (4), we can draw the R-I plot, which represent the laser intensity distribution on Linearity CCD, as Figure 2(a) shows. Based on it, we expand the one-dimensional laser intensity distribution to two-dimensional laser intensity distribution, which is the interference pattern on focal plane of convex lens, as Figure 2(b) shows.

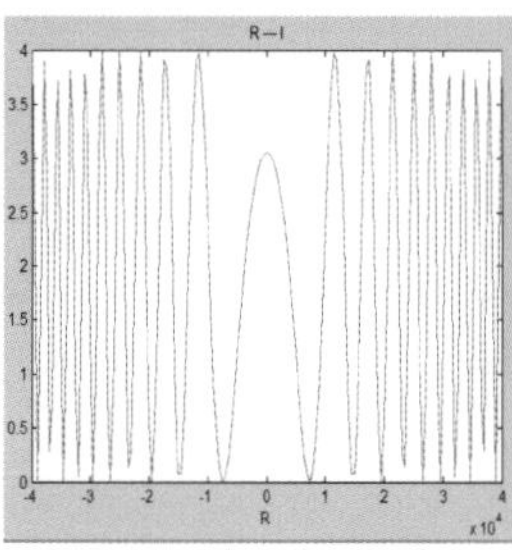

Fig. 2(a) Intensity distribution

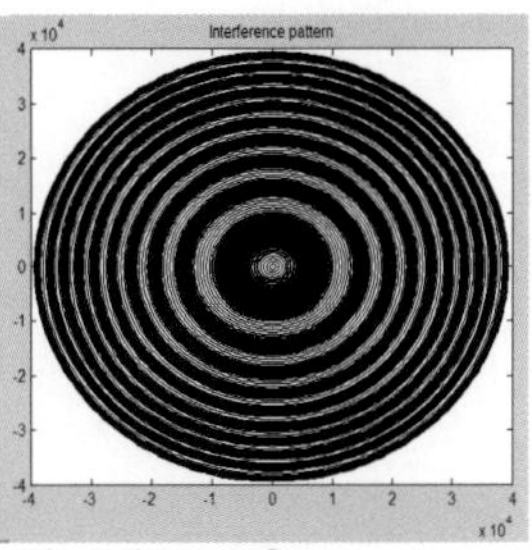

Fig. 2(b) Interference pattern

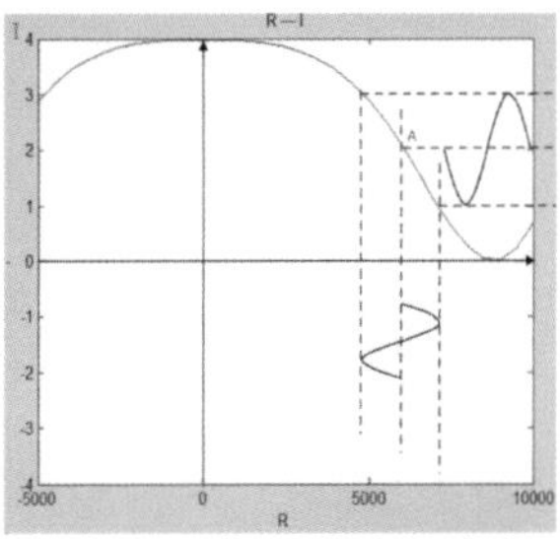

Fig. 3. Laser intensity's variation as the movement of fringes

3.1. ***Sound restore algorithm***

3.1.1. *Principle*

As shown in Figure 3, point A is in the middle of the intensity maximum and the intensity minimum, second derivative of point A approximately equal to zero, and the R-I function achieve best linear degree in point A. So the system's distortion is minimized, and we call point A static working point.

Due to the laser intensity of interference pattern is linear with the analog voltage that linear CCD collecting, the voltage of static working point corresponds to the laser intensity of this point, namely the vibration of glass.

3.1.2. *Realizing*

Since the optical path difference and the laser intensity in center is unknown at first, for the convenient of finding static working point, we use the main controller to produce pulse signal, thereby control piezoelectric ceramic aim to adjust the reference arm.

By this way, we can realize changing the optical path difference, make the center voltage V_a of the interference pattern that linear CCD collecting reach maximum, which is the initial state of system, as shown in Figure 4. The linear CCD will record center bright ring voltage value V_a, and find the point that have lowest voltage value V_b nearest to center. Calculate V_s via formula (5).

$$V_s = \frac{(V_a + V_b)}{2}. \qquad (5)$$

Vs is the point that voltage value is between V_a and V_b. Find the pixel, whose voltage is equal to V_s, and This pixel is the static working point. Continuously sampling the voltage of static working and using digital filter, sound waves will be recovered.

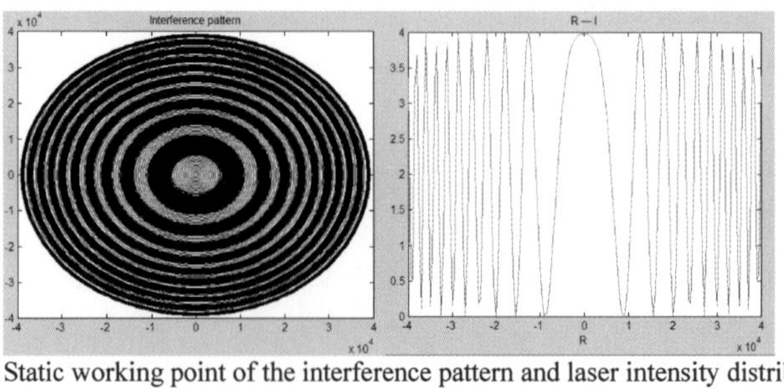

Fig. 3. Static working point of the interference pattern and laser intensity distribution

3.2. *Analysis and improvement of system*

3.2.1. *Single-frequency sinusoidal signal simulation and restore*

Through the actual experiment, we conclude that vibration of glass caused by sound waves cause is very little, the amplitude is only nanometer level. Suppose the vibration glass is the sum of input signal and input noise： input signal=50sin($w_o t$), w_o=2π×800*hz*, input noise =20sin($w_n t$), w_n=2π×50*Hz*, as shown in Figure 4(a). Put this vibration into sound restore algorithm, we can get the restored signal, as Figure 4(b) shows. Do FFT analyze to the restored signal, as Figure 4(c) shows, we can see the Total Harmonic Distortion (THD) of restored signal is very small, which approve the feasibility of our sound restore algorithm.

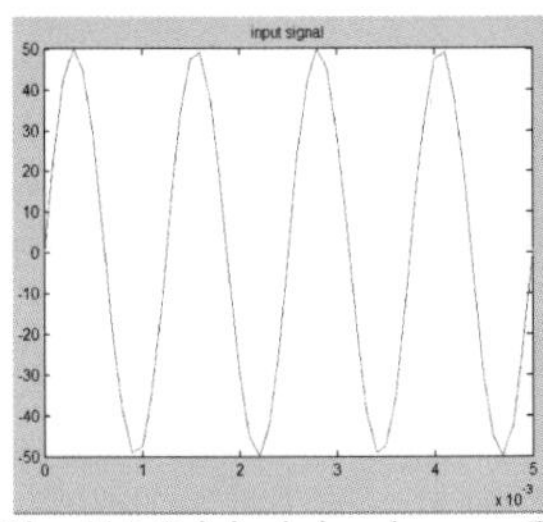

Fig. 4(a) Original signals

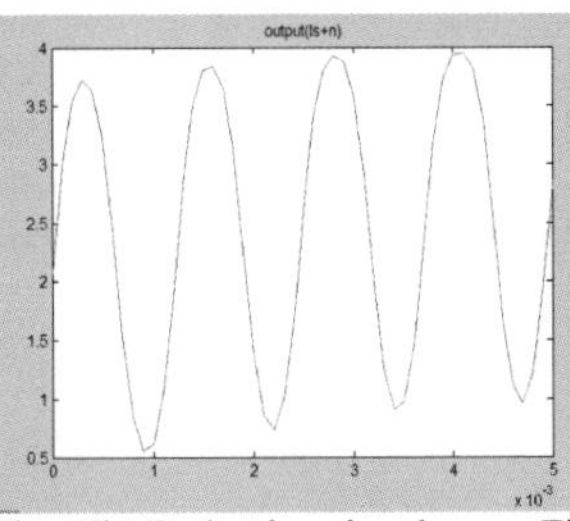
Fig. 4(b) Reductive signal

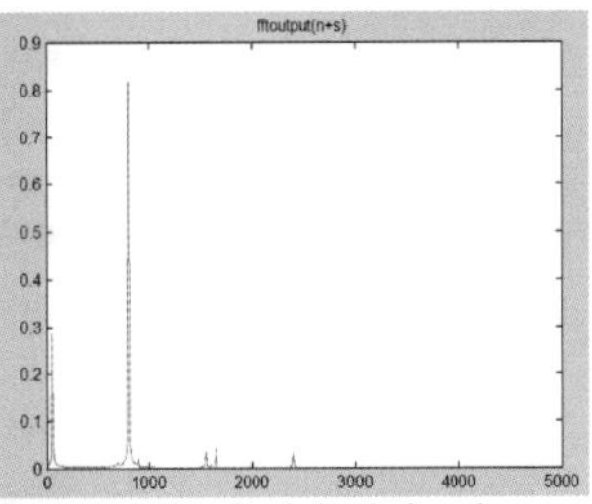
Fig. 4(c) FFT of original signal

3.2.2. *Sound waves signal simulation and restore*

Next, we use audio signal to test the system's performance. We first use wavread function reading a section of audio signal store in computer previously, as Figure 6(a) shows. Considering there is noise in reality, rand function is applied to generate a noise signal. We add the audio signal and noise signal together as the vibration of glass, put this vibration into sound restore algorithm, and get the recovered signal, shown in Figure 5(b). As shown in Figure 5(b), although the sound waves are recovered, there are much noise combined with signal. To solve this problem, we do some improvements: Due to attenuation of vibration that caused by sound waves is obvious, but the noise is in the whole glass [10], so we use two laser systems. Supposed that amplitude of signal that caused by laser 1 is 1, amplitude of signal that caused by laser 2 is 0.5. Subtract signal that recovered by CCD1 from signal that recovered by CCD2, then use 200-3000Hz bandpass filter, we obtain the recovered signal, shown in Figure 6(c). Play this signal by sound function, the sound is good.

This system is proved feasible by simulation, and can be tested by real experiment.

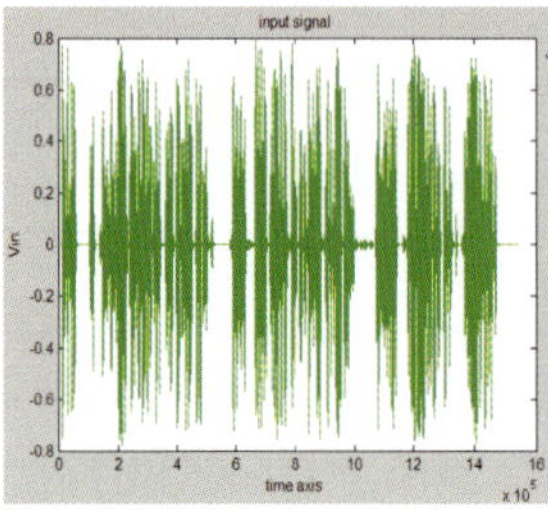

Fig. 6(a). Original audio signal

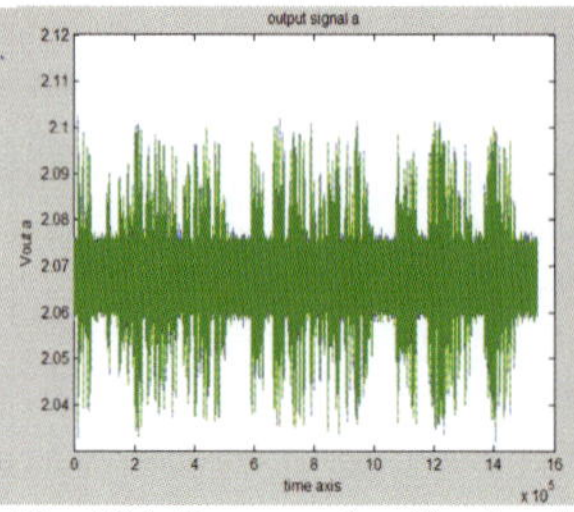

Fig. 6(b). Recovered audio signal

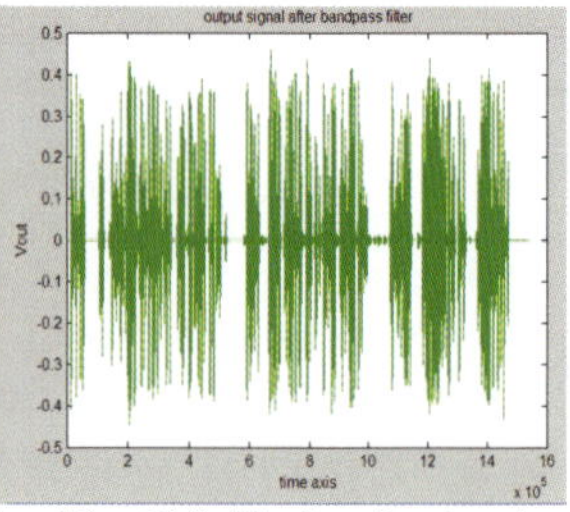

Fig. 6(c). Final audio signal

4. The Design of Laser Monitoring Experiment

4.1. *The Establishment of Light Structure*

The light structure established on the optical platform is illustrated in the following Figure 7. In the structure diagram, the glass is monitoring object. A laser beam, which comes from the laser, is divided into two exactly same laser beams by the light splitter. Then the beam expander expands two laser beams. The two lasers go through the beam splitter, which is inclined to optical axis at angle of 45, so the two lasers are divided into four lasers: two transmission (reference) lasers will reflect back on the original way after touching the glass, whose reflectivity is 0.04, and then they will be reflected by the beam splitter. Finally, the convex lens will focus these laser beams on the two linear CCDs; two reflection (signal) lasers is vertically incident on the monitored glass after and then reflect back on the original way. These two laser beams focused on the two former linear CCDs too. These two laser beams and the former two are focused on the same place together to produce interference. In the actual optical structure, 650*nm* and 1550*nm* DFB lasers are used to produce laser. The 650*nm* one is used to adjust optical path, and the 1550*nm* one is for actual monitoring. What's more, these two laser heads are installed on the precise motor interlinkage system for convenient and fast switching. Piezoelectric ceramic is used to fine tune the optical path difference between the reference laser and the signal laser, polaroid is used to adjust the intensity of the reference laser, to insure the equal intensity of the reference and signal lasers. CCD1 and CCD2 respectively collects the information of two interference pattern, which are used for the differential amplifying proceeding.

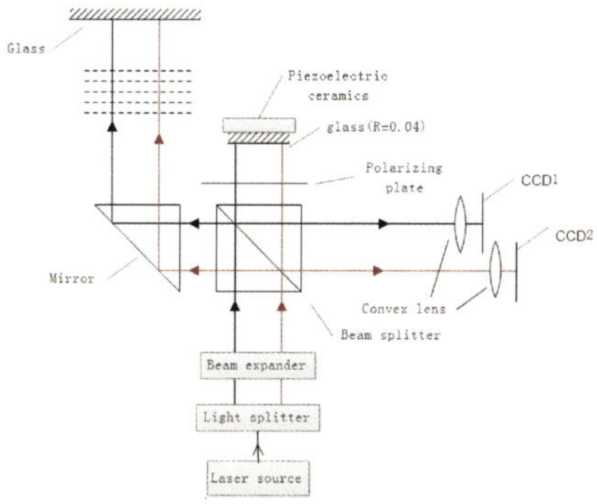

Fig. 7. Schematic diagram of optical structure

4.2. *Result Analysis*

As shown in Figure 8, there are two interference patterns which is produced by 650*nm* laser. Due to the lack of linear CCD that has high precision and high response speed, Sound restore experiment has not been done. But through the observation of the interference fringes, we verified that the vibrations, which are caused by sound waves, are small and the center interference stripe order changes within one level. The observation proved that the vibration amplitude of glass caused by sound wave is smaller than one λ. Meanwhile, we found that if the optical structure was placed on the table, the interference fringes kept moving, which has a negative effect on experiment. There, we placed the optical structure in the optical platform, so that interference fringes were stabilized and better experiment effect were achieved.

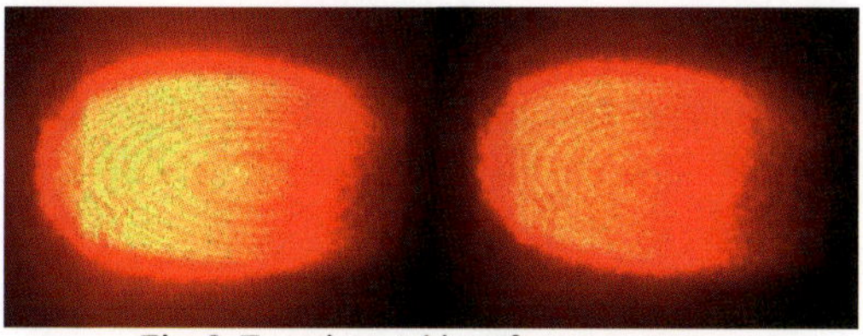
Fig. 8. Experimental interference pattern

5. Conclusions

Based on the basic principle of the Michelson interferometer, we obtained the connection between the sound waves and interference fringes. We verified a new laser monitoring system based on linear CCD mainly through simulation that tested the system respond of single-frequency sine signal and audio signal, and good results have been achieved. Besides, we also constructed the actual optical structure, completed some parts of experiment and verified the correctness of some conclusions. The next step we will buy suitable linear CCD as soon as possible, set up the experimental platform and conduct further study and improvement on the monitoring system.

Acknowledgment

Zhi-Yong Dai provides all the funds in the project.

References

1. Mims, Forrest M. Beware of laser eavesdropping [J]. American Journal of Physics, 1987, 55(10): 871-872.
2. Joel Solomon and Robert Prigo. Eavesdropping with a laser [J]. American Journal of Physics, 1987, 55(4): 381.
3. Luo Hai-jun and Zhu Xiao. Reach on Laser Eavesdropping Technique [J]. Laser & Optoelectronics Progress. 2003, 40(12): 53-56.
4. Zhang Chao-fan. The Improvement and Realization of Laser Eavesdropping [J]. Laser & Infrared. 2008,38(2)： 145-148.
5. Ning Qing-song, Ma Xiao-long and Yang Zhen. Experimental research on laser eavesdropping [J]. Physics Experimentation. 2009,29(12)： 38-41.
6. Qiao Jun-qing, Zhao Yan and Wan Chuan-xu. Detecting sound information from long distance by laser interference [J]. Physics Experimentation. 2013,33(5)： 5-8.
7. Yang Yi and Lv Xiao-gang. Design of voice recovery circuit system based on PSD sensor [J]. Application of Integrated Circuits. 2014， 40(1):50-52.
8. Tang Li-hong, Lai Yong-kuan and Lu Jie-fa. Detecting ting vibration using laser reflection method [J]. Physics Experimentation. 2013， 33(9):6-8.
9. Ye Yu-tang, Xiao Jun, Rao Jian-zhen. The second edition of Optics Course [M]. Beijing: Tsinghua University Press. 2011.9
10. Huang Jing, He Ya-ling, Li Ru. Numerical simulation of sound wave attenuation[J]. Journal of Engineering Thermophysics. 2008, 29(2):202-204.

Quantum Chemistry Calculation of the SOF_4 Product Formation Mechanism in SF_6 Discharge Gas

Zhen Li†, Shao-yi Chen, Shang-kun Gong, Bing Feng and Zhou Zhou

Hunan Electric Power Corporation Research Institute,

Changsha, 410007, China

†*E-mail: lizhenluke@163.com*

Due to its excellent insulating and arc-suppression properties, sulfur hexafluoride (SF_6) is widely used in gas-insulated switchgear. Under high voltage discharge condition, SF_6 decomposes into various by-products according to discharge scenarios. As a product of greater occurrence probability, the formation mechanism of SOF_4 is so far unclear. In this paper, high level quantum chemical method is used to research the mechanism of decomposition reaciton and products, in the presence of trace air or H_2O. For each reaction product, a detailed generatation pathway is provided, including the data and energy structure of reaction intermediates and transition states. The simulation results explain the experimental results accurately, and provide new understanding and support for the judgment of the discharge reaction state and subsequent electrical aging. These results are very helpful to build $SF_6/H_2O/O_2$ reaction model, and demonstrate the electrical aging process from microcosmic angle.

Keywords: Sulfur Hexafluoride; Thionyl tetrafluoride; Quantum Calculation; Decomposition Mechanism.

1. Introduction

SF_6 gas can significantly reduce the equipment size and has been widely used in gas-insulated equipment, such as Gas-Insulated Switchgear (GIS), Gas-Insulated Transformer (GIT), and Gas-Insulated circuit Breaker (GIB)[1][2]. Under electric arc, spark, corona or overheating conditions, SF_6 molecule may decompose into fragments of SF_x. In the presence of trace air or H_2O, the fragments of SF_x will react with O_2, H_2O, or some other fragments (H, O, HO, and HO_2), and then produce some SFOH molecules or radicals[3]. So it is possible to monitor the content of decomposition components and to forecast the level of discharge, which has great significance to master insulation condition and guide maintenance work of gas-insulated equipment.

Zhang and cooperaters have investigated the emergence probability of SF_6 decomposition components[4]. Statistical results show that SOF_4 are compounds with large probability in partial discharge and arc discharge. But the formation mechanism of SOF_4 is so far not clear. At present, the research on the mechanism of the reaction is much from the experimental measurement[5]. Those research is postulated with great arbitrariness. The decomposition products measurement is only a superficial phenomenon, but the discharge status should be corresponded with microcosmic mechanism.

Quantum chemical calculation by using high precision simulation theory is a very effective method to reveal the microcosmic mechanism of complex chemical reaction, and it can provide all possible change of the structure and the chemical reaction energy[6][7]. Therefore, in this paper, we used high level quantum chemical method to study the SOF_4 generation mechanism from SF_6. We provide a detailed generatation pathway, including the data and energy structure of reaction intermediates and transition states. The simulation results can explain the experimental results accurately, and provide new understanding and support for the judgment of the discharge reaction state and subsequent electrical aging.

2. Methodology

In general, for the molecular system of sulfur, advanced calculation method is usually required to meet the accuracy of the experimental results[8]. However, in this paper, the molecular contains multiple atoms,which will take a long time by using advanced calculation method. Therefore, the key of this study is finding a simple and accurate method.

Many reports have found that DFT-B3P86 method is very effective and accurate for the simulation of sulfur containing molecules[9][10]. Its accuracy can be similar to MP2 theory level, but the calculation efficiency is much higher. The calculation of sulfur atom usually need a large basis set, just like 6-311+G(3df). But at B3P86 level, satisfactory results can be obtained only by the basis set of DZP quality[11]. Therefore, in order to get systematic calculation results for compare, this paper uses B3P86/6-31G** method to simulate all the possible chemical reactions.

3. Results and Discussion

3.1. *Calculation accuracy verification*

In this paper, the theoretical level is DFT-B3P86/6-31G**, which should be considered in the medium level from the theoretical point of view. Although the effectiveness of the DFT method have been confirmed by many literature work [12], but for the complex reaction system of sulfur studied in this paper, the data calculated in B3P86/6-31G** level should be validated, to ensure its reliability.

Single molecule decomposition of SF_6 is the initial step of the reactions. Under high voltage discharge or high temperature condition, SF_6 can be decomposed into SF_5, SF_4, SF_3, SF_2, SF, S, F fragments. And these fragments will be the starting point for all subsequent reactions. Therefore, the study of the single molecule dissociation of SF_6 is a crucial step.

The decomposition of SF_6 has carried on the detailed calculation in literature, and its highest level is calculated by RCCSD (T)/AVQZ. In general, the accuracy of the results are close to the limits of theoretical calculation. In this paper, we firstly calculate the SF_6 decomposition on the level of B3P86/6-31G**. The calculation results will reveal the generation of SF_x (x = 5 - 0) fragments, and on the other hand, through the comparison to the calculation results of the literature, the reliability of the theoretical calculation will be validated. The calculation results are shown in Figure 1.

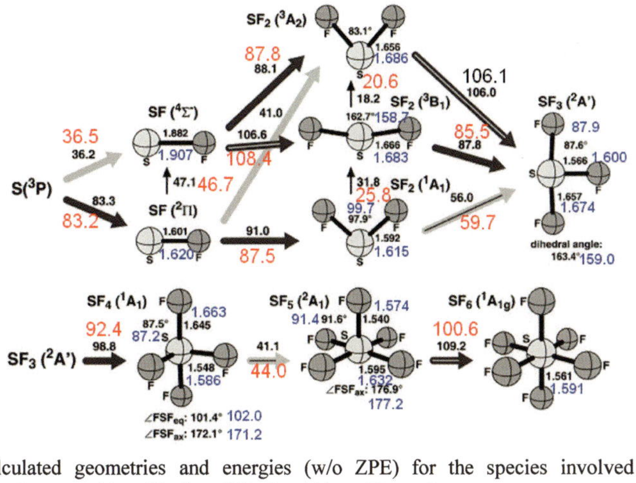

Fig. 1. Calculated geometries and energies (w/o ZPE) for the species involved in the SF_6 unimolecular decomposition. Black: ref. Red and blue: this work.

Figure 1 indicates that the calculation results in this paper are in good agreement with the literature in both the molecular structure and electronic energy[13]. The error of the reaction heat is controlled in the range of 0 - 10 kcal/mol, and the change of the molecular structure parameters are less than 10%. The comparison results fully verify the reliability of B3P86/6-31G** method selected in this paper. Therefore, we will use B3P86/6-31G** method to study all subsequent reactions.

3.2. *SOF_4 generation mechanism*

SOF_4 molecular can be generated from reaction between SF_4 and O molecular. Apart from this pathway, SOF_4 molecular can also be generated from SF_5 molecular, just like $SF_5+O \rightarrow SOF_4 + F$ or $SF_5+OH \rightarrow SOF_4 + HF$. The reactions associated with SOF_4 are summarized in Figure 2.

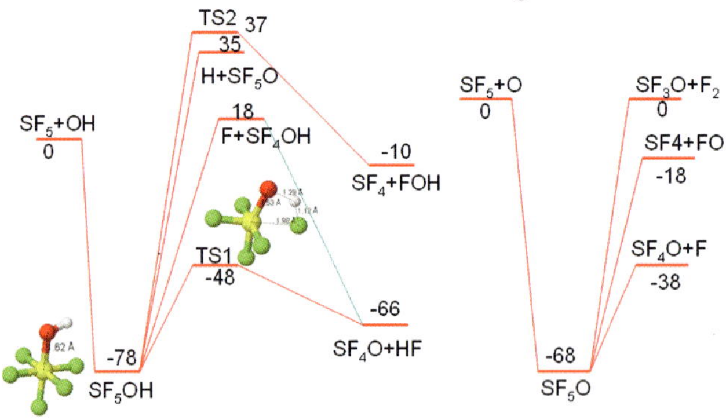

Fig. 2. Reaction mechanisms for the SF_5+OH and SF_5+O reactions. Energies are in kcal/mol.

As can be seen from Figure 2, the reactions with SF_5 of OH or O mainly produced SOF_4, instead of SF_4. Even generate SF_4 is exothermic, but its barrier is much higher than that of SOF_4 generation. Therefore, SF_5 should be one of the main sources of SOF_4 generation.

It is worth mentioning that the free radical fragments SF_5 and SF_4 can directly react with O_2, generating peroxide components. And these non-barrier composite reactions are slightly exothermic reaction process. But SF_xO_2 do not contribute to SF_xO generation. Breaking of OO bond have to overcome a very high energy barrier, and it is not feasible in thermodynamics.

In addition to the neutral reaction, SF_5 and SOF_4 can also be produced by the reaction of SF_6^- anion in the initial stage of discharge, as shown in Figure 3. SOF_4 can be efficiently generated by these negative ion - molecule reactions. These processes will yield an important contribution to SOF_4 generation.

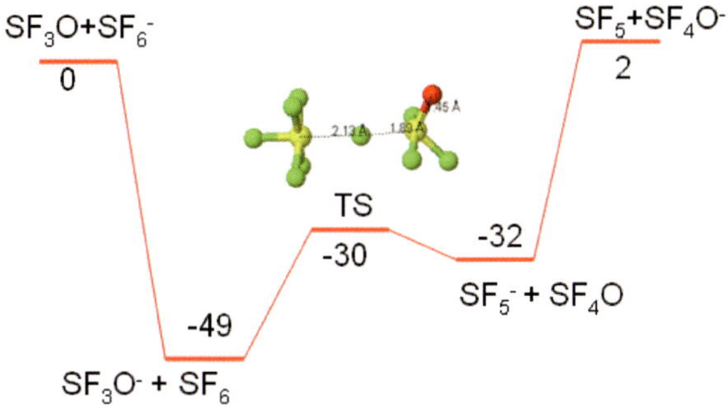

Fig. 3. Reaction mechanisms for the reactions of SF_6^- with SF_3O. Energies are in kcal/mol.

4. Conclusion

In this paper, the quantum chemical calculation method is used to study the reaction mechanism of SOF_4 generation from SF_6. A clear generation mechanism is given for important experimental observations parts of SOF_4. The feasibility of the reaction pathway is discussed from two aspects of thermodynamics and kinetics. These results are very helpful to establish the $SF_6/H_2O/O_2$ reaction model, and indicate the electrical aging process of SF_6 electrical insulation equipment from microcosmic angle.

References

1. Y. Martin, Z. Li, T. Tsutsumi, R. S. M. Nakano, and J. Suehiro, Detection of SF_6 decomposition products generated by DC corona discharge using a carbon nanotube gas sensor, *IEEE Trans. Dielectr. Electr. Insul.* **19**, 671 (2012).
2. W. Ding, W. Zhou, X. Ren, G. Li, C. Fan, W. Duan, and G. Song, Decomposition characteristics of SF_6 under partial discharges with point-to-plane electrode defect, *IEEE Trans. Dielectr. Electr. Insul.* **20**, 1727 (2013).
3. X. Zhang, F. Meng, and B. Yang, Use of hydroxyl-modified carbon nanotubes for detecting SF_6 decomposition products under partial discharge in gas insulated switchgear, *IEEE Trans. Dielectr. Electr. Insul.* **20**, 2246 (2013).
4. X. Zhang, E. Gockenbach, C. Y. Wang, G. Song, X. L. Yan, Y. S. Ji, R. Yang, Z. X. Su, Q. Yao, H. L. Chen, and G. Lu, SF_6 Decomposition and Fault Type in SF_6 Gas Insulated Equipment, *High Voltage Engineering and Application (ICHVE), 2012 International Conference on*, 514 (2012).

5. D. H. Shi, J. F. Sun, Y. F. Liu, Z. L. Zhu, and H. Ma, Total cross sections of electron scattering by several sulfur-containing molecules OCS, SO_2, SF_4, SF_6, SF_5CF_3, SO_2Cl_2 and SO_2ClF at 30–5000 eV, *Eur. Phys. J.* **54**, 43 (2009).
6. F. Y. Chu, SF_6 decomposition in gas-insulted equipment, *IEEE Trans. Electr. Insul.* **EI-21**, 693 (1986).
7. A. N. Rudenko, F. J. Keil, I. Katsnelson, and A. I. Lichtenstein, Adsorption of cobalt on graphene: electron correlation effects from a quantum chemical perspective, *Phys. Rev. B*, **86**, 075422 (2012).
8. X. Ma, K. Sakanishi, T. Isoda, and I. Mochida, Quantum chemical calculation on the desulfurization reactivities of heterocyclic sulfur compounds, *Energ. Fuel.* **9**, 33 (1995).
9. K. Somasundram, and N. C. Handy, Density functional theory studies on Sulfur–Nitrogen species, *J. Phys. Chem.* **100**, 17485 (1996).
10. J. Zhao, D. Xue, K. Zhang, and X. Cheng, Quantum chemical study of C–SH bond dissociation energies for some thiol compounds, *J. Mol. Struc.-THEOCHEM.* **909**, 9 (2009).
11. B. J. Smith, and L. Radom, Gas-phase acidities: a comparison of density functional, MP2, MP4, F4, G2(MP2, SVP), G2(MP2) and G2 procedures, *Chem. Phys. Lett.* **245**, 123 (1995).
12. B. J. Lynch, Y. Zhao, and D. G. Truhlar, Effectiveness of diffuse basis functions for calculating relative energies by density functional theory, *J. Phys. Chem. A*, **107**, 1384 (2003).
13. D. E. Woon, and T. H. Dunning Jr., Theory of hypervalency: recoupled pair bonding in SF_n (n = 1-6), *J. Phys. Chem. A*, **113**, 7915 (2009).

A Power Estimation Method of Coarse Grained Reconfigurable Pipelined Processor Based on Macro-model

Wang Qi † and Si-Kun Li

Computer Academy, National University of Defense Technology,

Chang Sha, Hu Nan, China

†*E-mail: 592963206@qq.com*

Most power estimation methods of the traditional coarse-grained reconfigurable processors use simulation method, although the accuracy is higher, but the efficiency is very low. According to the characteristics of coarse grained reconfigurable computing processor, this paper proposes a new method of power estimation based on the macro power model. This method builds a multiple linear equation which is based on the theory of multiple linear regression, and estimates the total power of the processor through the equation. Macro power model consists of variable and regression coefficients. Data word width, working frequency and intensive calculation amount of application algorithm as parameter variable are chosen. They have important impact on power. Regression coefficients express the contribution degree of variable to power. It could be solved by low level gate simulation under different input vectors. The macro power model of reconfigurable pipelined processor is constructed by using the architecture of coarse grained reconfigurable pipelined processor as an example. Matrix multiplication is used to test the different combinations of macro power model variables, and the average error between the estimated and the actual analog values is 14.12%, which speed is hundred times faster than the traditional simulation method. The experimental results demonstrate that the macro power model is an efficient and accurate power estimation method.

Keywords: Power Estimation; Macro Model; Multiple Linear Regression; Coarse Grained Reconfigurable Processor; Reconfigurable Pipeline.

1. Introduction

Analysis and estimation of power aim at to estimate the power at different design stages of the circuit, so as to ensure that the power in each phase of the circuit is required. On the one hand, it lays the foundation for low power design and power optimization. On the other hand, the number of returning last stage can be avoided and the design cost can be saved. Power estimation can find problems of power in the early phase of design. And programmer could adjust the design structure and coding style, reduce the failure probability due to power indicators effectively. So power evaluation is a necessary and important part in the process of processor design.

The reconfigurable processor is different from the traditional general purpose processor and application specific integrated circuits. It is a kind of new processor with dual programming of hardware and software, which is realized by the reconfigurable hardware structure. The reconfigurable hardware structure can be divided into fine structure and coarse grained structure. The data width of the fine granularity reconfigurable processing unit is from 1 to 3 bits, and the data width of the coarse grained reconfigurable processing unit is generally 4 to 64 bits[1]. At present, the widely used field programmable gate array FPGA is a fine grained reconfigurable chip[2], FPGA technology development has become mature, power evaluation method usually based on performance simulator and power statistical model[3]. Commonly used performance simulator is SimpleScalar and Gem5, power statistical models are Wattch, SimplePower, Pets[4] and so on. Most FPGA chip development platform embeds these two kinds of tools into the EDA system to achieve the power of FPGA.

The coarse grained reconfigurable processor is different from FPGA. It is a kind of processor with coarse granularity reconfigurable hardware structure, which is more efficient than FPGA on computing ability and reconstruction speed, but also in the development stage. At present, the traditional analog method is still used to evaluate the power of the coarse grained reconfigurable processor[5]. Although the estimation accuracy is high, but the large amount of configuration information makes the error prone and the time cost huge, so the efficiency of simulation is very low.

In 2005 Chan S. and Parameswaran J. proposed a macro model method for power estimation of network router on chip[6]. The macro model method combines the traditional simulation method with the probability method. On the basis of analyzing the key factors that affect the power of the circuit, using the multiple linear regression theory to establish the macro power model, the quantitative relationship between these factors and the power is expressed clearly. Macro model method is used to estimate the power of the circuit with the advantages of accuracy and efficiency. It mainly be used to estimate the power of network on-chip[7] and on-chip memory system [8] at present.

Research shows that the macro model method is suitable for power estimation of circuit system which is regular and extensible. Reconfigurable processing unit sets and reconfigurable interconnection network of coarse granularity reconfigurable processors have regular structure and good expansibility. So it is suitable to use the macro model to estimate the power. According to the characteristics of coarse grained reconfigurable computing processor, this paper proposes a new method of power estimation based on the macro power model. Macro power model consists of variable and regression coefficients. In this paper, we choose data word width, working frequency and intensive calculation amount of application algorithm as parameter variable. They are all important impact on power. Regression coefficients express the contribution of variable to power. It could be solved by low level gate simulation under different input vectors. In this paper, the macro power model of reconfigurable pipelined processor is constructed by using the architecture of coarse grained reconfigurable pipelined processor as an example. Matrix multiplication is used to test the different combinations of macro power model variables, and the average error between the estimated and the actual analog values is 14.12%, which speed is hundred times faster than the traditional simulation method. The experimental results show that the macro power model is an efficient and accurate power estimation method.

2. The architecture of coarse grained reconfigurable pipelined processor

2.1. *The architecture of coarse grained reconfigurable processor*

Coarse grained reconfigurable processor is usually composed of reconfigurable computing processing unit sets, reconfigurable data channel (reconfigurable interconnection network), reconfigurable controller and data buffer [9]. The structure of the graph is shown in Figure 1.

The number of arithmetic processing unit, instruction set and data width usually decided by the application demand of custom design. Switching network or router architecture is generally used in the reconfigurable interconnection network. The main function of the data buffer is to cache the data, so as to improve the efficiency of data access. The basic working principle of the coarse grained reconfigurable processor is that controller read the configuration information and complete the dynamic control of the function of computing processing units and the structure of the data channel under the driving of the configuration information flow. Different architectures have different implementation details.

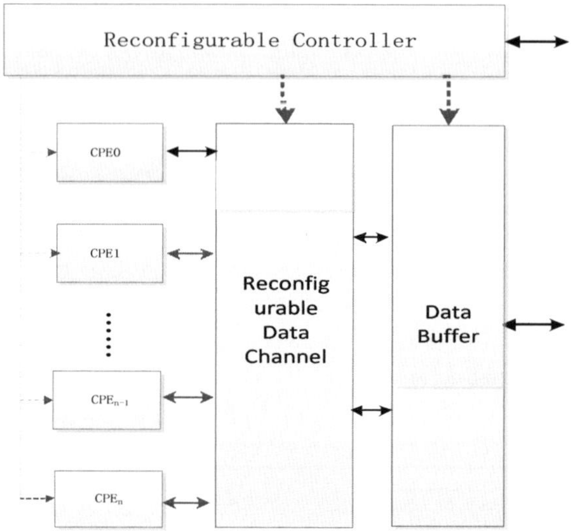

Fig. 1. Composition structure diagram of coarse grain reconfigurable processor

2.2. *The architecture of coarse grained reconfigurable pipelined processor*

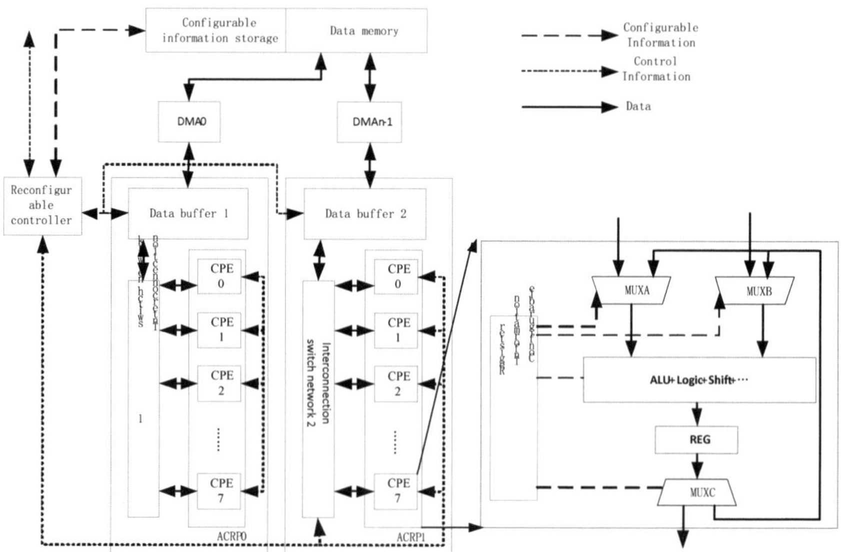

Fig. 2. System structure diagram of coarse grained reconfigurable pipelined processor

Figure 2 is a coarse grained reconfigurable pipelined processor architecture for speech enhancement processing application requirements[10]. This architecture has two reconfigurable pipelines and either of them has its own data buffer. Every pipeline has eight reconfigurable processing units, each of which contains 16 basic arithmetic and logical instruction sets. The reconfigurable data path uses a multiplexer switch network, which owns fast reconstruction speed and small delay. During the process of system working, DMA and reconfigurable controller read calculation data and configurable information from data memory and configurable information storage respectively. The reconfigurable controller sends control instruction to the reconfigurable switching network and the reconfigurable processing unit array according to the configuration information flow. After reading the instruction, they will complete data path reconstruction and functional reconstruction respectively. Until the action is over, the system begins to calculate. Configuration information flow can be changed in real time with the external program, so the function of the reconfigurable processing units and the reconstruction of the reconfigurable data path always changing constantly. This makes it to be highly flexibility and efficiency.

3. Modeling method of macro power model

3.1. *Modeling process of macro power model*

The macro model is a commonly used method for power evaluation of SoC based on IP core reuse. The macro power model based on multiple linear regression theory[11] is not limited by the circuit structure, similar to a "black box", only care about the input and output results of the circuit. Mapping the idea on the power evaluation of coarse grained reconfigurable computing processors concludes that macro power model is composed of the variables which have important influence on the power and regression coefficients that reflect contribution degree of variable to power. So the general expression for the macro power model is Eq. (1).

$$P = \beta_0 + \alpha x_1 + b x_2 + ... + n x_n . \quad (1)$$

Among them, x_1, $x_2 ... x_n$ is the variable parameters which have important influence on power. The choice of these parameters must be careful, otherwise it will lead to the failure of modeling. a, b...n is the regression coefficients of multiple linear regression equation. β_0 is a value which is no relationship with power, it is often regarded as a correction factor of modifying the model. The method of solving the regression coefficient is to simulate gate-level netlist which has completed placement and routing and to get the power of different input vectors, finally solve the matrix equation. The modeling idea of the macro model makes up for the disadvantages of the traditional power evaluation method, which has low efficiency and low accuracy. Macro power model is simple in structure, clear objectives, parameter selection with high flexibility. It is effective to abstract the power estimation problem of complex reconfigurable systems, is a combination of probability and simulation analysis method, so it has a double advantage in speed and accuracy. Figure 3 is a complete modeling process of the macro power model.

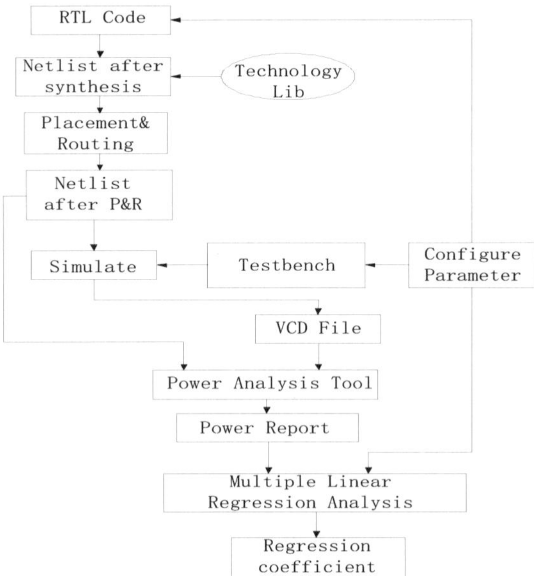

Fig. 3. Modeling process of the macro power model

3.2. *Power macro model of the coarse granularity reconfigurable pipelined processor*

Coarse grained reconfigurable pipelined macro power estimation model can be discussed from the following aspects:

(1) Parametric model, it's also called parameter extraction. This phase is to analyze and extract the variables that have a great impact on the power of the coarse grained reconfigurable pipelined processor. Through careful research and discussion, we give two parameter types which have a great impact on the power. They are structural parameter and application parameter. Structural parameters are closely related to the processor architecture and design variables, such like data width, voltage, frequency, temperature etc. Application parameters are mainly related to the application program which driving the processor to work. This kind of variable is mainly used to simulate the

performance of the design under different kinds of test vectors, designers set it according to specific requirements, the most typical is switch active factor α.

(2) Capture data, simulate the test design at gate level and record the power of each test type. The reason why use gate simulation is that gate level is not as long time as transistor level but as larger error as structural level. It is a tradeoff between speed and accuracy.

(3) Identify model, establish macro power model based on the significance of multiple linear regression model and extracted parameters.

(4) Solve and verify model parameters, fitting model coefficients using Matlab or other software with linear fitting function. To verify the authenticity of the model according to significant checking rules and calculate error through comparing data of estimation and simulation.

(5) Model repeatedly, in view of the large error model, we need to return to the first stage, check whether the parameters are right, and then extract parameters again, correct model until the values are satisfactory.

Combined with the structural characteristics of coarse grained reconfigurable pipelined processor based on the customization function of bone conduction speech enhancement application and the application characteristics of speech enhancement algorithm, data width and frequency are selected as the structural parameters, intensive calculation amount as the application parameters. It is pointed out that the intensive calculation amount is a numerical value between 0~1, which reflects the complexity of the different operations in the speech enhancement algorithm. So the dynamic power macro model for coarse grained reconfigurable pipelined processor is Eq. (2).

$$P_{dyn} = \beta_0 + \alpha x_1 + b x_2 + c x_3 . \qquad (2)$$

x_1 represents the data width, x_2 is on behalf of the working frequency, x_3 is the intensive calculation amount. The specific solution process will be described in detail in the next section.

4. Experimental analysis major headings

The most basic operation of the bone conduction speech enhancement algorithm is matrix multiplication. In order to make the simulation experiments close to the actual situation as far as possible and obtain the more accurate model coefficients, we use matrix multiplication as the application program to drive the processor working. Simulation environment is TSMC40nm process, tcbn45gsbwp12t cell library, fixed voltage 0.9V, temperature 25°C. Change the parameter configuration and observe the changes of power. Test different combinations of data width is 16bit, 24ibt, 32bit; clock frequency is 50MHz, 100MHz, 200MHz and intensive calculation amount is 0.25, 0.33, 0.4, 0.44. Finally, the regression coefficients of the dynamic power macro model obtained by the LINEST fitting is a=0.24, b=0.02, c=2.59, and β_0= -5.34 respectively. So the dynamic power macro model of coarse granularity reconfigurable pipelined processor is Eq. (3).

$$P_{dyn} = 0.24x_1 + 0.02x_2 + 2.59x_3 - 5.34 \quad (3)$$

It has experienced coding, simulation, synthesis, placement and routing, power simulation and so on with such auxiliary tools like Design Compiler, IC Compiler, NC_Verilog and Redhawk etc. during the process of solving these coefficients. Figure 4 is the line chart of true values and estimation values comparing.

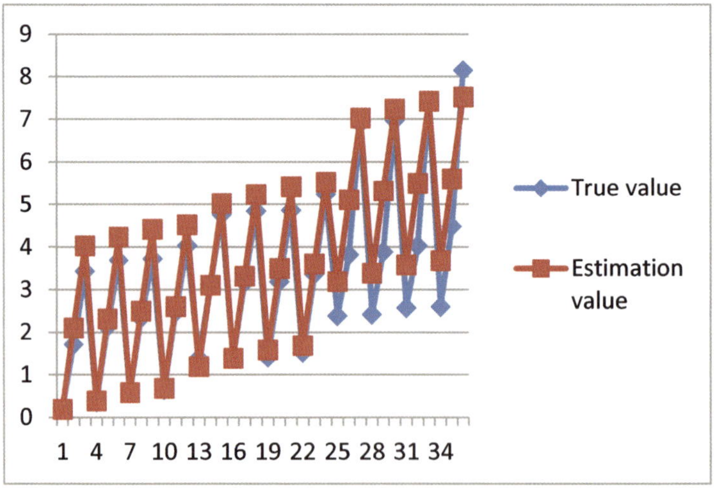

Fig. 4. Line chart of true values and estimation values comparing

The graph shows that the estimated value is approximately equal to the true value. Statistics indicate that the average error of the power of the reconfigurable processor is 14.12% in the total 36 sets of data. This conclusion proves the accuracy and reliability of the power macro model effectively. Power estimation with macro model is more than a hundred times faster than simulation in the same accuracy, this proves the great advantage of macro model in the evaluation speed.

5. Conclusion

As the standard of measuring processor power indicators and reference of optimizing processor structure, power evaluation technology will develop and progress with each processor revolution. Each progress will have a new improvement in the analysis speed and accuracy. A good and accurate power estimation method can effectively guide the designers to design more efficient products with lower power. In this paper, we further explore the characteristics of the macro model based on multiple linear regression, and map the characteristics on the coarse granularity reconfigurable computing processor for power evaluation after learning the macro model of network on-chip power analysis. The paper analyzes and summaries the modeling method and parameters extraction rules of coarse grained reconfigurable computing processor. Finally, the experimental results show that the macro model method can estimate the dynamic power of coarse grained reconfigurable processor quickly and accurately, and it has a great application value.

Acknowledgment

This work is supported by National Natural Science Foundation of China(61133007, 61076020). Work partially supported by grant 2-4570.5 of the Swiss National Science Foundation.

References

1. Singh M P, Jain M K. *A Survey of Reconfigurable Architectures[J]. International Journal of Computer Applications*, 2014, 98(14).
2. Xilinx Inc. All Programmable FPGAs [EB/OL]. http:/ / www. xilinx. com/products/silicon-devices/fpga/index.html[2014-6-7].
3. Goeders J B, Wilton S J E. *VersaPower: Power estimation for diverse FPGA architectures[C]//Field-Programmable Technology (FPT), 2012 International Conference on. IEEE*, 2012: 229-234.
4. Rethinagiri S K, Palomar O, Unsal O, et al. *Pets: power and energy estimation tool at system-level[C]//Quality Electronic Design (ISQED), 2014 15th International Symposium on. IEEE*, 2014: 535-542.
5. XiaoWei, Binyu Zang, Chuanqi Zhu. *Architecture level power model and analysis of reconfigurable processor[J]. Computer Engineering and Applications*, 2007, 43(26): 34-37.
6. Chan J, Parameswaran S. *NoCEE: energy macro-model extraction methodology for network on chip routers[C]//Proceedings of the 2005 IEEE/ACM International conference on Computer-aided design. IEEE Computer Society*, 2005: 254-259.
7. Zhou F, Wu N, Zhang Y, et al. *NoC router power macro-modeling at high level[M]//Advances in Computer Science and Engineering. Springer Berlin Heidelberg*, 2012: 199-206.
8. Gu Ming. *Performance model and optimization of embedded SRAM[D]. Dong Nan University*, 2006.
9. De Sutter B, Raghavan P, Lambrechts. *A Coarse-grained reconfigurable array architectures [M]//Handbook of signal processing systems. Springer New York*, 2013: 553-592.

10. Kongfei Du. *Research and Implementation on Reconfigurable Pipeling Co-processor for Bone-conduction Speech Enhancement Algorithm[D]. National University of Defense Technology, Chang Sha*, 2014.
11. Montgomery D C, Peck E A, Vining G G. *Introduction to linear regression analysis[M]*. John Wiley & Sons, 2015.

The Analytical Solutions of Two-Photon Spatial Solitons in Biased Photovoltaic Photorefractive Crystal*

Guang-Yong Zhang†, Guo-Zhen Chen, Mi-Zhuo Zhao and Xiao-Qing Zhang

Department of Physics, China University of Geosciences,

Wuhan, 430074, China

†E-mail: zhgyong@163.com

The analytical solutions of Two-Photon Screening Photovoltaic (TPSP) spatial solitons are obtained in biased two-photon photovoltaic photorefractive crystals. Under low-amplitude conditions, the dynamical evolution equations degenerate into Manakov equations, the analytical soliton solutions of these Manakov equations are investigated in detail. The analytical expressions of the full width at half maximum are obtained for bright and dark TPSP solitons. The results demonstrate that Manakov soliton can exist in biased two-photon photorefractive crystals. The bright TPSP solitons have analytical expressions of secant profile and the dark ones have analytical expressions of tangent profile. The results also demonstrate that the widths of bright and dark TPDP soltions are inversely proportional to the square root of the combination of the external applied electric field and the photovoltaic field.

Keywords: Nonlinear Optics; Photorefractive Effect, Spatial Soliton, Photovoltaic Effect, Two-Photon.

1. Introduction

Photorefractive spatial solitons observed experimentally by Duree et. al.[1] just a year after its theoretical prediction by Segev et. al.[2] and have attracted much special interest, for these photorefractive spatial solitons can be formed

---

at low light intensity and are potentially useful for all-optical switching, beam steering, and optical interconnects. At present, three different types of steady-state photorefractive spatial solitons have been investigated, namely, screening solitons [3,4], PhotoVoltaic (PV) solitons [5,6], and Screening-Photovoltaic (SP) solitons [7,8].

All of the above mentioned photorefractive solitons rely on single-photon photorefractive effect. Two-photon process, which is used in photorefractive holographic data storage system to avoid the problem that optical fields erase the information during readout, can enhance the photorefractive effect greatly[9]. Recently, Castro-Camus et. al. [10] provided a theoretical model for the Two-Photon PhotoRefractive (TPPR) effect, which includes a Valence Band (VB), a Conduction Band (CB), and an Intermediate allowed Level (IL). The intermediate allowed level is used to maintain a quantity of excited electrons from the valence band by photons with energy $\hbar\omega_1$ (gating beam). These electrons are then excited again to the conduction band by another photon with energy $\hbar\omega_2$ (signal beam). The pattern of the signal beam can induce a spatial dependent charge distribution that gives rise to nonlinear changes of refractive index in the medium. Very recently, based on Castro-Camus model, it have been predicted that Two-Photon Screening (TPS) solitons can be formed in biased non-photovoltaic photorefractive crystal [11], Two-Photon Photovoltaic (TPP) solitons be formed in photovoltaic crystal under open circuit conditions [12], and Two-Photon Screening Photovoltaic (TPSP) solitons can also be formed in biased two-photon photovoltaic crystal [13].

In this paper, the analytical soliton solutions of one dimensional spatial solitons in biased two-photon photovoltaic crystal are investigated under low-amplitude conditions. The analytical expressions of the full width at half maximum for the low-amplitude TPSP bright solitons and dark ones are obtained, respectively.

2. The dynamical evolution equation of the soliton in biased two-photon photorefractive crystals

Let us consider an optical beam that propagates in a TPPR crystal along the z-axis and is permitted to diffract only along the x-direction. The crystal is placed with its optical axis along the x coordinate and is illuminated by the gating beam I_1, which can be considered as a constant. Moreover, let us assume that the incident optical beam is linearly polarized along x and the external bias field is applied in the same direction. As usual, we express the optical field of the incident beam in terms of slowly varying envelope ϕ i.e., $E = \hat{x}\phi(x,z)\exp(ikz)$, where $k = k_0 n_e = (2\pi / \lambda_0) n_e$, λ_0 is the free-space wavelength of the lightwave employed, and n_e is the unperturbed extraordinary index of refraction. Under these conditions the optical beam satisfies the following envelope evolution equation [7]:

$$i\phi_z + \frac{1}{2k}\phi_{xx} - \frac{k_0}{2}(n_e^{\,3} r_{33} E_{sc})\phi = 0\,, \tag{1}$$

where $\phi_z = \partial\phi / \partial z$, $\phi_{xx} = \partial^2 \phi / \partial x^2$, r_{33} is the electro-optic coefficient of the TPPR crystal, and E_{sc} is the induced space-charge field. When the diffusion effect is neglected, the space-charge field can approximately given by[13]

$$E_{sc} = E_0 \frac{(I_\infty + I_{2d})(I + I_{2d} + \gamma_1 N_A / s_2)}{(I + I_{2d})(I_\infty + I_{2d} + \gamma_1 N_A / s_2)} + E_P \frac{s_2 (I_\infty - I)(I + I_{2d} + \gamma_1 N_A / s_2)}{(s_1 I_1 + \beta_1)(I + I_{2d})} \tag{2}$$

where $I = I(x,z)$ is the intensity of the input optical beams, $I_\infty = I(x \to \pm\infty, z)$ is the intensity in the area far from the centre of the two photon PR media; I_{2d} is the so called dark irradiance; E_0 is the external biased electric field, and $E_P = \kappa\gamma N_A / e\mu$ is the photovoltaic field constant; γ and γ_1 are the recombination factors of the CB-VB, IL-VB; κ, μ and e are, respectively, the photovoltaic constant, the electron mobility and the charge; s_1 and s_2 are photoexcitation crosses; β_1 and $I_{2d} = \beta_2 / s_2$ are the thermoionization probability constants for the transitions of VB-IL and IL-CB; N_A is the acceptor or trap density. According to Poynting's theorem, the intensity of the optical beams can be obtained by $I(x,z) = (n_e / 2\eta_0)|\phi|^2$, where $\eta_0 = (\mu_0 / \varepsilon_0)^{1/2}$.

The envelope evolution equation can now be established by insertion of Eq. (2) into Eq. (1). It proves more convenient, however, to study this equation in a normalized fashion. In doing so, let us adopt the following dimensionless coordinates and variables, i.e., $\xi = z / (k x_0^2)$, $s = x / x_0$, $\phi = (2\eta_0 I_{2d} / n_e)^{1/2} U$, Here x_0 is an arbitrary spatial width. Under these conditions, the normalized envelope U satisfies the following dynamical evolution equation:

$$iU_\xi + \frac{1}{2}U_{ss} - \beta \frac{(\rho+1)(|U|^2 + 1 + \sigma)}{(|U|^2 + 1)(\rho + 1 + \sigma)} U - \alpha \frac{\eta(\rho - |U|^2)(|U|^2 + 1 + \sigma)}{|U|^2 + 1} U = 0. \tag{3}$$

where $\rho = I_\infty / I_{2d}$, $\beta = (k_0 x_0)^2 (n_e^4 r_{33} / 2) E_0$, $\alpha = (k_0 x_0)^2 (n_e^4 r_{33} / 2) E_p$, $\eta = \beta_2 / (s_1 I_1 + \beta_1)$ and $\sigma = \gamma_1 N_A / s_2 I_{2d} = \gamma_1 N_A / \beta_2$.

In the case of low intensity ratio (i.e., $|U|^2 \ll 1$), by expanding the term due to the nonlinearity, Eq. (3) can be reduced to

$$iU_\xi + \frac{1}{2}U_{ss} - \frac{\beta(\rho+1)}{(\rho+1+\sigma)}[(1+\sigma) - \sigma(|U|^2)]U - \alpha\eta\rho(1+\sigma)U + [\alpha\eta\rho\sigma + \alpha\eta(1+\sigma)](|U|^2)U = 0. \tag{4}$$

Let $U = u(s,\xi)\exp(i\omega\xi)$ submitting it to the above equations, we get the following transformation

$$U(s,\xi) = u(s,\xi)\exp\{-i[\beta(1+\rho)(1+\sigma)/(1+\rho+\sigma)+\alpha\eta\rho(1+\sigma)]\xi\} \qquad (5)$$

Thus Eq. (5) can be rewritten to

$$iu_{\xi} + u_{ss}/2 + [\beta(\rho+1)\sigma/(\rho+1+\sigma)+\alpha\eta\rho\sigma+\alpha\eta(1+\sigma)](|u|^2)u = 0. \qquad (6)$$

Equation 6 is known as the manakov equation, which has bright and dark soliton analytical solutions, and can be solved exactly. In what follows, the soliton analytical solutions of Eq. (6) will be given.

3. The analytical solution of TPSP solitons in biased two-photon photorefractive crystals

3.1. *The analytical solution of the dark TPSP solitons*

We can derive the dark spatial solitons solution from Eq. (6), which can be written directly:

$$\begin{aligned} u(s,\xi) = \rho^{1/2}\tanh\{[-\frac{\beta(\rho+1)\sigma}{(\rho+1+\sigma)}\rho-\alpha\eta\rho^2\sigma-\alpha\eta(1+\sigma)\rho]^{1/2}s\} \\ \cdot\exp\{i[\frac{\beta(\rho+1)\sigma}{(\rho+1+\sigma)}\rho+\alpha\eta\rho^2\sigma+\alpha\eta(1+\sigma)\rho]\xi\} \end{aligned} \qquad (7)$$

We can see that dark manakov TPSP soliton can exist in biased two-photon photorefractive crystal if we properly chose the values and signs of α and β .

The full width at half maximum(FWHM) of the dark soltions is given by

$$W_d = 2\ln(1+\sqrt{2})/\sqrt{\Delta} \qquad (8)$$

Where $\Delta = -\beta\rho(\rho+1)\sigma/(\rho+1+\sigma)-\alpha\eta\rho^2\sigma-\alpha\eta(1+\sigma)\rho$, which is the combination of the external applied electric field and the photovoltaic field. We can see that the width of dark TPSP solitons in inversely proportional to the root of Δ.

3.2. *The analytical solution of the bright TPSP solitons*

For bright soliton case, the optical energy concentrates in the area near the beam centre, and thus the intensity is expected to vanish at infinity and thus $I_{\infty} = \rho = 0$. In this case, Eq. (6) take the forms:

$$iu_{\xi} + u_{ss}/2 + [\beta\sigma/(1+\sigma)+\alpha\eta(1+\sigma)](|u|^2)u = 0. \qquad (9)$$

The bright soliton solution is given by

$$u(s,\xi) = r_1^{1/2} \sec h\{[\frac{\beta\sigma}{1+\sigma} r + \alpha\eta(1+\sigma)r]^{1/2} s\}$$
$$\cdot \exp(i\{[\frac{\beta\sigma}{1+\sigma} r + \alpha\eta(1+\sigma)r]/2\}\xi) \tag{10}$$

And the FWHM of the bright soltions is obtained as

$$W_b = 2\ln(1+\sqrt{2})/\sqrt{\Lambda} \tag{11}$$

Where r represents the rations of the maximum intensity of the bright soliton to the dark irradiance, respectively; i.e. $r = I_{\max}/I_{2d} = I(0)/I_{2d}$, $\Lambda = \beta\sigma r/(1+\sigma) + \alpha\eta(1+\sigma)r$, which also is the combination of the external applied electric field and the photovoltaic field. It is shown that the width of dark TPSP solitons is inversely proportional to the root of Λ.

4. Conclusion

The analytical solutions of two-photon screening photovoltaic spatial solitons in biased two-photon photovoltaic photorefractive crystals are investigated theoretically under low-amplitude conditions. Under low-amplitude conditions, the dynamical evolution equations degenerate into Manakov equations, the analytical soliton solutions of these Manakov equations are obtained. The analytical expressions of the full width at half maximum for bright and dark TPSP solitons are also investigated. It is shown that Manakov bright and dark TPSP soliton can exist in biased two-photon photorefractive crystals, and the bright solitons have analytical expressions of secant profile and the dark ones have analytical expressions of tangent profile. It is also found that the widths of bright and dark TPSP soltions are related to the external applied electric field and the photovoltaic field, which are found inversely proportional to the square root of the combination of the external applied electric field and the photovoltaic field.

Acknowledgments

This work is supported by the National Basic Research Program of China (973 Program) (Grant No. 2011CB710606) and the Earth biology fund from State Key Laboratory of Biogeology and Environmental Geology (Grant No. GKZ15Y638).

References

1. G. C. Duree, J. L. Shultz, G. J. Salamo, M. Segev, A. Yariv, B. Crosignani, P. Di Porto, E. J. Sharp, and R. R. Neurgaonkar, "Observation of self-trapping of an optical beam due to the photorefractive effect," Phys. Rev. Lett. **71**, 533–536 (1993).
2. A. G. Grandpierre, D. N. Christodoulides, T. H. Coskun, M. Segev, and Y. S. Kivshar, "Gray spatial solitons in biased photorefractive media," J. Opt. Soc. Am. B **18**, 55–63 (2001).
3. M. Segev, G. C. Valley, B. Crosignani, P. Di Porto, and A. Yariv, "Steady-state spatial soliton in photorefractive material with external field," Phys. Rev. Lett. **73,** 3211– 3214 (1994)
4. A. G. Grandpierre, D. N. Christodoulides, T. H. Coskun, M. Segev, and Y. S. Kivshar, "Gray spatial solitons in biased photorefractive media," J. Opt. Soc. Am. B **18**, 55–63 (2001).
5. G. C. Valley, M. Segev, B. Crosignani, A. Yariv, M. M. Fejer, and M. C. Bashaw, "Dark and bright photovoltaic spatial solitons," Phys. Rev. A **50**, R4457–R4460 (1994).
6. W. L. She, K. K. Lee, and W. K. Lee, "Observation of Two-Dimensional Bright Photovoltaic Spatial Solitons," Phys. Rev. Lett. **83**, 3182–3185 (1999).
7. J. S. Liu, and K. Q. Lu, "Screening-photovoltaic spatial solitons in biased photovoltaic photorefractive crystals and their self-deflection," J. Opt. Soc. Am. B **16,** 550–555 (1999).
8. E. Fazio, F. Renzi, R. Rinaldi, M. Bertolotti, M. Chauvet, W. Ramadan, A. Petris, and V. I. Vlad, "Screening-photovoltaic bright solitons in lithium niobate and associated single-mode waveguides," Appl. Phys. Lett. **85,** 2193–2195 (2004).
9. Y. S. Bai and R. Kachru, "Nonvolatile Holographic Storage with Two-Step Recording in Lithium Niobate using cw Lasers," Phys. Rev. Lett. 78, 2944–2947 (1997).
10. E. Castro-Camus and L. F. Magana, "Prediction of the physical response for the two-photon photorefractive effect," Opt. Lett. **28**, 1129–1131 (2003)..
11. C. F. Hou, Y. B. Pei, Z. X. Zhou, and X. D. Sun, "Spatial solitons in two-photon photorefractive media," Phys. Rev. A **71**, 053817-1–053817-6 (2005).
12. C. F. Hou, Y. Zhang, Y. Y. Jiang, Y. B. Pei, "Photovoltaic solitons in two-photon photorefractive materials under open-circuit conditions," Opt. Commun. **273**, 544–548 (2007).
13. G. Y. Zhang, and J. S. Liu, "Screening-photovoltaic spatial solitons in biased two-photon photovoltaic photorefractive crystals." J. Opt. Soc. Am. B **26**, 113-120 (2009).

Intellectualized large diameter aspherical reflector manufacture technology

Bao-Wei Qiu, Wen Guo and Ting-Ting Hao

Beijing Institute of Space and Mechanics & Electricity, 104 Youyi Road Haidian District Beijing, China

E-mail: shylockqbw@163.com

Large diameter aspherical reflector machining technology is a kind of technology which has big difficulty, long process cycle and high production cost. Therefore, the machining efficiency and accuracy are reduced. But now the intelligent large diameter aspherical reflector machining technology greatly increases the machining efficiency and accuracy. This paper introduces intelligent large diameter aspherical reflector machining technologies. Intelligent large diameter aspherical reflector machining technology is a technology to make the skill of the advanced optical machining person digitization and quantization. Compared with traditional optical machining technology, intelligent optical machining technology makes the judgement of the optical shape more accurate, control of the process more reliable, machining efficiency and machining accuracy greatly improved. Take 1.3m diameter aspherical reflector as an example. Before manufacture, PV of the mirror is 8722nm and RMS is 1230nm. After simulation and manufacture, PV of the mirror is 4948nm and RMS is 358nm and PV of the mirror is 5390nm and RMS is 356nm respectively. The PV error between simulation and manufacture is +8.93%. The RMS error between simulation and manufacture is -0.56%. These results prove that intellectualized large diameter aspherical reflector manufacture technology is highly efficient and accurate.

Keywords: Intelligent; Optical machining; Large diameter; Aspherical; Reflector.

1. Introduction

In the early 1970s to 80s, the computer control technology is still in its initial stage. Because at that time the computer control technology which is restricted by the development of computer technology and precision measurement technology has not be totally practical application. During this time, Itek company which has done a lot of work in the computer simulation and optimization completed the world's first computer controlled optical machining machine. The computer control technology is also called intellectualized technology[1] [2].

Intellectualized large diameter aspherical reflector manufacture technology successfully resolved the problem that the simulation results are not consistent with the processing actual result. It truly achieve the prediction and controllability of the production route and cycle. Due to the integrated processing software and test software and its application in the mutual cooperation, it fully implement the intelligent technology, standardization and routinization.

2. Intellectualized reflector manufacture technology

Princeton put forward the famous Princeton hypothesis in 1972[3] [4]. He pointed that, in a wide range of values, polishing can be described as a linear equation, as in Eq. (1)

$$dz / dt = KVP \tag{1}$$

In this equation, K is proportionality constant, it is composed of all other factors except the velocity and pressure; V is speed about position and time function, V=V(x,y,t); P is polishing pressure about position and time function, P=P(x,y,t). In this hypothesis, Princeton attributed all other factors except the speed and pressure[5] [6] [7]. In this way, Princeton set up a linear relationship about material removal volume, pressure and instantaneous velocity. In this way, we can calculate the material removal remount Δz in this period of time according to the tool machining position, the relative velocity and pressure and the processing time t:

$$\Delta z(x,y) = K\int_0^t V(x,y,t)P(x,y,t)dt \tag{2}$$

$$\Delta z(x,y) = z_0(x,y) - z(x,y) \tag{3}$$

In Eq. (2) and Eq. (3), $z_0(x,y)$ is the surface height when the time is t=0; $z(x,y)$ is the surface height when the time is t.

After Princeton hypothesis, optical polishing process successfully simplified[8] [9] [10]. For process control, it is very favorable. The development of optical processing theory is mainly based on the Princeton hypothesis. Intelligent tobot digital machining technology is also based on the Princeton hypothesis.

Fig. 1 is the smart robot:

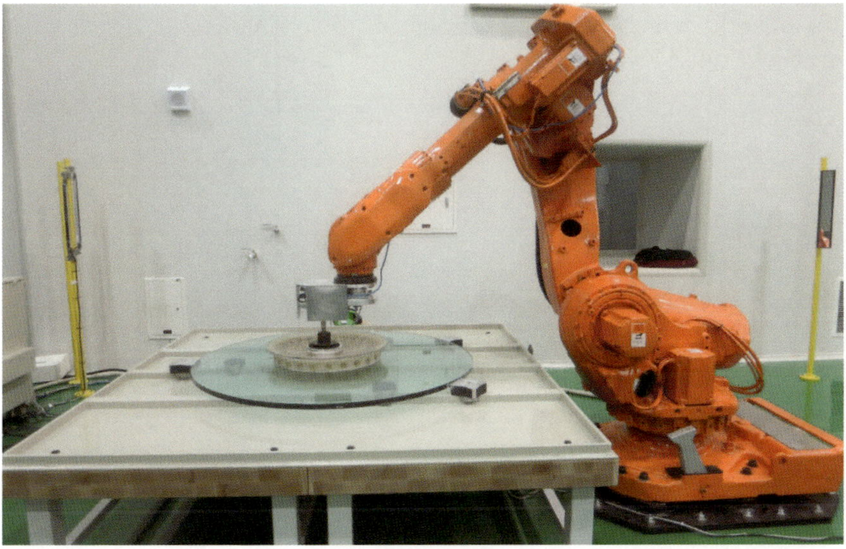

Fig. 1. Smart robot

3. Simulation and experimental result

We can use smart robot manufacture simulation software to simulate optical polishing process as Fig. 2:

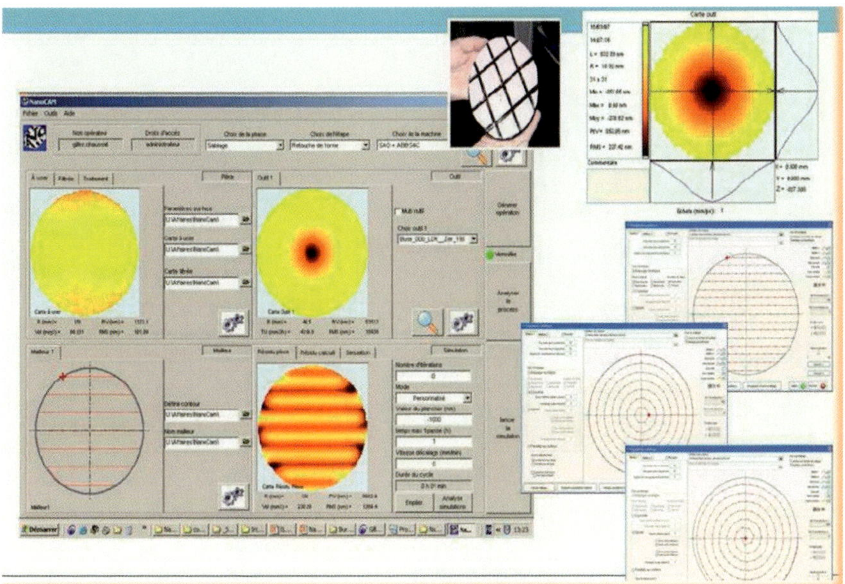

Fig. 2. Smart robot simulation software

Take 1.3m diameter mirror for example. Before manufacture the shape of the mirror is like Fig. 3:

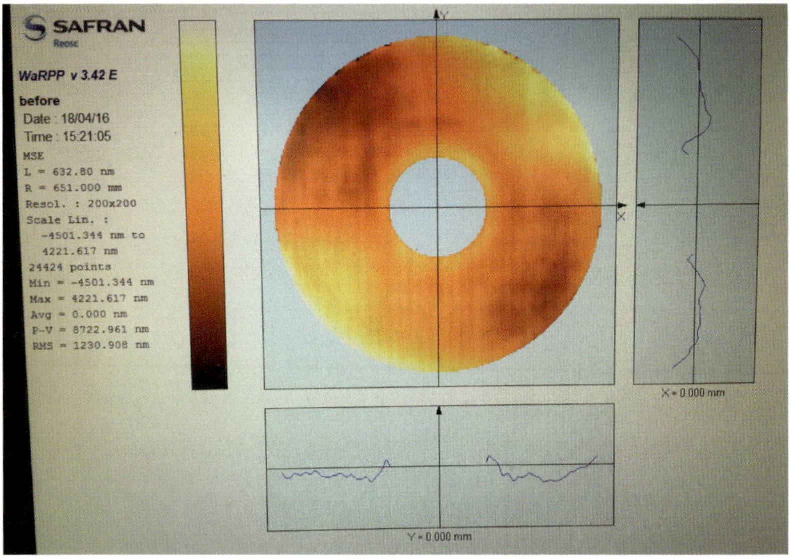

Fig. 3. Shape before manufacture

After simulation the shape of the mirror is like Fig. 4:

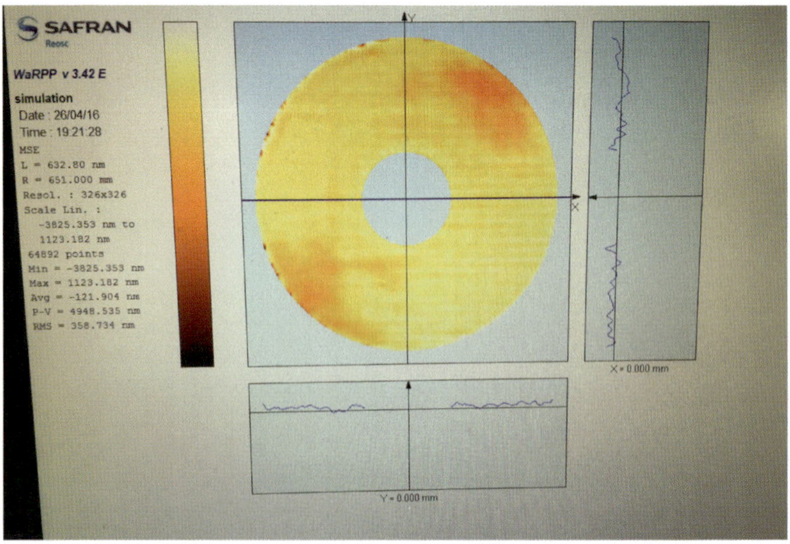

Fig. 4. Shape after simulation

After manufacture the shape of the mirror is like Fig. 5:

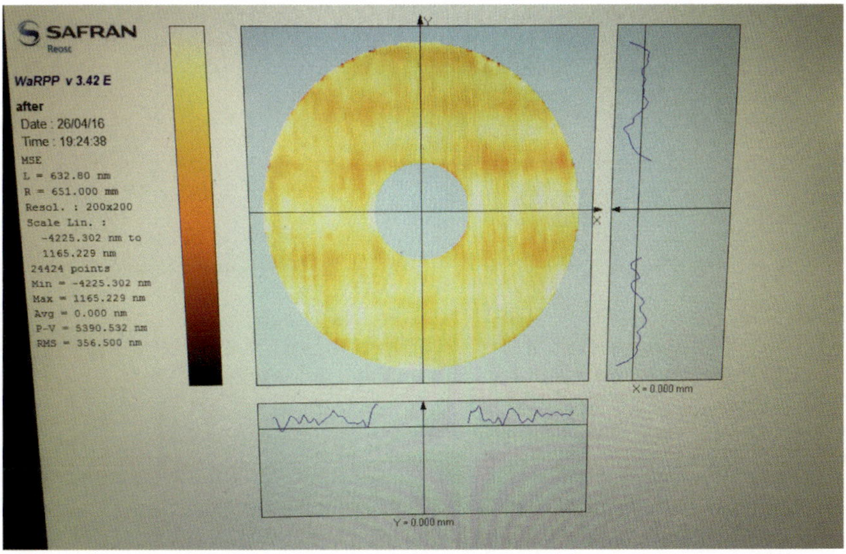

Fig. 5. Shape after manufacture

Error between simulation and manufacture is like table 1 below:

Table 1. Error between simulation and manufacture

	PV	RMS
Before manufacture	8722nm	1230nm
After simulation	4948nm	358nm
After manufacture	5390nm	356nm
Error between simulation and manufacture	+8.93%	-0.56%

We can see that all the errors are under 10%, so these results prove that the smart robot is efficient and accurate.

4. Conclusion

This paper proved that intellectualized large diameter aspherical reflector manufacture technology is efficient and accurate. Through the investigation of smart robot optical manufacture and the analysis of relevant industries, this paper introduced the result of intellectualized large diameter aspherical reflector manufacture technology. We did an experiment on 1.3m large diameter aspherical reflector through optical manufacture software. We proved that the efficiency and accuracy of smart robot optical manufacture through the results of simulation and the results of manufacture. And we made a solid foundation for the large scale batch manufacture of large diameter aspherical reflector in the future.

References

1. Li Cai, *Optical element manufacture technology*, 2nd edition.
2. Max Born and Emil Wolf, *Principles of Optics*,7th edition.
3. Daniel Malacara, *Optical Shop Testing,*3rd edition.
4. Bemhard Braunecker, Rudiger Hentschel and Hans Tiziani, *Advanced optics using aspherical elements.*
5. Wen-Kui Wang ,To Suet and Chi-Fai Cheung, A new method to test the photometric characteristics of lamps for motor vehicles, *Optik-International Journal for Light and Electron Optics.* **120**, 11 (2009).
6. D Gao, To S and W-B Lee, Tool path generation for machining of optical freedom surfaces by an ultra-precision multiaxis machine tool, *Proceedings of the Institution of Mechanical Engineers Part B-Journal of Engineers Manufacture.* **220**, 12 (2006).
7. Chi-Fai Cheung, H-F Li and W-B Lee, An integrated form characterization method for measuring ultra-precision freeform surfaces, *International Journal of Machine Tools and Manufacture.* **47**, 1 (2007).
8. Chi-Fai Cheung, H-F Li and L-B Kong, Measuring ultra-precision freeform surfaces using a robust form characterization method, *Measurement Science and Technology.* **127** (2006).
9. Chi-Fai Cheung, To S and X-Q Jiang, A study of pattern and feature analysis of surface generation in fast tool servo machining of optical microstructures, *Proceedings of ASPE's 21st Annual Meeting.* **321-324**(2006).
10. Chi-Fai Cheung, To S and T-C Kwok, Measurement and characterization of surface quality in fast tool servo machining of optical microstructures. *Proceedings of The 8th International Symposium on Measurement Technology and Intelligent Instruments* .**359-362**(2007).

Research on Gas Chromatography-Mass Spectrometer Method Measuring Corrosive Sulfur in Mineral Insulating Oil

Tao Wan†, Jun Hu, Bing Feng, Zhou Zhou, Hui Qian and Ying-Jun Qiu

Hunan Electric Power Corporation Research Institute,
Changsha, Hunan 410007, China
†*E-mail: 94525873@qq.com*

In recent years, many transformer failures occurred. In fact, the main reason was that the DiBenzyl DiSulfide (DBDS) in the transformer oil caused the corrosion of copper wind. In order to solve the problem, the content of corrosive sulfur DiBenzyl DiSulfide (DBDS) in insulating oils was tested by Gas Chromatography-Mass Spectrometer (GC-MS). The quantitative detection methods of DBDS, which use internal standard method, were built by exploring the experimental condition. Experimental results demonstrate that utilization of GC-MS method for measuring contents of DBDS in insulating oil is of good veracity and with recovery of 85.6%～100.6%. This means good repeatability, relative standard deviation of five times paralleled measuring results of DBDS is within 0.72%.

Keywords: Template; corrosive sulfur; GC-MS; dibenzyl disulfide; quantitative detection; internal standard method.

1. Introduction

Currently, corrosive sulfur in insulation oil lead to transformer faults. The threats to the safe and stable of power grid caused by corrosive sulfide have gained increased attention [1–3]. Corrosive sulfur is capable of direct reaction with silver, copper, iron and other metal under certain conditions, such as sulfur, hydrogen sulfide, mercaptan and some small molecule organic sulfide [4]. Disulfide dibenzyl (DBDS) is considered to be the main corrosive sulfide[5][6].

Disulfide dibenzyl is a kind of sulfur antioxidant, and molecular formula is $C_{14}H_{16}S_2$. The total sulfur content in transformer oil is mainly determined by the sulfur content of crude oil and the processing technology of transformer oil. Because of the improvement to the insulation oil oxidation stability with DBDS, some manufacturers added DBDS or homologues, instead of 2, 6-di-tert-butyl-p-cresol (DBPC). Cu_2S is considered to be a product of corrosive sulfide and copper windings under certain conditions[7]. Cu_2S is a semi-conductor, which penetrate and pollutes paper on copper wind[8]. The wire insulation strength weakened, caused by the corrosion of corrosive sulfur. The destruction of equipment insulation system, resulting in damage to the equipment greatly, influences the safety and stability of the transformer operation [9]. So it is very important to detect the presence of corrosive sulfur DBDS in transformer oil.

Most of the existing corrosive sulfide test method is qualitative method, through the color and shape of the test piece or insulating paper observed after chemical reaction, to identify the presence of corrosive sulfur in oil[10]. In 2012 the International Electrotechnical Commission released a new transformer oil Standard IEC 60296-2012, in which added the DBDS requirement in the oil of less than 5 mg/kg. and visible detection of DBDS content in transformer oil is very important. In this paper, the DBDS content was determined by gas chromatography-mass spectrometry (GC-MS).

2. Experimental

2.1. *Apparatus and reagent*

GC-MS: Thermo Fisher Trace1300-ISQ, configuration of split/splitless inlet;
Iso-octane: AR;
DiBenzyl DiSulfide (DBDS): purity greater than 99.95% (mass fraction);
DiPhenyl DiSulfide (DPDS): purity greater than 99.95% (mass fraction);

The base oil of mineral insulating oil without sulfide: the viscosity is less than 12mm^2/s (40°C).

2.2. *Gas chromatography*

Carrier gas: purity of helium >99.99%, constant flow rate of 1.2mL/min.
Split flow inlet port: inlet temperature of 275 °C, the split ratio is 20:1.

Hand sample: sample volume 1μL.

Capillary chromatographic column: 30m * 0.25 mm * 0.25um, 5% phenyl and 95% methyl polysiloxane Stationary phase, heating program of column temperature parameters are shown in Table 1.

Table 1 Heating program

Program	Parameters
Initial temperature/°C	90
Initial hold/min	0
Ramp rate/(°C·min^{-1})	10
Final temperature /°C	275
Final hold/min	10

2.3. *MS detection*

Electron Ionization (EI) Source：electron energy 70eV.
Interphase line temperature: 270°C
Source temperature: 200°C

Mass analyzer: set to transmit ions with DBDS and DPDS， the characteristic ions are shown in Table 2.

Table 2 Mass spectrometer parameters

DBDS Ions	Dwell time ms
246	100
DPDS Ions	
218	100

3. Method

3.1. *Stock solution*

A certain amount of DBDS and DPDS dissolved in blank mineral insulation oil (free from DBDS and DPDS) for the preparation of 1000mg/kg DBDS and 500mg/kg DPDS reserve solution. The reserve solution should be kept in a brown bottle and placed in a dark place. The solution should be replaced for no more than 3 months.

3.2. *Standard solution*

5mg/kg, 50mg/kg, 100mg/kg, 300mg/kg and 600mg/kg standard DBDS solution were prepared with DBDS stock solution and blank oil shown in Table 3.

Table 3 Standard solution preparation

Number	C_{DBDS}（mg/kg）	DBDS stock solution /g	Blank mineral insulating oil /g
1	5.000	0.025	4.975
2	50.000	0.250	4.750
3	100.000	0.500	4.500

4	300.000	1.500	3.000
5	600.000	3.000	2.000

3.3. *Analysis*

Weigh out a 0.25 g aliquot of homogenized oil sample in a centrifuge tube. Record the weight of sample as W_{oil}. Add a known amount of DPDS reserve solution, so that the concentration of internal standard DPDS in the oil sample is 50mg/kg. Dilute to 5mL with isooctane. Mix the solution obtained by shaking it briefly by hand, and takes an aliquot for the analysis. Inject 1 μL of sample solution into the GC-MS analyzer by means of a micro spyringe. Applicable scope of this method is 5mg/kg ~ 600 mg/kg.

3.4. *Calibration procedure*

The standard solution of different concentrations of 1 μL was injected into the GC-MS, and the response value (peak area) was recorded. The standard calibration curve was illustrated with C_{DBDS}/C_{DPDS} versus A_{DBDS}/A_{DPDS}.

4. Results and discussion

4.1. *Calibration*

The concentration of 50 mg/kg of DPDS and 600 mg/kg DBDS chromatographic diagram is shown in Figure 1. The ratios of C_{DBDS}/C_{DPDS} and A_{DBDS}/A_{DPDS} are listed in Table 4. The standard calibration curve of the concentration 5~600 mg/kg is shown in Figure 2.

Table 4 Different ratios of concentration and response with standard solution

Number	Concentration ratios (C_{DBDS}/C_{DPDS})	Response ratios (R_{DBDS}/R_{DPDS})
1	0.10	0.025
2	1.00	0.273
3	2.00	0.530
4	6.00	1.645
5	12.00	3.294
6	Correlation coefficient	0.9999
7	Linear equation	Y=0.27506x-0.00734

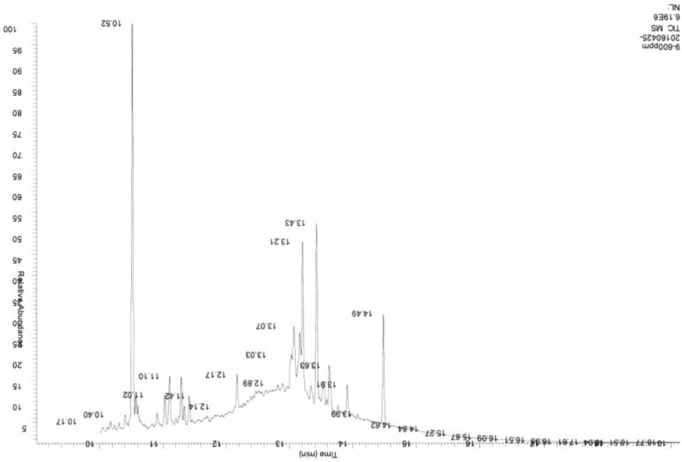

Figure 1 DBDS GC-MS chromatograms（12.17min: DPDS; 14.49min: DBDS）

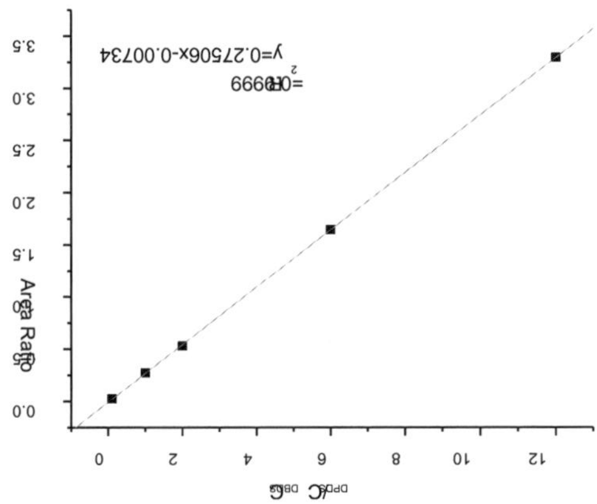

Figure 2 Standard calibration curve

4.2. *Repeatability*

Standard oil sample of 20mg/kg, 150mg/kg and 400mg/kg were used for Repeatability test. Each sample was measured in parallel 5 times. In Table 5, the results showed that the method has a good repeatability.

Table 5 DBDS repeatability test

Number and result	DBDS concentrations（mg/kg）		
	20	150	400
1	19.7	140.7	398.8
2	19.5	141.1	399.4
3	19.5	140.3	398.9

4	19.8	140.8	398.4
5	19.5	140.4	398.8
Average value	19.60	140.66	398.86
Standard deviation	0.14	0.32	0.36
Relative standard deviation %	0.72	0.23	0.09

4.3. *Recovery test*

By adding DBDS to the unknown used oil samples determine the recovery rates. The addition of DBDS in oil samples to achieve the concentration of 5mg/kg, 50mg/kg, 100mg/kg, 300mg/kg and 600mg/kg. Parallels were measured three times (n = 3), and the results are shown in Table 6.

Table 6 Recovery test

Addition DBDS solution	Test results of DBDS concentration / （mg/kg)				Recovery
concentration / （mg/kg)	1	2	3	Average value	rates /%
5.0	6.4	6.5	6.5	6.43	128.6
50.0	50.1	50.2	49.8	50.33	100.7
100.0	97.8	97.0	97.8	97.53	97.5
300.0	303.4	302.4	302.6	302.80	100.9
600.0	604.3	604.9	605.9	605.30	100.9

4.4. *Sample analysis*

The DBDS contents of 14 used insulating oil samples were tested, and the results are shown in Table 7.

Table 7 Used insulating oil samples test

	DBDS concentration / （mg/kg)		
Number	1	2	Average value
1	4.2	4.1	4.15
2	4.3	4.3	4.30
3	3.3	3.4	3.35
4	4.0	4.1	4.15
5	NA	NA	NA
6	3.3	3.4	3.35
7	3.8	3.8	3.80
8	NA	NA	NA
9	NA	NA	NA
10	NA	NA	NA
11	2.4	2.4	2.40
12	2.8	2.8	2.80
13	2.2	2.2	2.20
14	4.8	4.8	4.80

5. Conclusion

A method for quantitative determination of corrosive sulfur DBDS in unused and used mineral insulating oil was established by using internal standard method. The concentration range of the method was 5 to 600mg/kg. At the same time, the recovery test and repeatability test were carried out. The results show that the method has good recovery and repeatability.

This method detected the DBDS contents of 14 used transformer oils. The results shown 4 oils are detected no DBDS, and 9 oils are detected DBDS from 2.2mg/kg to 4.8mg/kg.

References

1. Scatiggio F, Tumiatti V, Maina R, et al. Corrosive Sulfur in Insulating Oils: Its Detection Correlated Power Apparatus Failures[J]. *IEEE Transactions on Power Delivery*, 2008, 23(1):508-509.
2. Cao Shunan, Li Rui, Qian Yihua. Overview of Research on Corrosive Sulfur in Transformer Oil [J]. *Power Plant Chemistry*, 2008, 10(8): 481-486
3. CIGRE WG A2-32, "Copper sulphide in transformer insulation," Final Report Brochure 378, 2009.
4. Qian Yihua. Corrosive sulfur protection of mineral insulating oil in electrical power equipment [M]. Beijing: China electric power press, 2013.
5. Hajek J, Bennstam G, Dahlund H, et al. Quality of Oil Makes the Difference, ABB Discovers the Solution to Transformer Breakdowns[J]. *ABB Review*.
6. F. Scatiggio, V. Tumiatti, R. Maina, et al, "Corrosive Sulfur Induced Failures in Oil-Filled Electrical Power Transformers and Shunt Reactors," *IEEE Trans. Power Del.*, Vol. 24, pp. 1240-1248, 2009.
7. T. Wan, H. Qian, Z. Zhou, et al, "Suppressive Mechanism of the Passivator Irgamet 39 on the Corrosion of Copper Conductors in Transformers," *IEEE Trans. Dielectr. Electr. Insul.*, Vol. 19, pp. 454–459, 2012.
8. H. Kawarai, Y. Fujita, J. Tanimura, et al, "Role of Dissolved Copper and Oxygen on Copper Sulfide Generation in Insulating Oil," *IEEE Trans. Dielectr. Electr. Insul.*, Vol. 16, pp. 1430–1435, 2009.
9. P. Wiklund, M. Levin, and B. Pahlavanpour, "Copper Dissolution and Metal Passivators in Insulating Oil," *IEEE Electr. Insul. Mag.*, Vol. 23, No. 4, pp. 6-14, 2007.
10. S. Toyama, J. Taninmra, N. Yamada, E. Nagao, and T. Amimoto, "Highly sensitive detection method of dibenzyl disulfide and the elucidation of the mechanism," *IEEE Trans. Dielectr. Electr. Insul.*, Vol.16, No. 2, pp. 509-515, 2009.

Lens' Design and Verification Used for Terahertz Beam Collimation

Qiu-Jie Yang†, Jing-Guo Huang, Zhi-ping He, Zhi-Ming Huang and Rong Shu
Key Laboratory of Space Active Opto-Electronics Technology, Shanghai Institute of Technical Physics, Chinese Academy of Sciences
Shanghai, 200083, P. R. China
†E-mail: yqj488112gxx@163.com

The free space transmission is an important research content in terahertz technology. The beam collimation is the key to the free space transmission. Nowadays, positive lenses are widely used in terahertz beam collimation. Compared with visible laser, the wave front of terahertz beam is neither plane wave nor spherical wave but mixed. From theoretical derivation, these negative lenses are more appropriate for terahertz beam collimation than positive lenses. The positive lenses with big F number can improve the energy utilization ration which are good for the space transmission but do little to improve the beam collimation. The parameter for negative lens depends on terahertz laser and propagation distance. Experiments demonstrate that the negative lens with f' = -101.6mm can improve the terahertz beam collimation from 6°to 0.1°, while the positive lens with f' = 101.6mm at the best position can make beam convergence with an angle of 2°. The discovery will make a big difference in the free space transmission and terahertz imaging.

Keywords: Terahertz Laser; Beam Collimation; Negative Lens; Divergence Angle.

1. Introduction

Terahertz (THz) is a kind of electromagnetic wave with the frequency between 0.1THz and 10THz. Both optics and electric method can generate terahertz radiation, but limited by the device size, the terahertz radiation based on electric method cannot achieve frequency above 1THz and the device is also fragile [1][2]. Based on DFG and QCLs, the terahertz radiation can cover full-wave band in terahertz. So the terahertz emission generated by optics is the mainstream direction for terahertz source [3][4]. Limited by the poor efficiency of DFG, the small effective emission cross section of QCLs [5][6], and also the high atmospheric absorptivity, to get efficient transmission of THZ waves must find the best transmission medium [7][8]. A lot of research shows that free space transmission is more advantage than any other transmission modes such as metallic waveguide and so on.

When solving the free space transmission of terahertz wave, many research institutions use positive lens to improve the divergence of the terahertz beam [9][10]. There is no doubt that positive lens can easily make full use of the terahertz energy, and which is good for the free space transmission. But positive lens has little effect in improving terahertz beam collimation. Limited by terahertz source and detector, this work has not been done by any other people. We find that those negative lenses are more appropriate for terahertz beam collimation. In this paper, the difference between terahertz laser and visible laser has been elucidated detailed. Theoretical derivation shows that the negative lenses are better for terahertz beam collimation, and the way to get the parameter for negative lens is also included in the paper. Finally, the theoretical derivation has been certificated by our experiment.

2. Theoretical analysis

The terahertz source based on optic mothed, in essence, is stimulated radiative transition of the molecule. So the terahertz is a kind of lasers [11]. It should be described by Gaussian optics as formula (1) and the terahertz wave front is showed as figure 1.

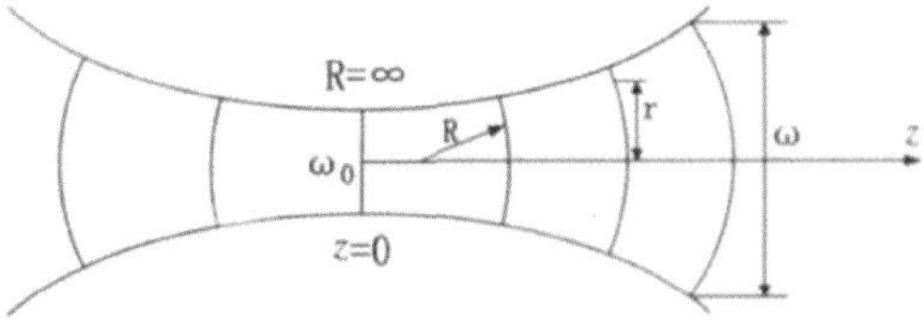

Fig. 1. Gaussian beam transmission along z-axis. In the figure, ω_0 is the beam waist, r is the distance between point (x,y) and z axis, R is the spherical wave radius, and ω is the spot radius transmission along z-axis.

$$\psi_{mn}(r,z) = A_{mn}\exp(-\frac{r^2}{w^2})\exp(-jkz)\exp(j\phi_{mn})\exp(-j\frac{kr^2}{2R}) \qquad (1)$$

In the formula (1), $\psi_{mn}(r,z)$ is the field distribution of mn order mode along z axis. It's defined that the laser waist ω_0 is at the position $z = 0$.A_{mn} is normalized amplitude. $r=\sqrt{x^2+y^2}$ is the distance between point (x,y) and z axis. z is the distance between the wave front and laser waist. The right side of formula (1) represents Gauss profile, plane wave phase, additional phase and spherical wave phase, respectively. Each parameter is calculated as follows.

$$\phi_{mn} = (m+n+1)\arctan\left(\frac{\lambda z}{\pi\omega_0^2}\right) \tag{2}$$

$$\omega^2 = \omega_0^2\left[1+\left(\frac{\lambda z}{\pi\omega_0^2}\right)^2\right] \tag{3}$$

$$R = z\left[1+\left(\frac{\pi\omega_0^2}{\lambda z}\right)^2\right] \tag{4}$$

For visible light, such as $\lambda = 587.6\text{nm}$, with laser waist $\omega_0 = 1\text{mm}$, it means $\frac{\lambda}{\omega_0{}^2} \ll 1$, so those formulas can be simplified as follows.

$$\phi_{mn} = (m+n+1)\arctan\left(\frac{\lambda z}{\pi\omega_0^2}\right) \approx 0 \tag{2-a}$$

$$\omega^2 = \omega_0^2\left[1+\left(\frac{\lambda z}{\pi\omega_0^2}\right)^2\right] \approx \omega_0^2 \tag{3-a}$$

$$R = z\left[1+\left(\frac{\pi\omega_0^2}{\lambda z}\right)^2\right] \approx \infty \tag{4-a}$$

$$\psi_{mn}(r,z) = A_{mn}\exp\left(-\frac{r^2}{w^2}\right)\exp(-jkz)\exp(-jkz) \tag{1-a}$$

It means that visible laser can be seemed as plane wave.

For terahertz bean, such as $\lambda_{1T} = 0.3\text{mm}$, with laser waist $\omega_0 = 1\text{mm}$, it means $\frac{\lambda}{\omega g^2} = 0.3$, so these formulas can be simplified as follows.

$$\phi_{mn} = (m+n+1)\arctan\left(\frac{\lambda z}{\pi\omega_0^2}\right) \approx (m+n+1)\arctan\left(\frac{z}{11}\right) \tag{2-b}$$

$$\omega^2 = \omega_0^2\left[1+\left(\frac{\lambda z}{\pi\omega_0^2}\right)^2\right] \approx \omega_0^2\left[1+\left(\frac{z}{11}\right)^2\right] \tag{3-b}$$

$$R = z\left[1+\left(\frac{\pi\omega_0^2}{\lambda z}\right)^2\right] \approx z\left[1+\left(\frac{z}{11}\right)^2\right] \tag{4-b}$$

$$\psi_{mn}(r,z) = A_{mn}\exp\left(-\frac{r^2}{w^2}\right)\exp(-jkz)\exp(-j\phi_{mn})\exp\left(-j\frac{kr^2}{2R}\right) \tag{1-b}$$

It means that terahertz bean cannot be seemed as plane wave or spherical wave which is quite different from visible lasers.

In order to describe easily, we select m=0 and n=0 which mean that the laser is fundamental mode. For λ = 500nm and λ = 0.3mm, the relation between ω, R and transmission distance z is showed in figure 2, respectively. From figure 2(a), we can easily find that the beam waist for visible laser just change a little along z axis, while terahertz beam waist changes fast along z axis which means terahertz beam has a big divergence angle about 5.7°which is bad for the free space transmission. More details, if z < *3mm*, the terahertz beam waist change also a little along z axis meaning its divergence angle no more than 1°. From figure 2(b), we can easily find that the spherical wave radium for visible laser is thousands times bigger than its beam waist which is another reason why visible laser can be seamed as plane wave. But for terahertz beam, the spherical wave radium is just the same order as its beam waist except the zone z < *3mm*. So, formula 1-b can be rewritten as follows.

$$\psi_{mn}(r,z) = \begin{cases} A_{mn}\exp\left(-\frac{r^2}{w^2}\right)\exp(-j(kz-\phi mn)) & 0 < z \le 3 \\ A_{mn}\exp\left(-\frac{r^2}{w^2}\right)\exp(-j(kz-\phi mn))\exp\left(-j\frac{kr^2}{2R}\right) & z > 3 \end{cases} \tag{1-c}$$

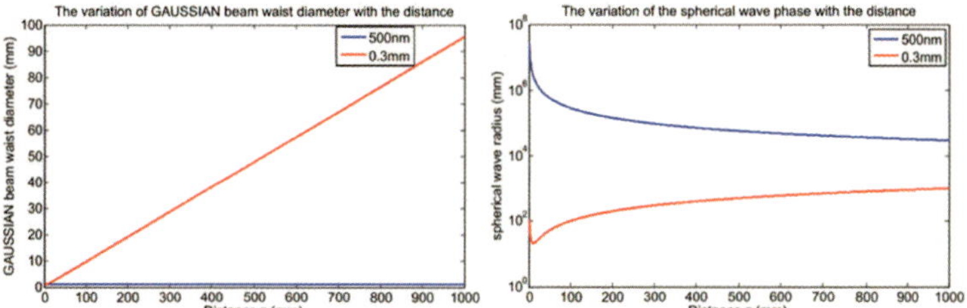

Fig. 2. The relation between Gaussian beam waist diameter, spherical wave radius and transmission distance z.

In order to achieve far distance transmission, lens must be used for terahertz beam to improve the beam collimation. From formula (1-c), different zone along z axis, the beam is quite different. In the zone $z < 3mm$, the terahertz beam can be seemed as plane wave, so a positive lens with big F number can makes beam converge in a long distance which could be full use of the beam energy. This has been used in some commercial terahertz source. In the zone $z < 3mm$, the terahertz beam can't be seemed as plane wave. From formula (1-c), it's obvious that the plane wave front adds a spherical wave phase. So to get plane wave front, we must eliminate the spherical wave phase $\exp\left(-j\frac{kr^2}{2R}\right)$ from the terahertz wave front. In Fourier optics Lens can be used as phase converter which can be described as formula (5). It means putting a negative lens after terahertz beam can remove the spherical wave phase $\exp\left(-j\frac{kr^2}{2R}\right)$ from formula (1-c). In this way, far distance transmission can be achieved. It can be explained as figure 3.

$$t_1(x,y) = \exp[-j\frac{k}{2f}(x^2+y^2) \tag{5}$$

In the formula (5), $t_1(x,y)$ s phase converter caused by the lens. f is the focus length of the lens. k is the wave vector.

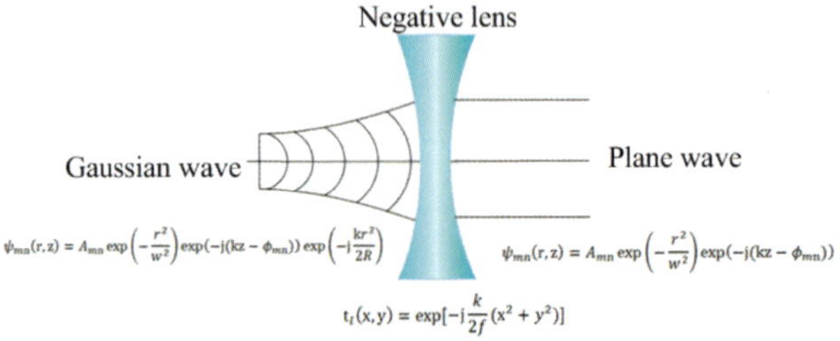

Fig. 3. Schematic diagram of wave front converting.

Compared figure 3 and formula (5), we can easily find that if formula (6) satisfied, we can converge the Gauss wave into plane wave. In order to get the parameter of the negative lens, we need to calculate the parameters of the Gauss wave carefully. To get far distance transmission, the spot diameter is the smaller the better which means the distance between lens and laser waist should be small.

$$f = -R = -z\left[1+\left(\frac{\pi\omega_0^2}{\lambda z}\right)^2\right] \tag{6}$$

3. Experiment and analyst

3.1. *The design and manufacture of the terahertz lens*

Lenses are used in so many terahertz systems to improve the beam collimation and energy utilization ration. Lens used in terahertz systems are divided into two categories depending on the material: lens made from high resistivity silicon and lens made from plastics, such as HDPE, PTFE, and Picairn. In our experiment, the lens is made from HDPE of which transmissivity curve in terahertz region is shown in figure 4(a).

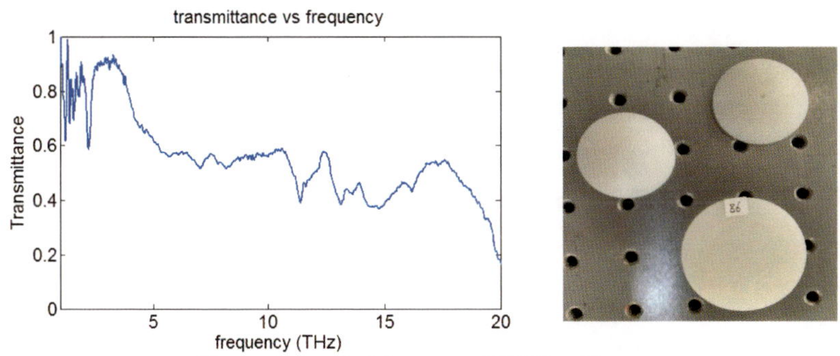

Fig. 4. Transmissivity curve of HDPE and lenses

In our lab, the 1T terahertz emission generated in GaSe is produced by pumped lasers with wave length 1068.1nm and Nd:YAG laser with wave length 1064nm from DFG. The terahertz beam has a beam waist of 1mm. From formula (3-b) and (4-b), we can calculate the spot diameter ω = 9.6mmand spherical radius R = 101.1mm at the distance z = 10mm. From formula (6), the focus length of the negative lens is 101.1mm. We don't use aspheric surface in the design because the wave length is long which is no sensitive to the surface roughness. The parameter of the negative lens optimized by ZEMAX is listed in table 1. The manufactured lenses are shown as figure 4(b).

Table 1. The parameter of terahertz lens

Surf	type	Radius	Thickness	Glass	Semidiameter	Conic
OBJ	Standard	Infinity	Infinity	--	0	0
STO	Standard	-51.832	3	HDPE	10	0
2	Standard	Infinity	-103.083	--	10.264	0
IMA	Standard	Infinity	--	--	0.133	0

3.2. *The measure and analysis of the divergence angle*

First, we measure the divergence angle without any lens. The experiment photo is shown as figure 5(a) and the detector we used in the experiment is bolometer working at liquid helium condition. We measure the spot diameter at place 1, place 2 and place 3 respectively. The relation between the signal intensity distribution and horizontal x at each place is shown in figure 5(b). From the figure, we can easily calculate the divergence angle of the terahertz source with 6°.

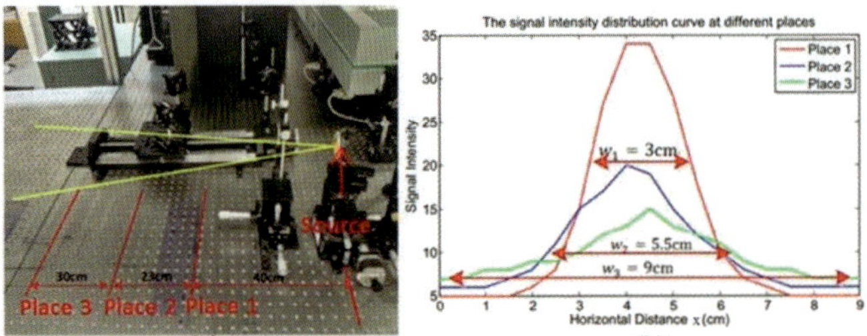

(a) (b)

Fig. 5. Schematic diagram of experiment and results.

Secondly, we measure the divergence angle with a positive lens of which the focus length is 101.1mm at the place $z = 10mm$ which is the best place for the lens. The experiment photo is shown as figure 6(a) and the detector we used in the experiment is bolometer working at liquid helium condition. We measure the spot diameter at place 1, place 2 and place 3 respectively. The relation between the signal intensity distribution and horizontal x at each place is shown in figure 6(b). From the figure, we can easily calculate the divergence angle of the terahertz beam after the positive lens is 2°. We also find that the positive lens converts the divergence beam into convergent beam, because after the lens we can easily find a place where the signal is bigger than any other places.

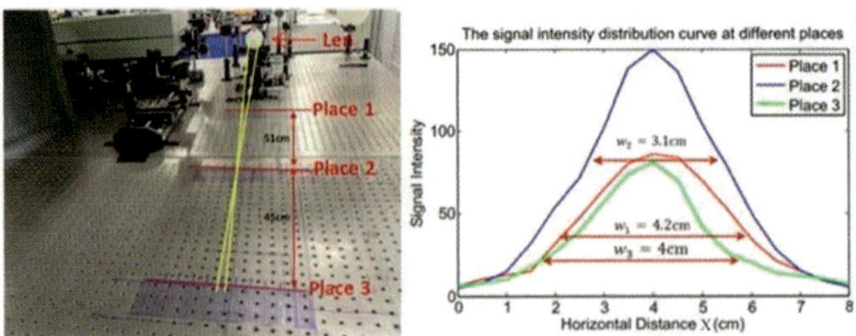

(a) (b)

Fig. 6. Schematic diagram of experiment and results.

Finally, we measure the divergence angle with a negative lens of which the focus length is 101.1mm at the place z = 10mm which is the best place for the lens. The experiment photo is shown as figure 7(a) and the detector we used in the experiment is bolometer working at liquid helium condition. We measure the spot diameter at place 1, place 2 and place 3 respectively. The relation between the signal intensity distribution and horizontal x at each place is shown in figure 7(b). From the figure, we can easily calculate the divergence angle of the terahertz beam after the positive lens is 0.1°.

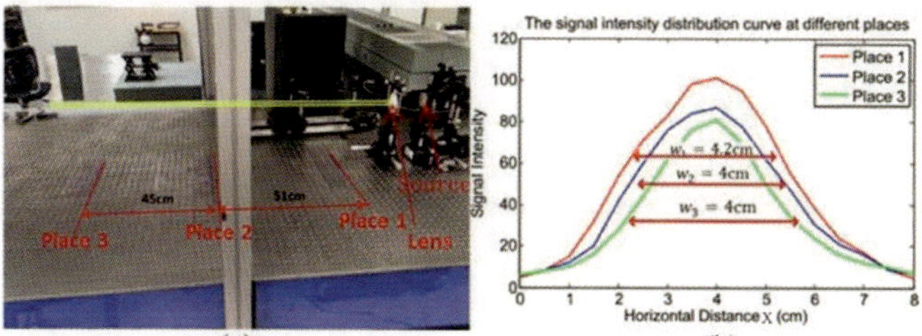

(a) (b)

Fig. 7. Schematic diagram of experiment and results.

From the experiment, we can get some conclusions as follows. Terahertz beam has a big divergence angle which is bad for the space transmission. The positive lens with big F number can improve the energy utilization ration which is good for the space transmission. But the positive lens can just do a little to improve the beam collimation. The negative lens can easily change the beam collimation which is important for the space transmission, terahertz imaging and so on.

4. Conclusion

Compared with visible laser, the wave front of terahertz beam is neither plane wave nor spherical wave but mixed. Terahertz beam has a big divergence angle which is bad for the space transmission. The positive lens with big F number can improve the energy utilization ration which is good for the space transmission. But the positive lens can just do a little to improve the beam collimation. The negative lens can easily change the beam collimation which is important for the space transmission, terahertz imaging and so on.

Reference

1. F. Nakajima, T. Furuta, and H. Ito, High-power continuous terahertz wave generation using resonant-antenna-integrated uni-travelling-carrier photodiode, Electron.Lett., 40,20, 1297-1299(2004).
2. R. W. Mcgowan, G. Gallot, and D. Grischkowsky, Propagation of ultrawideband short pulses of Thz radiation through submillimeter-diameter circular waveguides, Opt. Lett., 24, 1431-1433 (1999).

3. E. Schomburg, N. V. Demarina, and K. F. Renk, Amplification of a terahertz field in a semiconductor superlattice via phase-locked k-space bunches of bloch oscillating electrons, *Phys. Rev. B.* **67**, 155302 (2003).
4. M. Ingvarson, A. Ø. Olsen, and J. Stake, Design and analysis of 500GHz heterostructure barrier varactor quintuplers, in *14th International Symposium on Space TeraHertz Technology*, Tucson, AZ, (2002).
5. H. Willenberg. G. H. Dohler, and J. Faist, Intersubband gain in a bloch oscillator and quantum cascade laser, *Phys. Rev. B.* **67**, 085315 (2003).
6. M. Rochat, L. Ajili, H. Willenberg, J. Faist, H. Beere, G. Davies, E. Linfield and D. Ritchie, Low-threshold terahertz quantum-cascade lasers, *Appl. Phys. Lett.* 81, 1381-1383(2002).
7. B. S. Williams, H. Callebaut, S. K. Q. Hu, and J. L. Reno, 3.4-thz quantum cascade laser based on longitudinal-optocal-phonon scattering for depopulation, *Appl. Phys. Lett.* 82,1015-1017(2003)
8. R. Sachs and H. G. Roskos, Mode calculations for a terahertz quantum cascade laser, Opt. Express 12, 2062-2069(2004)..
9. D. Wvander Weide, J. Murakowski, and F. Keilmann, Gas-absorption spectroscopy with electronic terahertz techniques, IEEE Trans. Microw Theory Tech., 48(4, pt2), 740-743, (2000).
10. M. K. Choi, et al., Broadband 10-300 GHz stimulus-response sensing for chemical and biological entities, Physics in Med. and Bio., 47,3777-3787(2002).
11. R. Köhler, Terahertz semiconductor-heterostructure laser, Nature, 156-159(May 9, 2002).

Severity Diagnosis of Surface Partial Discharge Provoked by Immobilized Metal Wire on GIS Insulators

Xie Yaoheng[1,+], Zhou Weihua[1], Ye Huisheng[1], Liu Yun[1], Zhang Guoqi[2] and Ding Yuzhu[2]

[1]*Electric Power Research Institute, State Grid Hunan Electric Power Company, Changsha, China*

†*E-mail: xieyaoheng@163.com*

[2]*State Grid Hunan Electric Power Maintenance Company, Changsha, China*

The surface Partial Discharge (PD) triggered by immobilized metal wire on Gas Insulated Switchgear (GIS) insulators in terms of the discharge phenomena alongside its evaluation process was researched in this paper. With the application of gradually increased voltage, long-term tests were conducted on a well-established 252kV GIS experiment platform to observe the entire evolution process of PD from its very initiation till the eventual flashover. Ream-time measurement was undertaken during the tests to capture the trend curve as a result of test time. The results indicate that PD initiated by defects in GIS can be classified into three severity levels, namely, petty discharge, medium discharge, and threatening discharge. Moreover, a procedure based on k-means cluster analysis is proposed to diagnose and assess the severity of PD automatically.

Keywords: Partial discharge; diagnose and assessment; severity levels; insulation condition; defects.

1. Introduction

In complex high voltage components local insulation defects can cause partial discharges, especially in highly stressed GIS, these PD affected defects can lead to major blackouts. In the analysis of PD test results aspects, a single discharge frequency and amplitude analysis of the test results can not fully assess the dangers of discharge source, the need for more detailed and comprehensive PD source degree of risk assessment is necessary[1-4].

However, for the condition assessment of GIS equipment, in addition to the need to determine the type of defect and the discharge source position, also need to understand and obtain the severity of the discharge. At present, only according to the amplitude of PD it is difficult to determine the severity of the GIS discharge[5-8], the identification and assessment of the PD severity has always been a key point of the maintenance decision for GIS, which is particularly urgent to be solved.

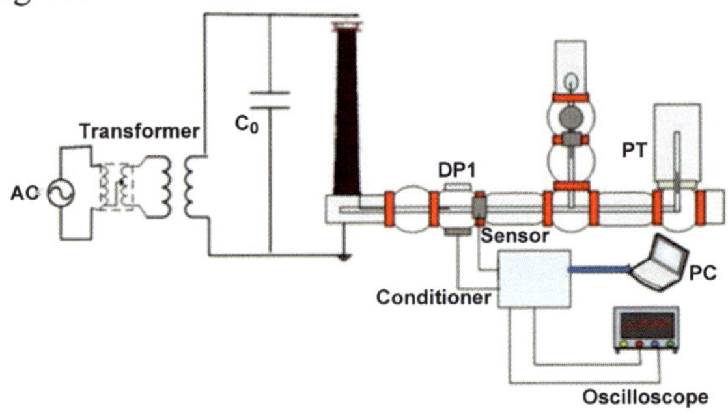

Fig. 1 The experimental setup

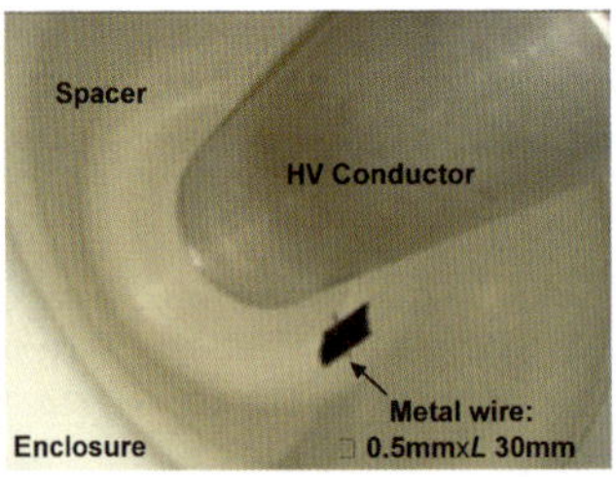

Fig. 2 The defect of metal wire on spacer surface

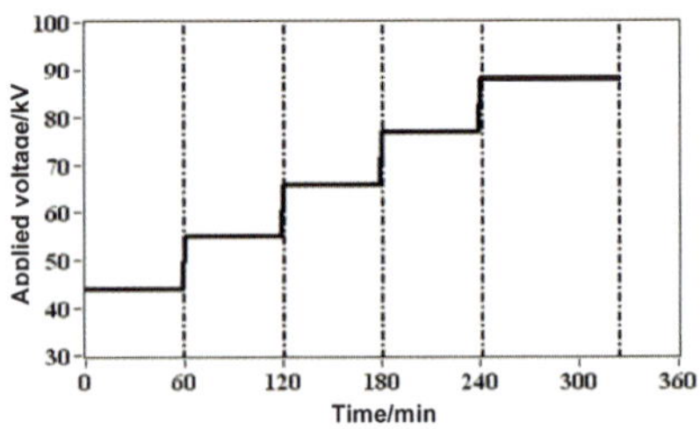

Fig. 3 The test voltage curve

2. Experimental setup and results

2.1. *Experimental setup*

PD measurement system for a 252 kV GIS is established in laboratory to acquire Ultra-High-Frequency (UHF) signals, as shown in Fig. 1. The GIS used in the test simulates the structure and operation of on-site GIS, and is composed of a defect chamber, bushing, Potential Transformer (PT), Current Transformer (CT) and dis-connectors, etc. The central conductor diameter and enclosure diameter are about 180 and 586 mm, respectively. Artificial defects are mounted in DP1 chamber, which is filled with SF_6 gas of 0.45 MPa. One internal disc UHF coupler and one external UHF sensor are assembled to receive PD-excited electromagnetic (EM) wave. The coupled PD signals are acquired and recorded with a high speed digital oscilloscope (Agilent DSO9404, 4 GHz). The sampling rate and record length of one UHF signal acquisition are 20 GS/s and 1000 ns, respectively. Moreover, the UHF Phase Resolved Partial Discharge (PRPD) and partial discharge pulse sequence (PRPS) patterns are recorded with a PD measurement system. A typical defect model of the surface discharge on GIS insulator is designed, as shown in Fig. 2. The surface as shown in Fig. 2 is a metal wire of 30 mm in length adhered on the insulator and the diameter of the wire is about 0.5 mm.

The stepped increasing voltage is used to accelerate the deterioration of the defect. After increasing the degree of deterioration, the constant voltage is applied until the end of the experiment to the test. The stable PD signal appears when the applied voltage reaches 45 kV, then progressively increase the voltage, held 60 minutes at 45, 55, 65, 75 kV respectively. In order to further accelerate the deterioration of the defect, the applied voltage reaches 85kV and maintained constant for 80 minutes until the end of the test. PD acquisition system uses real-time monitoring to record the discharge data every 10 s to analyze the PRPD and PRPS features.

2.2. *Results*

The development trend of maximum discharge amplitude V_{max}, average discharge amplitude V_{ave} and discharge number N over the test time are shown in Fig. 4. It can be seen that the discharge amplitude will appear the trend of decrease after the first increase in each of the stepped voltage in the test which is not monotone. Fig. 5 shows various kinds of PRPD and PRPS diagrams including Q–N–φ gray spectrum diagrams of PRPD, $\Delta u/\Delta t$ and Δu spectra of PRPS spectrum diagram.

According to the characteristics of the fingerprint data for the surface in each stage, the following six fingerprint parameters with the monotonous change characteristics are extracted to diagnose the severity of PD for GIS, as shown in Table 1. On the basis of the fingerprint data, the paper has given a kind of the severity diagnosis method of PD. The change trend of the six feature fingerprint parameters with test time for the surface defect on the insulator are shown in Fig. 6. The black curve is the original change trend with the time for each feature fingerprint, and the red solid line is the exponential smoothing curve of the black curve. It can be seen that all the six feature fingerprints present the monotonic increasing or monotonic decreasing trend with the test time, that is, φ_w, $\mu(N)$, $\sigma(N)$, $E_n(V)$, $\sigma(N_v)$ and $E_n(V_{max})$ are monotonic increasing before the insulator flashover, then there is monotonic decreasing.

3. Severity diagnosis and assessment

By using the principal components analysis(PCA) projection algorithm[9], the six-dimension features mentioned above in section 2 are reduced to two-dimension parameters $P1$ and $P2$, as shown in Fig. 7. It is observed that the features of the surface defect falls into distinct clusters in the map. In the paper,

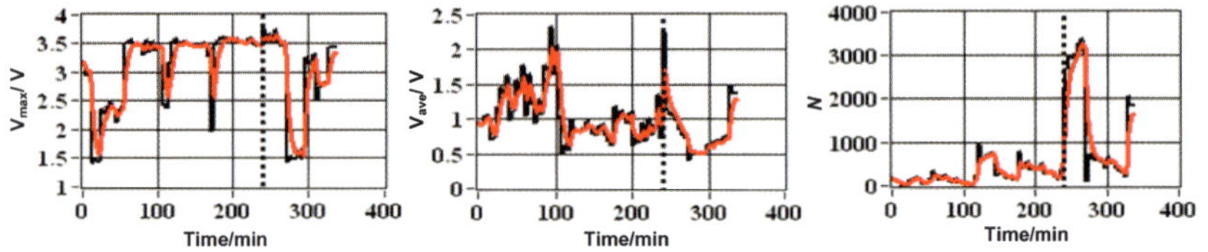

Fig. 4 The change trend of PD amplitudes with test time

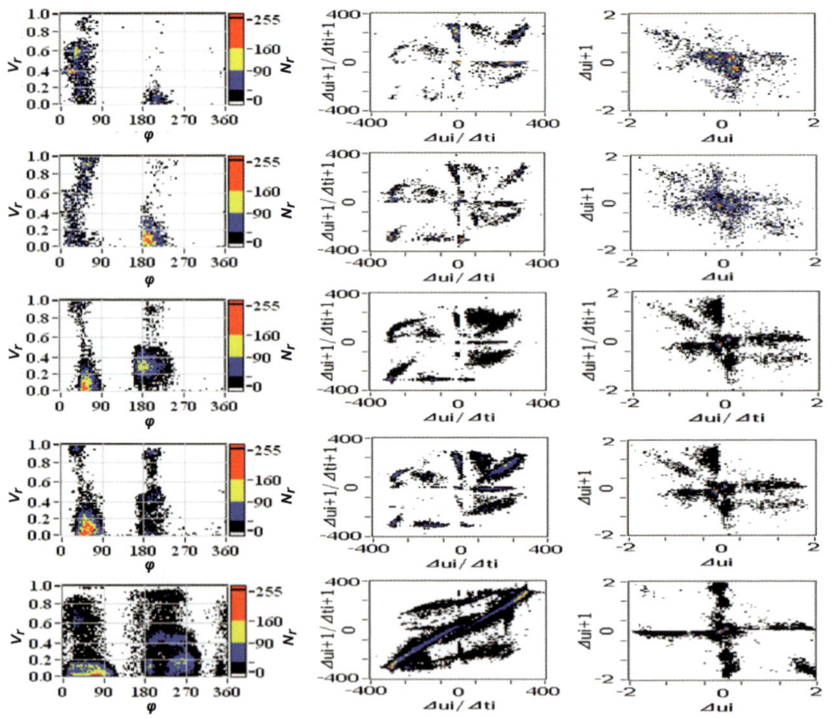

Fig. 5 The PRPD and PRPS spectrograms in different discharge stages

(V_r is the relative amplitude of PD, N_r is the relative number of PD, φ is the phase angle of PD, Δu is the voltage difference between two partial discharge pulses, Δt is the time interval between two partial discharge pulses)

Table 1 Severity characterization feature fingerprints of the surface defect

Notation	Feature fingerprints
φ_w	Phase width of PD
$E_n(V)$	The entropy of the discharge amplitude
$\mu(N)$	The average discharge times
$\sigma(N)$	The standard deviation of discharge times
$E_n(V_{max})$	The entropy of the maximum discharge amplitude
$\sigma(N_v)$	The standard deviation of the discharge amplitude

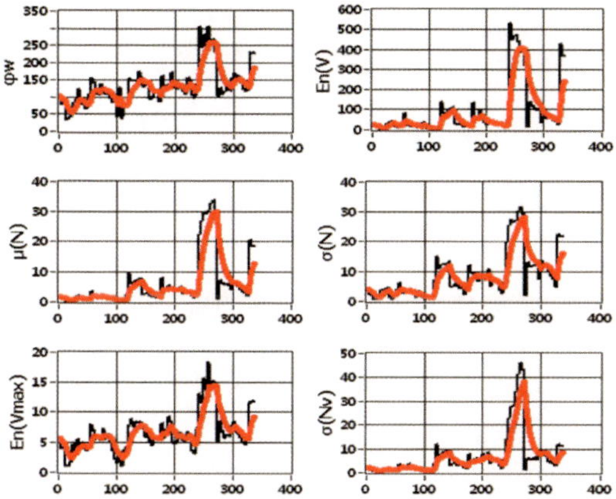

Fig. 6 The change trend of PD fingerprint features with test time

different discharge stages are used to assess the severity of the defect. The k-means clustering algorithm is adopted to classify the feature parameters of the surface defect (see Fig. 7d). Before performing k-means clustering analysis, the number of clusters needs to be predefined. The membership value of data points are calculated as a function of distance, then the cluster centroids can be updated during the optimization procedure with a function of data points and their membership value in each cluster.

In order to avoid the effect of subjective judgment on severity assessment, the six feature fingerprints by clustering analysis is used to define the discharge stage which represents the severity of the defect. Based on the experimental results, the discharge stage is divided into three parts, including initial stage S_{ini}, development stage S_{dev}, and severe stage S_{ser}. Then the k-means clustering method with three clusters is applied to classify the feature parameters of the surface defect, the fingerprint characteristics collected by the test are as severity diagnosis clustering samples. According to the principle of the minimum distance, the discharge phase of the PD defect can be judged automatically.

4. Conclusions

PD measurements are performed on a kind of typical defect of the surface in a 252 kV GIS. The extracted features are applied to diagnose severity which is in accord with the three discharge stages of PD sources. The discharge stage is successfully divided by clustering the extracted features with k-means algorithm and the proposed algorithm is proved effective.

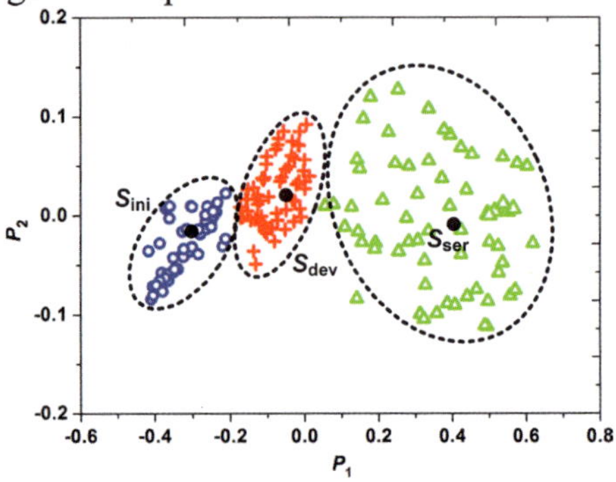

Fig. 7 Feature distribution after using PCA projection algorithm

References

1. A. Bargigia, Detection of Partial Discharges in Gas Insulated Substations, *IEEE Trans. Power Delivery*, 1992, 7(3): 1239-1249.
2. H.-G. Kranz, Fundamentals in Computer Aided PD Processing, PD Pattern Recognition and Automated Diagnosis in GIS, *IEEE Trans. Dielectr. Electr. Insul.*, 2000, 7(2): 12–20.
3. Judd M D, The excitation of UHF signals by partial discharges in GIS, *IEEE Trans. Dielectr. Electr. Insul.*, 1996, 3(2): 213-228.
4. Gulski E, Kreuger F H, Computer-aided recognition of discharge sources, *IEEE Trans. Dielectr. Electr. Insul.*, 1993, 27(1): 969-983.
5. Okabe S, Kaneko S, et al, Simulation of propagation characteristics of higher order mode electromagnetic waves in GIS, *IEEE Trans. Dielectr. Electr. Insul.*, 2006, 13(4): 855-861 .
6. Pompili M, Mazzetti C and Bartnikas R, PD pulse burst characteristics of transformer oils, *IEEE Trans. Power Delivery*, 2006, 21(2): 689-698.
7. Candela R, PD recognition by means of statistical and fractal parameters and a neural network, *IEEE Trans. Dielectr. Electr. Insul.*, 2000, 7(1): 87-94.
8. L.E. Lundgaard, Particles in GIS, Behavior and Possibilities for Characterization from Acoustic Signatures, *IEEE Trans. Dielectr. Electr. Insul.*, 2001, 8(6): 1064–1074 .
9. K. Dreisbusch, Determination of a Failure Probability Prognosis based on PD-Diagnostics in GIS, *IEEE Trans. Dielectr. Electr. Insul.*, 2008, 15(6): 1707-1714.
10. CIGRE WG D1.03, Risk assessment on defects in GIS based on PD diagnostics, *IEEE Trans. Dielectr. Electr. Insul.*, 2013, 20(6): 2165-2172.

Design of a Vertically Stacked Substrate Integrated Circular Cavity (SICC) Filter in LTCC Technology

Huan Wu, Jian-Rong Chen, Wen-Qiang Jia and Xiao-Yang Zhang

Institute of Space Radio Technology, Xi'an, China

†*E-mail: yimimomo@126.com*

A new K-band fourth order bandpass filter is proposed in this paper based on Substrate Integrated Circular Cavity(SICC) resonators. Taking advantage of the multilayer substrate wiring merit of the Low-Temperature Co-fired Ceramic(LTCC) technology, four resonant cavities which are stacked vertically in proper order are coupled through inductive circular arc slot or capacitive hole. The novel three-dimensional structure makes the filter more compact compared to the same type of traditional planar filter, and it still maintains a good frequency selection performance. TM_{010} mode is chosen as the dominant mode of the SICC, which makes the design more flexible. I/O CoPlanar Waveguide(CPW) feeds are applied for planar circuit integration. The measured results show that insertion loss of the designed SICC filter is less than 1.55 dB and return loss of the filter is greater than 13.04 dB in the passband centered at 18.45 GHz with a fractional bandwidth of 10.8%. The stopband of the designed filter is from 16.16 GHz to 21.5 GHz with a rejection over 30 dB. Measured results demonstrate good agreements with the simulation results.

Keywords: bandpass filter; LTCC; substrate integrated circular cavity; vertically stacked; TM_{010} mode; K-band.

1. Introduction

With the constant improvement of the integration densities of microwave and millimeter wave circuits and the prosperous development of multilayer circuit technologies, the use of conventional waveguide or planar transmission line is no longer sufficient for the miniaturization demand of the circuit and system. In this context, there is a growing interest in a novel transmission structure called Substrate Integrated Waveguide(SIW). It not only inherits the advantages of the traditional metallic waveguide like low loss, complete shielding, high quality factor and high power handling capability, but also unites the merits of planar transmission lines such as easy integration, easy fabrication, compact size, low weight and so on[1–3]. Hence, emerging SIW technology is considered to combine the best and complementary features of both planar transmission lines and traditional waveguides[4]. And SIW is effective in the integration of planner circuits using a multilayered circuit fabrication process such as LTCC technology.

LTCC, as a promising and fast development implementation form of package structure with circuits, has served to realize some passive components with three-dimensional construction which can substantially decrease the package size and improve the design flexibility[5–6]. Recently, LTCC-based SIW cavity filter has been widely studied because of its excellent performance [7–9]. At present, conventional SIW filters are predominantly based on rectangular cavities. However, substrate integrated circular cavity(SICC) filters have not only the same attractiveness as classical SIW filters in terms of quality factors, low losses and planar compatibility, but they also present a total design flexibility as long as the chosen mode is revolution-invariant [10–12]. But most of these designs are based on the planar structures.

In this paper, a novel fourth order SICC bandpass filter is proposed and implemented using LTCC technology. Four resonant cavities are vertically stacked in proper order and coupled through inductive circular arc slot or capacitive hole. Owing to the multilayer capability of the LTCC technology, the structure of the designed filter with high performance successfully evolve into a compact three-dimensional structure which makes the total size greatly reduced compared to a planar structure.

2. SICC Resonator and Operating Mode

The structure of the SICC resonator is illustrated in Fig. 1(a). The disks in horizontal plane and the circular side walls consisted of via-holes are made of metals. Thus, a circular cavity is delimited into the substrate. Here d, p and R stand for the diameter of the via holes, the distance between two adjacent via holes and the radius of the SICC, respectively. To meet the constraint during the fabrication process, p is usually at least 2.5 times bigger than d.

TM_{mn} mode can be established and propagated in a round-shaped cavity. When the cavity radius-to-height ratio is less than 2.1, TM_{010} mode is the dominant mode of the SICC. The distributions of electromagnetic field in a SICC working on TM_{010} mode is shown in Fig. 1 (b). The mode is revolution-invariant, and resonant frequency is only fixed by the radius of the cavity[11]. That means we can adjust the height of cavity as needed without affecting its performance in a LTCC multilayer circuit. Besides, compared with the SIW rectangular cavity, the input and output ports of the SICC can be placed in an angle according to the layout need as shown in Fig. 1(c). These benefits can largely improve the design flexibility of passive circuits based on SICC resonant cavity. SICC resonant frequency is approximated by the following formula (1).

$$f_0 = \frac{2.45c_0}{2\pi R\sqrt{\varepsilon_r}} \tag{1}$$

Where c_0 is the light velocity in vacuum and ε_r is the dielectric constant of the substrate.

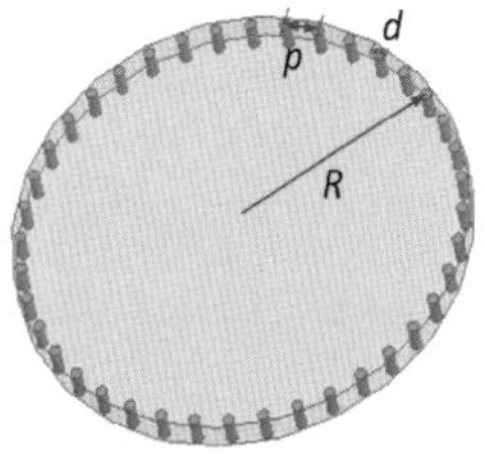

(a)

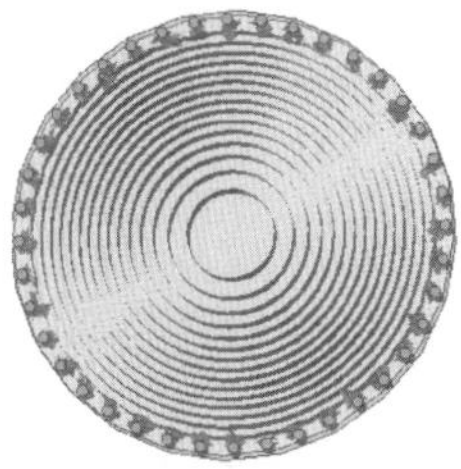

(b)

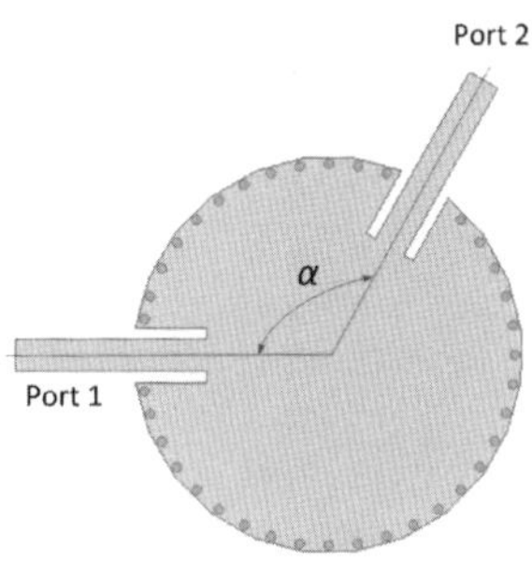

(c)

Fig. 1. (a) Structure of the SICC resonator (b) Electromagnetic field distributions in a SICC working on TM010 mode (c) SICC with an angle between the input and output ports.

The LTCC substrate in this design is Ferro-A6M (ε_r = 5.9, tanσ = 2×10^{-3}), with a thickness of 0.096 mm for every layer. According to the LTCC process rules, d and p are determined to be 0.13 mm and 0.41 mm in this design. The central frequency f_0 of the proposed filter is 18.45 GHz, so the initial radius of the SICC can be obtained by the formula (1). The calculated R is 2.61 mm.

3. Design of the SICC Filter

3.1. *Coupling Matrix*

According to the need, a fourth order directly coupled bandpass filter centered at 18.45 GHz with the fractional bandwidths of 10.8% is proposed. Based on the classical Chebyshev low-pass prototype, the external coupling Q_e value and coupling coefficients meeting the design requirements can be synthesized[8]. And the external coupling Q_e value is shown as follows:

$$Q_e=10.23 \tag{2}$$

The coupling matrix can be given as follows. Because the filter is directly coupled, so $M_{14}=0$

$$M=\begin{bmatrix} 0 & 0.0901 & 0 & 0 \\ 0.0901 & 0 & 0.0712 & 0 \\ 0 & 0.0712 & 0 & 0.0901 \\ 0 & 0 & 0.0901 & 0 \end{bmatrix} \tag{3}$$

3.2. *Structure of the Filter*

Fig. 2 shows the configuration of the proposed filter. The filter is composed of four SICC which are stacked vertically in proper order and each resonant cavity consists of three layers of dielectric substrate. The coupling between the cavity 1 and 2 is realized through inductive circular arc slots at the common metal surface of cavities, and the coupling manner between cavity 3 and 4 is the same to them. The two symmetrical slots are located at about 0.77R near the sidewalls, where the magnetic field is the strongest for the TM_{010} mode. Therefore this coupling method is magnetic coupling. The coupling between the cavities 2 and 3 is achieved through the capacitive circular hole at the center of common metal layer of the two cavities. Since the electric field in this position for the TM_{010} mode is the maximum, this coupling manner can be considered as electrical coupling. These magnetic/electrical coupling strength are controlled by the dimensions of the slot and the hole.

The initial dimensions of the resonator cavities with the above coupling manners can be determined in Ansoft full-wave simulation software HFSS according to the analytical equations. Due to the presence of coupling between two resonator cavities, the central frequency peak point of the single resonator cavity separates into two distinct peaks, and the two resonant frequency peaks can be obtained using the eigenmode solution solver of HFSS. Thus, The coupling coefficient M_{ij} can be extracted by using the following equation[2].

$$M_{ij}=\frac{f_1^2-f_2^2}{f_1^2+f_2^2} \tag{4}$$

Where f_1 and f_2 stand for the upper and lower resonant frequencies of the resonant cavity, respectively.

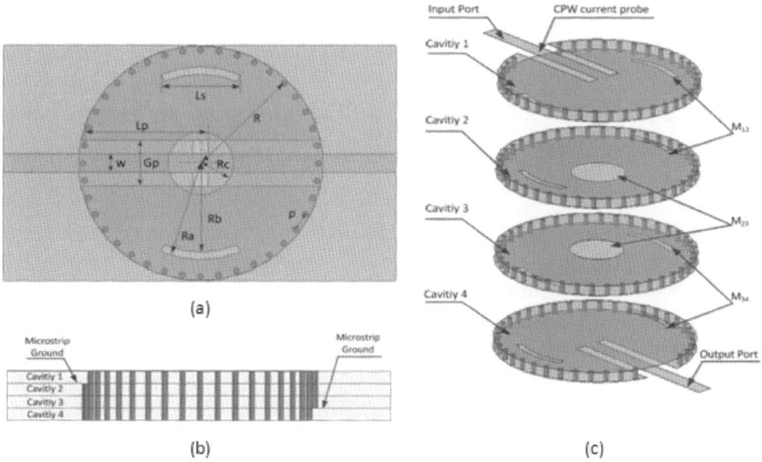

Fig. 2. The configuration of the filter (a) Top view (b) Side view (c) Anatomy view.

In the design, one feeding technique named CPW current probe is applied to excite the TM_{010} mode in the SICC filter, as shown in Fig. 3. Moreover, in order to enable easy measurement and integration of surface mount components, a 50 Ohms microstrip is connected with CPW current probe and the central metal of CPW is as wide as the microstrip. Previous researches showed that the microstrip to SIW transition realized by CPW current probe has a smaller size than a gradual microstrip transition and a lower insertion loss than a via coupling excitation[13].

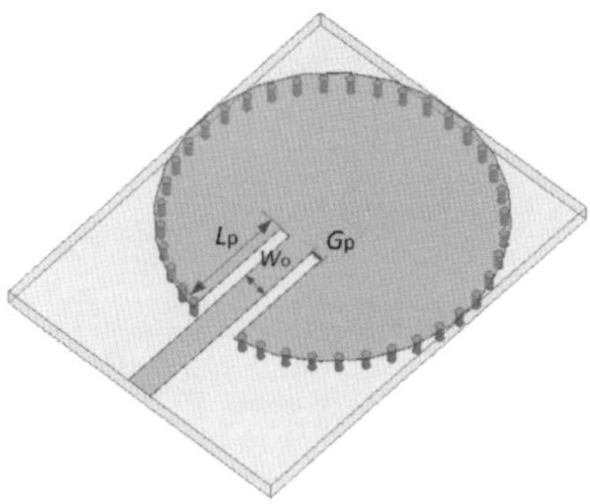

Fig. 3. The excitation structure of the SICC.

The coupling strength between the probe and the cavity changes as the length *L*p and width *G*p of the probe change. So the external coupling Q_e value can be determined by the dimensions of the CPW current probe. By simulating the phase of scattering parameter S11, the external coupling Q_e value can be obtained by the following formula[2].

$$Q_e = \frac{\omega_0}{\Delta\omega_{\pm 90^\circ}} \tag{5}$$

Where ω_0 is the resonant frequency of single-loaded resonant cavity, and $\Delta\omega_{\pm 90^\circ}$ is the relative frequency span of $\pm 90^\circ$ phase shift with respect to ω_0.

3.3. *Fabrication and Measurement of the Filter*

After simulation and optimization in HFSS, the geometry parameters of the proposed filter are as follows: R=2.58 mm, Ls=1.74 mm, Ra=2.2 mm, Rb=2.03 mm, Rc= 0.725 mm, Lp=2.71 mm, Gp=1.06 mm, w=0.43 mm. In order to validate our design, the filter is fabricated on the substrate Ferro-A6M with metalized via-hole with the diameter of 0.13 mm and via-to-via pitch of 0.41 mm. Fig. 4 shows the photograph of the fabricated SICC filter. The size of the filter is $5.4 \times 5.4 \times 1.152 \text{mm}^3$ after de-embedding the test leads. Because of the merit of the LTCC technology, the substrate area occupied by the designed filter is reduced to approximately 66.1% compared to the proposed filter centered at 25.7 GHz in[9], and is 41% of the proposed filter centered at 14.6 GHz in[11].

Fig. 4. Photograph of the fabricated SICC filter.

The filter was tested by the Agilent 5224A vector network analyzer with K joints connected to the microstrip line. The comparisons between simulated and measured S parameters of the proposed SICC filter are plotted in Fig. 5 by the dashed and solid line, respectively. As can be seen, the proposed filter has a center frequency of 18.45 GHz with a bandwidth of 2 GHz. The measured insertion loss of the filter within pass band is less than 1.55 dB which is 0.12 dB larger than the simulated result. And the return loss within pass band is greater than 13.04 dB which is less than the simulated result. The discrepancy should be due to the fabrication tolerance and the slight impedance mismatch between K joints and microstrip line. Comparing the measured and simulated stopband rejection results, good agreement is observed to a certain extent. The stopband of the designed filter is from 16.16 GHz to 21.5 GHz with a rejection over 30 dB.

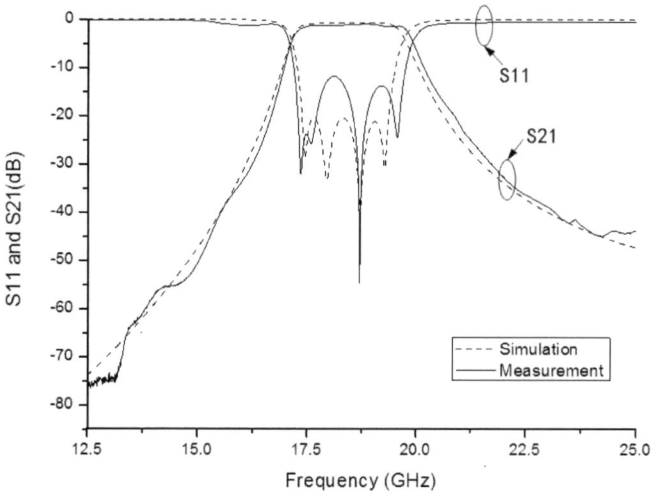

Fig. 5. Simulated and measured S parameters of the proposed SICC filter.

4. Conclusion

In this paper, a 18.45 GHz SICC-based fourth order bandpass filter was designed. Owing to the multilayer capability of the LTCC technology, four resonant cavities can be vertically stacked in proper order and coupled through inductive circular arc slot or capacitive hole. The proposed SICC filter with high performance successfully evolved into a compact three-dimensional structure which made the total size greatly reduced compared to a planar structure. The measured results show that insertion loss of the filter is less than 1.55 dB and return loss of the filter is greater than 13.04 dB in the passband centered at 18.45 GHz with a fractional bandwidth of 10.8%. Reasonable agreement were obtained between simulated and measured results.

Acknowledgments

The authors wish to acknowledge Fangqing Mu from CETC43 for his supporting in fabrications. Thanks to ZhiChao Li for his help in reviewing the paper.

References

1. N. H. Baba, M. I. Ismail, M. T. Ali, A. H. Awang and H. M. Hizan, "Substrate integrated waveguide radiating bandpass filter for c-band applications," *Applied*

Electromagnetics (APACE), 2014 IEEE Asia-Pacific Conference on, Johor Bahru, 2014, pp. 183-186.
2. L. Shaolan, Y. Bo and Z. Runqi, "A Ka-band FSIW band-pass filter with LTCC technology," *Microwave Technology & Computational Electromagn -etics (ICMTCE), 2011 IEEE International Conference on*, Beijing, 2011, pp. 196-199
3. X. P. Chen and K. Wu, "Substrate Integrated Waveguide Filters: Design Techniques and Structure Innovations," in *IEEE Microwave Magazine*, vol. 15, no. 6, pp. 121-133, Sept.-Oct. 2014.
4. F. Cheng, X. Q. Lin, M. Lancaster, K. Song and Y. Fan, "A Dual-Mode Substrate Integrated Waveguide Filter With Controllable Transmission Zeros," in *IEEE Microwave and Wireless Components Letters*, vol. 25, no. 9, pp. 576-578, Sept. 2015.
5. X. Zhang, C. Ma and F. Wang, "Design of compact dual-passband LTCC filter exploiting stacked QMSIW and EMSIW," in *Electronics Letters*, vol. 51, no. 12, pp. 912-914, 2015.
6. B. Wang, J. Liao and L. Zhu, "Design of LTCC multi-layer band-pass filter," *Microwave and Millimeter Wave Circuits and System Technology (MMWCST), 2012 International Workshop on*, Chengdu, 2012, pp. 1-3.
7. H. Grubinger, H. Barth and R. Vahldieck, "An LTCC-based 35-GHz substrate-integrated-waveguide bandpass filter," *Microwave Symposium Digest, 2009. MTT '09. IEEE MTT-S International*, Boston, MA, 2009, pp. 1605-1608
8. Bingyao Liu, Jianming Zhou, Ruobing Liu, Qiong Wu and Kunpeng Zhang, "A 35 GHz reduced-size bandpass filter based on SIW in LTCC technology," *Microwave Technology & Computational Electromagnetics (ICMTCE), 2013 IEEE International Conference on*, Qingdao, 2013, pp. 77-80.
9. R. Zhang, Z. Wang, B. Yan and R. Xu, "FSIW cavity filter and Derivative FSIW cavity and its filters with LTCC technology," *2009 Asia Pacific Microwave Conference*, Singapore, 2009, pp. 1360-1363.
10. Q. F. Wei, Z. F. Li, L. S. Wu, W. Y. Yin and J. F. Mao, "Investigation on some novel multi-layered cross-coupled substrate integrated waveguide (SIW) circular cavity filters," *2008 Asia-Pacific Microwave Conference*, Macau, 2008, pp. 1-4.
11. B. Potelon, J. C. Bohorquez, J. F. Favennec, C. Quendo, E. Rius and C. Person, "Design of Ku-Band Filter based on Substrate-Integrated Circular Cavities (SICCs)," *2006 IEEE MTT-S International Microwave Symposium Digest*, San Francisco, CA, 2006, pp. 1237-1240.
12. B. Potelon, J. F. Favennec, C. Quendo, E. Rius, C. Person and J. C. Bohorquez, "Design of a Substrate Integrated Waveguide (SIW) Filter Using a Novel Topology of Coupling," in *IEEE Microwave and Wireless Components Letters*, vol. 18, no. 9, pp. 596-598, Sept. 2008.
13. W. q. Jia and H. x. Yu, "A 27.5–29GHZ SIW cross-coupling filter in LTCC," *IET International Radar Conference 2015*, Hangzhou, 2015, pp.1-4.

Research on the Impact of Meteorological Factors on Electric Power Accidents

Chen Jia†, Chengrui Wang[a], Chao Zhang[b] and Jing Zhou[c]
NARI Group Corporation
Beijing, China
E-mail: †jiachen1@sgepri.sgcc.com.cn, [a]wangchengrui@sgepri.sgcc.com.cn,
[b]zhangchao5@sgepri.sgcc.com.cn, [c]zhoujing12@sgepri.sgcc.com.cn

Most of the electric power accidents occur because of the damage of the extreme weather. Meanwhile, the meteorological disaster has the greatest influence on the power system. This paper chooses a selected region in Southeast China as an example, and collects the electric power accident data continuously for 48 months and corresponding data of 15 meteorological factors of this region. Then, using logistic regression to establish the impact model of meteorological factors to research on the relations between electric power accidents and weather conditions. After that, use the test data collected of another year to check the accuracy of the established model. The model is mainly used to set up a warning system on protecting the security of power grid. The electric power corporation can choose relevant preventive measures to reduce the possible losses that are caused by the hazard weather in the future.

Keywords: Meteorological Factor; Power Accident; Power System; Logistic Regression; SPSS.

1. Introduction

In recent years, the global extreme weather occurs more frequently, which causes the domestic and foreign power grid accidents increasing. The major blackout caused by natural disasters can account for 53.64% of the total power outage, and the proportion is over half of the total[1]. It is easy to conclude that natural disaster is the largest hazard source of the large area blackout of the power system. Natural disasters include meteorological disaster, geological disaster, space weather disaster, flood disaster and drought disaster[2]. Among them, the meteorological disaster has the greatest influence on the power system. In particular, meteorological disaster can be divided into five kinds: storm disaster, snow and ice disaster, heavy fog disaster, thunder and lightning disaster and temperature disaster. Due to the different characteristics of the different natural disasters, the impacts on the power system are not same either. Generally, the failure of natural disasters can cause great damage to the power system infrastructure[3].

All the time, meteorological disaster has always been one of the main threats of the security of the power grid. In recent years, meteorological disasters occurred quite frequently and seriously, the destructive power and damage on the power grid increase year by year. Due to the severe climate changes in these years, extreme bad weather always occurs in summer, such as the torrential rain and thunderstorm. While, this kind of weather phenomenon always occurs in a short period and focus in one particular area, which gives a great damage of the region[4–7].

As our country has a vast territory, the energy is mostly distributed in the West while most of the power users are located in the East. Therefore, the long distance trans-regional power transmission mode is widely used in our country. The advantage of this transmission mode is economical and high efficiency, but the grid needs too long distant transmission and it will go through a wide variety of topographic features. This makes the probability of meeting the extreme adverse weather conditions become very high. As the lines in the long distance trans-regional transmission channel are mostly impact transmission lines, while multiple lines are erected on one pole then the transmission lines are more concentrated. Once the severe weather occurs, the transmission lines are more vulnerable to wind bias, lightning, ice flashover and other threats resulting in line trip and other accidents, which can even lead to the full loss of power transmission more seriously.

Currently, the scholars have already done related researches of the impacts of meteorological factors on electric power security. Literatures [8–11] emphasize that the traditional electric power defense system should be improved in order to resist the extreme natural disasters. Literatures [12–15] focus on research of the effects of seasonal variation on power grid loads and it shows that the extreme temperature changing in summer and winter can cause drastic change of grid load. This paper choose a region in Southeast China as an example, collect the data of recent years and test the effect of meteorological factors on electric power accidents by using metering method. The significance of the study is to establish the model on studying the impact of meteorological conditions on electric power accident by scientific methods, which also builds a solid foundation for the establishment of the warning system on electric power accident.

2. Analysis of Measurement Methods

Multiple linear regression analysis is a statistical analysis technique of determining the quantitative relation between two or more explanatory variable, which has been widely used in a variety of fields[16]. Some assumed conditions of the multiple linear regression are quite important, of which if being breached will lead to unreasonable estimate. Such as the multiple linear regression method requiring the values of variables to be continuous random variables, but according to the statistics of electric power system accidents, the values of variables are discrete random variables and using multiple linear regression analysis will lead regression estimation to a serious error. Logistic regression can supply the deficiency of the multiple linear regression, which has less assumed conditions on the variables[17]. According to the former researches, Logistic regression has better fitting and predicting results than the multiple linear regression. With that reason, this paper uses Logistic regression model to do the research on the variable correlation.

Logistic regression model is a widely used multivariate statistical analysis method in the field of doing regression analysis on binary dependent variables(y=1 or y=0). The main method of Logistic model is using the Maximum Likelihood Estimation to calculate the parameters through the sample data, while the integral equation is an expression of probabilities. There are two main purposes of the regression model: one is searching for the dangerous elements, such as the dangerous elements of one disease; the other one is prediction, if the Logistic regression model has been established, then it can be based on the model, predict the probability of a certain situation with different independent variables.

This paper uses Logistic regression model to calculate the influence of the meteorological factors on electric power accidents in a certain period of time. The mathematical expression of Logistic regression model is shown as in Eq. (1):

$$\log it(\mathrm{y}) = \ln\left(\frac{p}{1-p}\right) = a_0 + a_1F_1 + a_2F_2 + \cdots + a_nF_n \tag{1}$$

Or as in Eq. (2):

$$\mathrm{p}=\frac{\exp(\mathrm{a_0+a_1F_1+a_2F_2+\cdots+a_nF_n})}{1+\exp(\mathrm{a_0+a_1F_1+a_2F_2+\cdots+a_nF_n})} \tag{2}$$

In Eq. (1), y=(1,0) stands for the number of times that one particular event occurred; when y=1, it stands for occurred; when y=0, it stands for not occurred; p=P(y=1) stands for the probability of event occurred; a_i stands for parameters to be estimated; F_i stands for the independent variables, i=1,…,n. Like other multiple regression methods, Logistic regression model is very sensitive to multiple linear as well, if the correlation of variables is high, even a tiny change of the sample will still lead to a great change of the parameters estimation, which would reduce the accuracy of the model. In primitive variables, the meteorological elements like the monthly average wind speed, the monthly maximum wind speed, the monthly extreme wind speed have higher correlation among these variables, in order to effectively eliminate the effects of multi collinearity among primitive variables without affecting its information contents. In this paper, we firstly analyse the factors of meteorological indicators.

Factor analysis method is a multivariate statistical method, which researches the inner related dependency relation of variables and converses complex variables to several comprehensive indicators. The basic thought of the method is to classify the observed variables. It will classify variables with high correlation in one class, while the different classes of variables have lower correlation. Then, every class of variable is presenting one basic structure, and that is the common factor. Through factor analysis extracting common factors, can effectively overcome original variable multiple linear, and use common factors instead of original variables for Logistic regression.

3. Sample Data Selection

This paper collects the power accident statistical data from January 2010 to December 2013 of a region in Southeast China, and chooses the data of 12 months of 2014 as the test sample. According to the distribution characteristics of the monthly accident number in four consecutive years, the accident number of spring and summer is apparently much higher than that of autumn and winter (Fig. 1). There is a certain relationship between the accident number and the change of seasons.

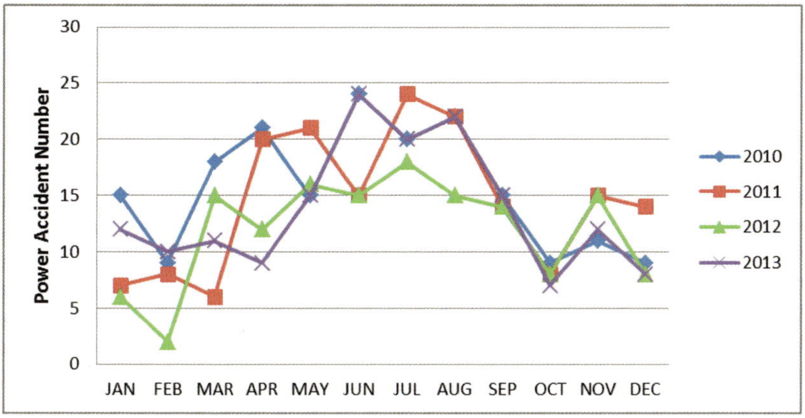

Fig. 1 The monthly illustration of accidents from 2012 to 2013

According to the histogram (Fig. 2) generated from the 48 electric power accident effective samples in four consecutive years of this region, the accident mode can be apparently decided as 15. Because this paper chooses the Logistic regression method, we define the samples greater or equal to 15 are high incidents of accident and assign $y=1$; the samples less than 15 are low incidents of accident and assign $y=0$.

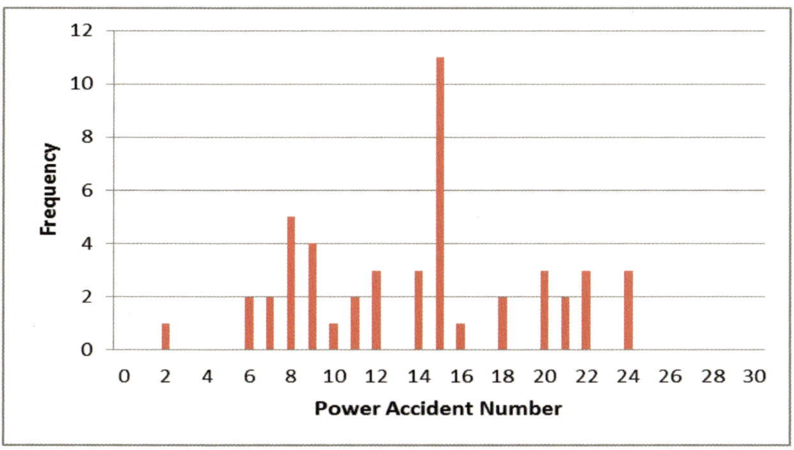

Fig. 2 Power accident frequency

Meanwhile, select the monthly meteorological observation data in this area corresponding to the year, and there are 15 kinds of meteorological factors: the monthly average wind speed X_1, the monthly maximum wind speed X_2, the monthly extreme wind speed X_3, the monthly mean relative humidity X_4, the monthly minimum relative humidity X_5, the monthly total precipitation X_6, the daily maximum total precipitation X_7, the monthly mean temperature X_8, the monthly maximum temperature X_9, the monthly minimum temperature X_{10}, the monthly extreme high temperature X_{11}, the monthly extreme low temperature X_{12}, the number of thunderstorm days per month X_{13}, the number of haze days per month X_{14}, the number of lightning days per month X_{15}. In the empirical analysis, use application factor analysis method to study the internal dependency relation among variables, then do further dimension decreasing and simplified analysis of the four years date, finally, the main common factor can be calculated.

4. Establish Econometric Model

4.1. *Factor Analysis*

Use SPSS18.0 to do the factor analysis, and choose principle component method to do the factor extraction. Choose the variance contribution rate of the selected factors to indicate the importance of the factor structure. The greater the variance contribution rate, the greater the percentage of the sum of the original variable variance explained by the factor, that is, the factor is more important. If the cumulative variance contribution rate of all the factors increases, the representativeness of the original variables by the factors will also increases. There are 15 independent variables in the original data, while it can get 15 principle components in theory, the total variance explained in Table 1. Generally, select the eigenvalue over or equal to 1 or the cumulative contribution rate of the eigenvalue over or equal to 85% to determine the number of common factors. Therefore, extract first 4 components, and can explain 85.673% of the original variables in total.

Table 1 Total variance explained

Component	Initial Eigenvalue		
	Total	Variance Contribution Rate	Cumulative Variance Contribution Rate
F_1	8.169	54.498	54.498
F_2	2.248	15.019	69.529
F_3	1.529	10.216	79.743
F_4	0.903	6.028	85.673
F_5	0.629	4.204	89.979
F_6	0.504	3.393	93.378
F_7	0.388	2.609	95.989
F_8	0.193	1.297	97.298
F_9	0.148	1.022	98.323

F_{10}	0.114	0.779	99.103
F_{11}	0.067	0.449	99.559
F_{12}	0.049	0.334	99.899
F_{13}	0.011	0.081	99.978
F_{14}	0.002	0.013	99.999
F_{15}	0.000	0.001	100.000

4.2. *Factor Explained*

In order to explain the common factors and eliminate the subjectivity of interpretation, using the factor rotation method could make the projection of each original vector on new coordinate axes polarize to 1 and 0 as far as possible, and then get the rotated component matrix (Table 2). Rotated component matrix illustrates the factor load of each meteorological element to these common factors. It is the correlation of common factor and variable, while the greater the value, the more closely the relationship is. Select the variables with larger load of each factor, define the factor an appropriate name with the meaning of the variables. From Table 2, the load to the original variable of each factor is shown below:

Table 2 Rotated component matrix

Meteorological Factor Index	F_1	F_2	F_3	F_4
Monthly Mean Temperature X_1	0.951	0.194	0.148	-0.148
Monthly Maximum Temperature X_2	0.962	0.171	0.100	-0.130
Monthly Minimum Temperature X_3	0.937	0.201	0.186	-0.161
Monthly Extreme High Temperature X_4	0.920	0.150	0.155	-0.094
Monthly Extreme Low Temperature X_5	0.901	0.214	0.176	-0.259
Monthly Mean Relative Humidity X_6	0.119	0.347	0.859	0.017
Monthly Minimum Relative Humidity X_7	0.437	-0.014	0.689	-0.086
Monthly Total Precipitation X_8	0.171	0.867	0.359	-0.117
Daily Maximum Total Precipitation X_9	0.142	0.842	0.202	-0.161
Monthly Mean Wind Speed X_{10}	-0.579	-0.477	-0.476	0.159

Monthly Maximum Wind Speed X_{11}	-0.242	-0.114	-0.114	0.912
Monthly Extreme Wind Speed X_{12}	-0.167	-0.128	0.052	0.937
Number Of Thunderstorm Days Per Month X_{13}	0.481	0.529	0.502	-0.040
Number Of Haze Days Per Month X_{14}	-0.033	-0.327	-0.772	0.010
Number Of Lightning Days Per Month X_{15}	0.453	0.614	0.132	-0.053

Common factor F_1 has greater load on X_1, X_2, X_3, X_4,X_5, so it is mainly determined by the monthly mean temperature, the monthly maximum temperature, the monthly minimum temperature, the monthly extreme high temperature and the monthly extreme low temperature these five meteorological elements. It can represent the effect of temperature on power system accident, define it as temperature factor.

Common factor F_2 has greater load on X_8, X_9, X_{13}, X_{15}, so it is mainly determined by the monthly total precipitation, the daily maximum total precipitation, the number of thunderstorm days per month and the number of lightning days per month these four meteorological elements. It can represent the effect of precipitation on power system accident, define it as precipitation factor.

Common factor F_3 has greater load on X_6, X_7, X_{14}, so it is mainly determined by the monthly mean relative humidity, the monthly minimum relative humidity and the number of haze days per month these three meteorological elements. It can represent the effect of humidity on power system accident, define it as humidity factor.

Common factor F_4 has greater load on X_{11}, X_{12}, so it is mainly determined by the monthly maximum wind speed and the monthly extreme wind speed these two meteorological elements. It can represent the effect of wind on power system accident, define it as wind factor.

After determining the significance of the meteorological information by each factor, it can obtain the linear expression of various factors from the original meteorological elements according to Table 3, the expression of F_1 is shown in Eq.(3):

$$\begin{aligned}F_1=&0.223X_1+0.236X_2+0.209X_3+0.221X_4+0.188X_5-0.094X_6+\\&0.041X_7-0.098X_8-0.095X_9-0.042X_{10}+0.077X_{11}+0.089X_{12}+\\&0.021X_{13}+0.114X_{14}+0.044X_{15}\end{aligned} \quad (3)$$

Table 3 Component score coefficient matrix

Meteorological Factor Index	F_1	F_2	F_3	F_4
Monthly Mean Temperature X_1	0.223	-0.037	-0.062	0.046
Monthly Maximum Temperature X_2	0.236	-0.033	-0.091	0.062
Monthly Minimum Temperature X_3	0.209	-0.047	-0.039	0.035
Monthly Extreme High Temperature X_4	0.221	-0.064	-0.041	0.068
Monthly Extreme Low Temperature X_5	0.188	-0.045	-0.038	-0.032
Monthly Mean Relative Humidity X_6	-0.094	-0.088	0.452	0.006
Monthly Minimum Relative Humidity X_7	0.041	-0.322	0.442	-0.046
Monthly Total Precipitation X_8	-0.098	0.437	-0.081	0.024
Daily Maximum Total Precipitation X_9	-0.095	0.492	-0.183	0.002
Monthly Mean Wind Speed X_{10}	-0.042	-0.075	-0.101	0.003
Monthly Maximum Wind Speed X_{11}	0.077	0.093	-0.078	0.546
Monthly Extreme Wind Speed X_{12}	0.089	0.017	0.037	0.562
Number Of Thunderstorm Days Per Month X_{13}	0.021	0.124	0.103	0.062
Number Of Haze Days Per Month X_{14}	0.114	0.065	-0.412	0.021
Number Of Lightning Days Per Month X_{15}	0.044	0.322	-0.185	0.087

4.3. *Logistic Model*

Select F_1, F_2, F_3, F_4 as regression variables and use SPSS software to do Logistic regression. Choose entering method when the variable entering the model, and all the independent variables will be put into the model.

Table 4 Variables in the Logistic equation

Variable	B Test	S.E.	Wald	Sig
F_1	4.408	2.035	4.687	0.031
F_2	2.219	1.074	4.291	0.039

F_3	1.776	0.901	3.926	0.048
F_4	2.801	1.327	4.446	0.036
Constant	-2.371	1.036	5.236	0.021

The result will be shown in Table 4. B represents the value of test statistic; S.E. is the asymptotic standard error of parameters, reflects the estimation of the reliability of the variables, and represents the discrete degree of the data; Wald is Chi-square test statistic, used to test for a significant degree of the partial regression coefficient, the independent variable shows more significant effect when the value becomes bigger; Sig represents the significance level, if given significance probability as α=0.05, then the confidence level will be 0.95 and indicates that the true value falls within a certain range by a reliability of 95%.

Through the Table 4, the four common factors all pass the test of Sig, which illustrates all the factors having significant relation with the dependent variables. Then, we can get the model of meteorological elements on power accident in Eq.(4):

$$\ln\left[\frac{p}{1-p}\right]=-2.371+4.408F_1+2.219F_2+1.776F_3+2.801F_4 \qquad (4)$$

That is also shown as in Eq.(5):

$$p=\frac{\exp(-2.371+4.408F_1+2.219F_2+1.776F_3+2.801F_4)}{1+\exp(-2.371+4.408F_1+2.219F_2+1.776F_3+2.801F_4)} \qquad (5)$$

In this equation, p indicates the probability of power accident occurrence by the effects of meteorological conditions, and F_i (i=1,2,3,4) indicates the score of common factor for fitted model.

5. Model Testing

According to the Logistic regression equation, set 0.5 as the primitive variable to do the determination. If the value of p is bigger than 0.5, then determine the accident will occur; if the value of p is less than 0.5, then determine the accident will not occur.

Select the power accident data and meteorological data of 12 months in 2014 as the test samples to verify the accuracy of the fitted model. Table 5 gives the test samples result of the model. From Table 5, it can conclude that the accurate rate of 6 non-accident events was 100%, the accurate rate of 6 accident events was 66.7% and the total accurate rate was 83.3%.

Table 5 Test results of the model to test samples

Month	January	February	March	April	May	June
P/%	0.8	27.5	11.3	40.5	42.8	99.3
Number of Accident	9	4	13	17	13	18
Accident Judgment	0	0	0	0	1	1
Month	July	August	September	October	November	December
P/%	98.7	59.4	64.7	9.8	18.4	2.3
Number of Accident	22	17	16	7	5	10
Accident Judgment	1	1	1	1	0	0

From this we can know, this model can precisely illustrate the relationship between the power accident and meteorological element of the chosen region in the statistical year. Besides, the temperature factor is the most influential factor to the electric power accident.

Especially in summer, on one side, the high temperature will increase the power grid load which will also increase the probability of occurrence of the electric power accident; on the other side, the high temperature will have bad effect on the insulation of the obsolete equipment and lines. Secondly, the impact of the wind factor on power accidents is also relatively large. The impact of the wind factor on power accidents are mainly reflected in power tower collapsed and transmission lines broken. The heavy wind will blow down or cut off the transmission lines, which can cause partial flashover accidents and lead to power supply interruption. Besides that, the precipitation factor is another important factor affecting the power accident. Heavy precipitation may ruin the power line pole or reduce the insulation performance of transformer and other electrical equipment, which will lead to various leakage accidents. While in the summer, the heavy rainfall always companied with lightning and the transmission and transformation equipment are mostly exposed to the air or set up in the upper air, which suffered a larger probability of hit by lightning.

Last, the humidity factor can also cause the power accidents. The influence of the humidity factor on electric power accident mainly depends on the contamination of the equipment. The moisture content will increase the conductivity of the contamination layer on insulator, which can reduce the insulation property of the equipment and the probability of pollution flashover on insulator will increase heavily.

Furthermore, the determination accuracy on occurring accident of this model is 66.7%，comparing with that of non-occur accident, the accuracy is quite low. Because the factors that lead to electric power accidents not confined to the bad weather conditions, the equipment failure, human operational error and some other reasons will also cause the electric power accidents.

6. Conclusion

According to the Logistic equation constructed, the meteorological factors that affect the electric power accidents sorted by the importance degree are: temperature factor, wind factor, precipitation factor and humidity factor. The test sample of 2014 gives the result that the accuracy rate of the occurrence of accident is 66.7% and the total determination accuracy can reach about 83.3%. The impact model has a good fitting precision to the influence of meteorological condition on electric power accident. This impact model quantifies the relationship between the meteorological conditions and electric power accident, which can help power corporations summarize the features of the power accidents. The power corporation can choose relevant preventive measures to reduce the possible losses of the future.

Acknowledgment

Science and Technology Project of State Grid Corporation of China (SGSHPD00ZSJS1406054).

References

1. Xue Yusheng, Fei Shengying, Bu Fanqiang. Upgrading the blackout defense scheme against extreme disasters-part i: new challenges and reflection[J]. Automation of Electric Power Systems,2008,32(9):1-6.
2. Lu Jiazheng, Jiang Zhenglong, Lei Hongcai, et al. Analysis of Hunan power grid ice disaster accident in 2008[J]. Automation of Electric Power Systems,2008,32(11):16-19.
3. Fu Guiqin, Cao Xin. The characteristic analysis of thunderstorm gale and electrical network disaster in Hebei[J]. Meteorological Monthly, 2012,38(3):353-357.
4. Lin Weifang, Sun Huadong, Tang Yong, et al. Analysis and lesson of the blackout in Brazil power grid on November 10,2009[J]. Automation of Election Power Systems,2010,34(7):1-2.
5. Zeng Ming, Li Na, Dong Jun, et al. Key technologies of power grid operation and management based on philosophy of general security a view of blackout in India[J]. Automation of Election Power System,2012,36(16):9-13.

6. Liao Feng, Xu Congying, Yao Jiangang, et al. Load characteristics of changed region and analysis on its influencing factors[J].Power System Technology,2012,36(7):117-125.
7. Liu Pingyuan, Luo Senbo, Zheng Xiaoguang, et. al. Research of meteorological warning system on anti-pollution flashover in Guangdong power grid[J].Southern Power System Technology, 2008, 2(5):67-70.
8. Zhang Jingwei, Zheng Lin, Yin Jianmin, et. al. The analysis of main meteorological factors on the impact of the power load of the Nanchang city in Poyang lake ecological economic region: based on Nanchang[J]. Journal of Jiangxi Normal University: Natural Science Edition, 2009,33(6):743-748.
9. Niu Dongxiao, Li Yuanyuan, Qi Jianxun, et al. A novel approach for load analysis based on empirical mode decomposition and influencing factors[J].Proceedings of the CSEE,2008,28(16):98-99.
10. Zhang Ziyin, Ma Jingjin, Lei Yangna. Beijing electric power load and relation with meteorological factors in summer[J]. Journal of Meteorological Science,2011,22(6):760-765.
11. Tripepia G, Jagerb K J, Stelb V S, et al. How to deal with continuous and dichotomic outcomes in epidemiological research: linear and logistic regression analyses[J].Nephron Clin Pract,2011(118):c399-c406.
12. He Yongxiu, Zhu Jiang, Luo Tao, et al. Risk assessment of natural disaster in urban electric power network planning[J]. Transactions of China Electrotechnical Society,2011,26(12):205-210.
13. Wang Shaohua, YE Ziqiang. Influence of atrocious weather on transmission lines of Zhejiang Electric Power Grid[J]. Electric Power, 2011, 44(2): 19-22, 57.
14. Billinton R, Kumar S, Chowdhury N, et al. A reliability test system for educational purposes-basic data[J]. IEEE Trans on Power Systems, 1989, 4(3):1238-1244.
15. Wei Zhenbo, Liu Junyong, Zhu Guojun, et al. Vulnerability evaluation model to power grid based on reliability-parameter-weighted topological model[J].Transactions of China Electrotechnical Society, 2010,25(8): 131-137.
16. Varela G, Novoa N, Jimenez M F, et al. Applicability of logistic regression risk modeling to decision making in lung cancer resection[J].Interactive Cardiovascular and Thoracic Surgery,2003(2):12-15.
17. Zou Zhihong, Yun Yi, Wang Huiwen, et. al. Research for prediction of water quality basing on logistic regression[J].Mathematics in Practice and Theory,2008,38(1):82-87.

Optical Analysis and Simulation of Spherical Vapor Cell

Yu-Ting Zhao†, Gang Liu, Wei Quan and Ming Ding

School of Instrument Science and Opto-Engineering, Beihang University, Beijing, China

†E-mail: zhaoyt@buaa.edu.cn

Accurate theoretical calculation of the outcome light that goes through an ideal spherical vapor cell and generated the matrix of transmissivity based on calculation results has been determined. The calculation demonstrates that the transmissivity of the spherical vapor cell is related to positions. For different incident light, the outcome light shows different character with the highest transmissivity for linearly polarized light is 88.43%. When the incident light is a uniform circularly polarized light field, the outcome light field is nearly a uniform field. Then simulating images and experiment are provided. The comparison between these two results has proved the validity of the calculation and simulation. The method proposed in this paper is useful to quality inspection on spherical vapor cell. The calculation and simulation method has provided possible ways to quality inspection for spherical surface and can also be improved to apply on inspection on other curve surfaces.

Keywords: Spherical Vapor Cell; Optical Analysis; Simulation; Transmissivity; Quality Inspection.

1. Introduction

Atom-spin instruments have been developing rapidly in recent years and have been widely used in variety areas such as atomic gyroscopes, atomic magnetometers and atomic clocks [1–4]. Atom-spin instruments take atoms as sensitive medium and measure physical parameters based on the spin of atoms [1]. The key component of these instruments is alkali vapor cell which is a glass cell filled with alkali-metal atoms and other working substances [5]. The limitation of sensitivity partly depends on the performance of the vapor cell.

Efforts have been made to improve the performance of vapor cells, such as coating, optimizing the parameter of the cells [6–8] and developing the fabrication methods. The two common methods of fabrication of vapor cells are glass-blowing technique [5, 9] and microfabrication technique [10–14] at present.

Among all of these cells, spherical cells are preferred in some specific applications to minimize the magnetic self-fields generated by the atoms inside the cell and it also provides more optical ports than other types of cells [9]. However, spherical cells have curved surface which makes it hard to achieve high quality. Besides, the refraction and reflection effects of curved surfaces on light beam also makes it hard to select quality cells. The analysis of these effects is urgent needed to achieve a better performance.

In this paper, we calculate relevant parameters of a spherical vapor cell based on an ideal model of a typical spherical vapor cell, especially the light tracing and transmissivity, to know the detailed information of the outcome light field and the effect of the cell on incident light field. Then, simulating images and experiment are proposed. We make simulating images under specific incident condition to make comparison to experimental result. This result is followed by a discussion of ways in which the quality of spherical vapor cells can be detected to select high quality ones in the future.

2. Modeling and theoretical calculation

Most of the spherical vapor cells are approximately a hollow sphere, and the Inner Diameter (ID) is close to the Outer Diameter (OD). It has been published in the past that a spherical shell can represent the vapor cell well [13, 15]. Furthermore, an ideal spherical model can simplify the calculation efficiently. Therefore, we take an ideal spherical shell as the model of the analysis.

When a uniform optical field is incident on an ideal spherical glass shell, the light beam will be refracted except for the optical axis. As a result of this effect, the emergent light field will show a character of divergence. Besides, the incident angle of individual points is relate to the position of the incident point, the transmissivity for individual points can be different for different point. Therefore, a theoretical calculation of an ideal model will contain two parts, the calculation of light tracing and the calculation of transmissivity.

2.1. *Calculation of the light tracing*

The incident light beam that goes through an entire spherical cell will meet four times of refractions, as is shown in Fig. 1. The incident light is represented by the black line and arrow; the first, the second and the third reflected light is represented by red, green and blue line and arrow respectively; the outcome light is represented by purple color. The geometric optical tracing can be calculated one by one. The first refraction can be described as following equations

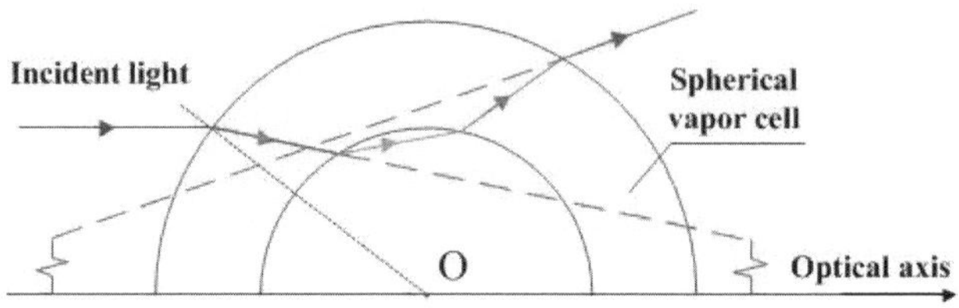

Fig. 1. The light tracing of an incident light beam that goes through an ideal spherical vapor cell. O is the center of the cell.

$$n \cdot sin\theta_1 = n' \cdot sin\theta_1' \tag{1}$$

$$sin\theta_1 = sin\varphi_1 = \frac{h_1}{r} \tag{2}$$

$$u_1 = \theta_1 - \theta_1' \tag{3}$$

where n and n' is the refractive index of air and cell material for specific wavelength respectively, $2r$ is the OD of the vapor cell, all θ and θ' in this paper represent the incident angles and the refracted angles for different refractions, u is the angle between optical axis and the emergent light, φ is the angle between optical axis and the normal, h is the height from the incident point to the optical axis. All the subscripts on the right side of these letters represent the number of refraction. Eq. (1) follows the refraction law of the first refraction; Eq. (2) and Eq. (3) show the geometrical relationship of angles that involved. Therefore, θ_1 and θ'_1 can be calculated.

Similarly, the second refraction can be described as follows

$$n' \cdot sin\theta_2 = n'' \cdot sin\theta_2' \tag{4}$$

$$r \cdot sin\theta_1' = r' \cdot sin\theta_2 \tag{5}$$

$$-u_2 = (\theta_2' - \theta_2) + (\theta_1' - \theta_1) \tag{6}$$

$$\varphi_2 = \theta_2 + \theta_1 - \theta_1' \tag{7}$$

where n'' is the refractive index of vapor inside the cell for specific wavelength, in normal conditions, n'' is equal to n, since the selection of vapor cell usually happened before the cell is filled and sealed, and the vapor inside and outside the chamber are the same; $2r'$ is the ID of the vapor cell. Generally, $2r'$ cannot be measured directly, we use the empirical value or the simulation result necessarily.

The third and the forth refraction can be calculated easily since an ideal spherical cell is a central symmetry geometry. Therefore, we can draw the conclusion that

$$\theta_3 = \theta_2', \ \theta_3' = \theta_2, \ \theta_4 = \theta_1', \ \theta_4' = \theta_1 \tag{8}$$

Other angles that involved in these two refractions can also be calculated. The results are represented as follows

$$-u_4 = 2 \cdot (\theta_2' - \theta_2) + 2 \cdot (\theta_1' - \theta_1) \tag{9}$$

$$-\varphi_4 = 2\theta_2' - 2\theta_2 + 2\theta_1' - \theta_1 \tag{10}$$

$$h_4 = r \cdot sin(-\varphi_4) \tag{11}$$

Therefore, if parameters of the spherical vapor cell and the wavelength of the incident light are fixed, the light tracing of every point can be calculated by Eq. (9) and Eq. (11), where Eq. (9) is the angle of the outcome beam and Eq. (11) is the height of the outcome point.

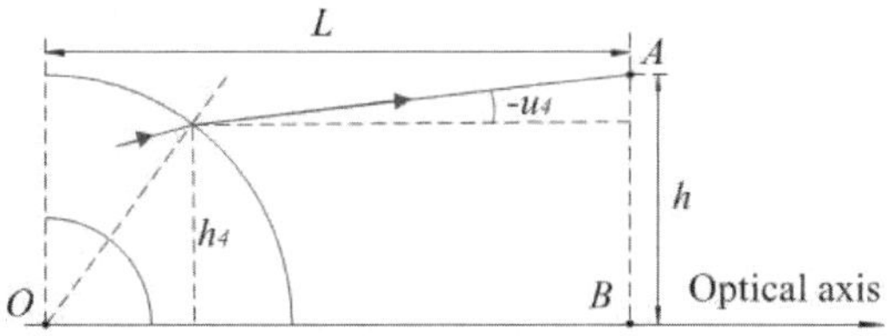

Fig. 2. The emergent light of the vapor cell. OB is the Optical axis and AB shows the position of a screen.

However, to generate the ideal image, we have to follow the light tracing to a receiving screen. As is shown in Fig. 2, AB indicated the receiving screen and it is in a distance of L from the center of the vapor cell, then the accurate position of emergent light can be represented as follows

$$h = h_4 + tan(-u_4) \cdot [L - cos(-\varphi_4) \cdot r] \tag{12}$$

where h is the distance from the recorded point on the screen to the optical axis which means that the original detect light of height h_1 will be refracted four times and eventually emergent on the position of height h on receiving screen. Therefore, if the parameters of the cell are known, the tracing of the detecting light that goes through every point on the surface of the vapor cell can be calculated accurately.

In fact, most of the vapor cells have a very thin glass wall which means that r' and r are considered to be close to each other. Besides, the calculation above has a limitation on h_1 that $h_1 \le r' \cdot \frac{n''}{n} \approx r'$. When h_1 is larger than r' and smaller than r, the light beam will be totally reflected and no energy can travel to the screen.

2.2. *Calculation of transmissivity*

To a spherical vapor cell, different point on the surface has different incident angle and only the center point of the surface is incident vertically. The transmissivity is related to these angles and these angles are depended on h_l.

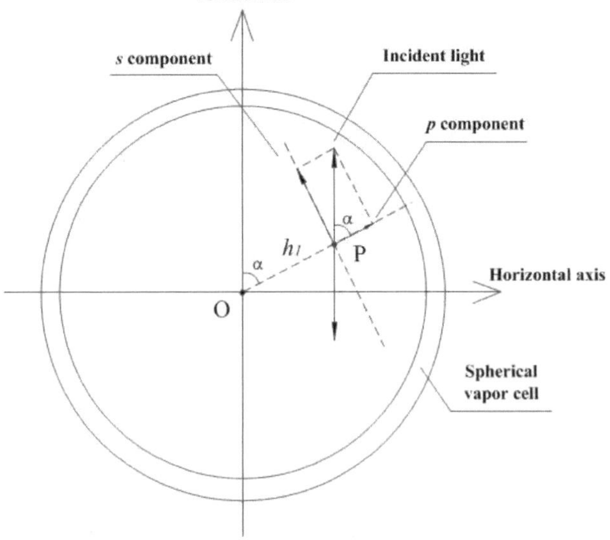

Fig. 3. The sketch of decomposition of a linearly polarized incident light.

Here, we take a linearly polarized incident light as an example. As is shown in Fig. 3, to an arbitrary point P (P is the subpoint), when a linearly polarized beam is used to detect the vapor cell and its direction of vibration is along the vertical axis, the optical beam can be decomposition into two independent orthogonal components, s and p component. As we can know from Fresnel formula, the transmissivity of p component and s component can be written as

$$\Gamma_{sl} = [1-\frac{sin^2(\theta_l-\theta_l')}{sin^2(\theta_l+\theta_l')}]^2\cdot[1-\frac{sin^2(\theta_2-\theta_2')}{sin^2(\theta_2+\theta_2')}]^2 \tag{13a}$$

$$\Gamma_{pl} = [1-\frac{tan^2(\theta_l-\theta_l')}{tan^2(\theta_l+\theta_l')}]^2\cdot[1-\frac{tan^2(\theta_2-\theta_2')}{tan^2(\theta_2+\theta_2')}]^2 \tag{13b}$$

As is shown in Fig. 3, in this case, the decomposition of s and p components follows Marlus principle, and the intensity of the outcome light can be written as

$$I_t = I_p \cdot \Gamma_{pt} + I_s \cdot \Gamma_{st} \tag{14}$$

where I_t is the intensity of the emergent light on receiving screen, I_p and I_s is the intensity of p component and s component respectively. We assume that I is the intensity of the incident light, then the transmissivity K of this point can be calculate as follows

$$K = \frac{I_t}{I} \tag{15}$$

For linearly polarized incident light with a horizontal vibrational direction or circularly polarized light, the decomposition is similar to equations above. Therefore, the transmissivity can be calculated without knowing the light intensity.

3. Simulation and analysis

The image analyzed by the computer is basically a series of data. The data is set as a format of matrix, therefore, when we are simulating the outcome light field, it means to generate a matric that describe the distribution of light intensity. This matrix should include the position information and the light intensity that is corresponding to this position. We describe this matrix as $K(a,b)$, while a and b represent the position and value $K(a,b)$ is the light intensity of this point. The individual position in this matrix can be regarded as pixels.

In order to generate a simulating image, we need an original light field and then calculate the transmissivity for all different pixels. Meanwhile, the average of the light intensity should also be counted. We assume that the center point O of the vapor cell is $K(a_0,b_0)$, then for a random pixel $K(a_m,b_n)$ on the simulating image, parameters of this point can be written as follows

$$\alpha = arctan[\frac{L_b \cdot (a_m - a_0)}{L_a \cdot (b_n - b_0)}] \tag{16}$$

$$h = \sqrt{[L_b \cdot (a_m - a_0)]^2 + [L_a \cdot (b_n - b_0)]^2} \tag{17}$$

where α is the angle between p component and the vertical axis, L_a and L_b (micron/pixel) is the pixel scale of horizontal and vertical axis respectively. L_a and L_b are only related to the parameters of the recording system.

For conditions that the incident light field is unknown and all parameters of vapor cell are known, we can solve Eq. (12) to gain the corresponding value h_1 for different h, and the transmissivity is then calculated by Eq. (15) for each point. Transmissivity of individual point show differences and can generate a matrix, as is shown in Fig. 4, all the image here are generated in the condition of $2r = 12mm, 2r' = 11.6mm,\quad n = n'' = 1, n' = 1.4714,\quad La = 0.086mm, Lb = 0.07885mm,$ $L = 300mm$, which is quite common in experiments. All images are generated by Matlab with 141*153 elements.

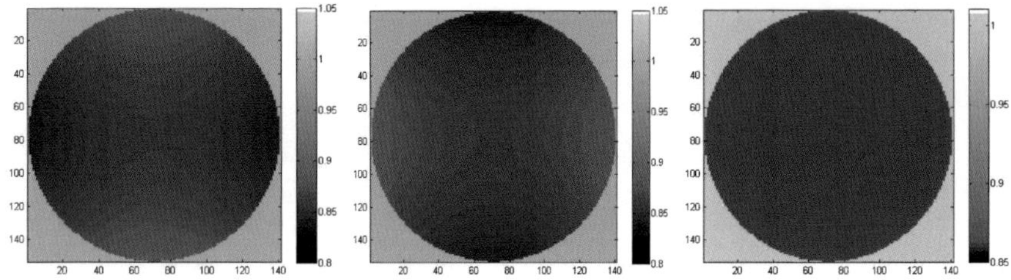

(a) (b) (c)

Fig. 4. The images of matrix of transmissivity for different incident light: (a) in condition of a linearly polarized incident light with vertically vibrational direction (b) in condition of a linearly polarized incident light with horizontally vibrational direction (c)in condition of a natural or circularly polarized light.

Figs. 5(a) and (b) are the images of transmissivity matrix with a linearly polarized incident light with a vertical direction and a horizontal direction respectively. As is shown in these images, vapor cells do affect the distribution of the light intensity, especially on the position that near the edge. The transmissivity is basically a symmetrical pattern and fluctuates from 83.80% to 88.43% in both Figs. 5(a) and (b). The effect of vapor cell on the transmissivity of circularly polarized light field is relatively small, as is shown in Fig. 5(c) The maximum transmissivity is 86.21% and the minimum transmissivity is 86.14%. This means that if a vapor cell is detected by an optical method and the incident light is a uniform circularly polarized light field, the outcome should also be a nearly uniform light field. However, as we can see from Figs. 5(a) and (b), the transmissivity of the two components on the same point is different which means that the transmissivity of individual point is nearly equal but the state of polarization is different. To be accurate, the outcome beam is elliptically polarized light except for the light beam that goes through point O.

The matrix of transmissivity can help to generate the simulation image. For conditions that the incident light field is known, the simulating image is the product of the two matrix.

4. Experiment

A spherical vapor cell has been inspected to prove the calculation and simulation results. The vapor cell is made by glass blowing method with Schott 8436 tube. The OD is 12mm with a thickness of 0.9mm. Commonly, vapor cells used on atomic instruments have a thickness which is no more than 0.2mm to minimums the effect of refraction. To make a contrast of the simulation image and the experimental image, we use a CCD camera to record images, and the parameters of the CCD camera is L_a=0.0172 and $L_b = 0.015$ mm/pixel. The incident light is 770nm. Three images are demonstrated below in Fig. 7.

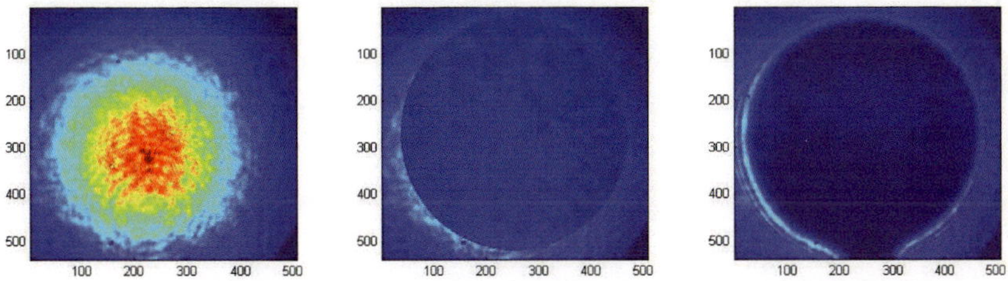

(a) (b) (c)

Fig. 5. Three images of quality inspection (a) the original image of the detect laser beam; (b) the simulation image of the outcome; (c) recorded image by a CCD camera.

The light that is used to detect the vapor cell is a Gauss beam with a diameter of 8mm as is shown in Fig. 5(a). Fig. 5(b) is the simulation image and (c) is the experimental image. Figs. 5(b) and (c) show the same character which means that the calculation and simulation are correct. The light intensity in simulating image is higher than experimental image. This is because of the absorption of cell material.

We envisioned that this method of calculation and simulation of outcome light field can be applied on quality inspection of spherical glass cell. Generally speaking, optical quality inspections involve image recording system. A photo of the incident light field is recorded as a matrix I, then simulating result of the outcome light field should be I_t as is shown in Eq. (14) if the cell is of an ideal shape. Therefore, the experimental image can be compared to I_t to show defects on the surface of the cell. The advantage of this method is that requirement of the quality of incident light field is relatively low.

5. Conclusion

We made theoretical calculation of the emergent light that goes through an ideal spherical vapor cell and generated the matrix of transmissivity based on calculation results. The calculation shows that the transmissivity of the spherical vapor cell is related to positions. For different incident light, the outcome light shows different character. When the incident light is a uniform circularly polarized light field, the outcome light field is nearly a uniform field. The simulating images and experiment has proved the validity of the calculation and simulation. The method proposed is useful to quality inspection on spherical vapor cell. The calculation and simulation method has provided possible ways to quality inspection for spherical surface and can also be improved to apply on inspection on other curve surfaces.

Acknowledgments
This work is supported by National Science Foundation of China (NSFC) (61227902, 61374210, and 61121003).

References

1.Dmitry Budker and Michael Romalis, Nature Physics, 3, 227–234 (2007).
2.Filler A. G, Internet Journal of Neurosurgery, 7(1), 1-85 (2009).
3.Rice H. F. and Benischek V., 2008 IEEE/ION, Monterey, CA, IEEE: 933-939 (2009).
4.Groeger S., Bison G, and Schenker J. L., European Physical Journal D, V38 (2): 239-242 (2006).
5.Bell W. E., Bloom A. L. and Lynch J., The Review of Scientific Instruments, V32 (6): 688-691 (1961).
6.M. A. Bouchiat, J. Brossel, Physical Review, V147(1):41-54 (1966).
7.M. V. Balabas, D. Budker, J. Kitching, P. D. D. Schwindt, J. E. Stalnaker, Journal of Optical Society of America B, 23(6):1001-1006 (2005).
8.Wei Quan, Yang Liu, Yao Chen, Chinese Physics Letters, 31(3), 030701, 2014.
9.E. Jesper Eklund, Andrei M. Shkel, Svenja Knappeb, Elizabeth Donley, John Kitching, Sensors and Actuators A Physical, 143(1):175-180 (2008).
10.L. A. Liew, S. Knappe, J. Moreland, H.Robinson, Applied Physics Letters, 84(14):2694-2696 (2004).
11.Y. P′etremand, C. Affolderbach, R. Straessle, M. Pellaton, D. Briand, G Mileti, N. F. de Rooij, Journal of Micromechanics and Microengineering, 22(2):25013-25020 (2012).
12.E. J. Eklund, A. M. Shkel, Journal of Microelectromechanical Systems, 16(2):232-239 (2007).
13.T. Baluktsian, C. Urban, T. Bublat, H. Giessen, Quantum Physics, 35(12):1950-1952 (2010).
14.Nezih Dural and Michael V. Romalisa, Applied Physics Letter Materials, 2(8):641-644 (2014).
15.R. Straessle, M. Pellaton, C. Affolderbach, Y. Petremand, Journal of Applied Physics, 113, 064501 (2013).

Author Index

An, 377
An, Jian, 377
Bai, Yan-Li, 356
Bao, Li-Hong, 85
Bian, Hai-Feng, 364
Cai, Heng-Bin, 247
Cai, Qing-Qiu, 73
Cao, Ming-hua, 310
Chen, Guo-Zhen, 152, 401
Chen, Jian-Rong, 435
Chen, Jing, 138
Chen, Jun, 180
Chen, Shao-yi, 385
Chen, Ya-zhou, 278
Dai, Zhi-Yong, 377
Ding, Ming, 456
Ding, Xue-Yong, 270
Dong, Pin-Hua, 197
Dong, Zhi, 19, 225
Fan, Yang-Yu, 340
Fang, Zhu, 321
Feng, Bing, 385, 413
Gao, Fang, 216
Gong, Shang-kun, 385
Guo, Quantong, 105
Guo, Wen, 407
Guo, Xinju, 239
Guoqi, Zhang, 429
Hai-tao, SUN, 210
Han, Jun, 286
Han, Kun, 113
Han, Li-Hong, 35
Hao, Ting-Ting, 407
He, Zhi-ping, 420
Hong, Bao-Li, 85
Hu, Jun, 413
Hu, Yin-Ping, 42
Huang, Jing-Guo, 420
Huang, Shen-Hai, 42
Huang, Yuan-Sheng, 42
Huang, Zhi-Ming, 420
Huang, Zhi-Ping, 19, 225
Huisheng, Ye, 429
Hui-yong, DENG, 210
Jia, Chen, 443
Jia, Ke-jun, 310
Jia, Wen-Qiang, 435
Jian, 377
Jiang, Xin, 105
Jin, Rong-Jiang, 364
Jiu-he, MA, 210
Kuang, Hu, 138
Lan, Qing-Ping, 180
Lei, Ding, 321
Lei, Yanjun, 105
Li, Jie-Jun, 73, 79
Li, Qiang, 295, 317
Li, Rong-Ping, 3
Li, Si-Kun, 391
Li, Xian, 239
Li, Zhen, 385
Li, Zhong-Guang, 171
Li, Zong-Ying, 146, 158
Li. Yun-Qing, 138
Liang, Wang-Feng, 247
Liu , Xin, 56
Liu, Gang, 456
Liu, Hong-Wen, 233
Liu, Wen-Xiang, 119
Liu, Xiang, 146, 158
Lu, Jian-hao, 301
Lv, Yao-Ping, 204
Mao, Qian-Ren, 63
Mao, Yubin, 239

Nie, Yi-xiong, 301
Pan, 332
Pan, Xian-Jun, 204
Pan, Yan-Qing, 332
Peng, Shengliang, 349
Qi, Wang, 391
Qian, Hui, 413
Qin, Jiang-Yi, 19, 225
Qin, Zhou, 119
Qing, Huang, 129, 164
Qing, Yan, 332
Qiu, Bao-Wei, 407
Qiu, Ying-Jun, 413
Quan, Wei, 456
RUAN, Wei, 9
Sang, Bo, 233
Sen, Zhao, 3
Shan, Ning, 113
Shao, Shuai, 197
Shu, Rong, 420
Shu, Zheng-Ping, 171
Si-Ying Wang, 3
Song, Jia-Hao, 377
Song, Ziyu, 349
Su, Ying-Xue, 119
Sui, Hui-Bin, 332
Sun, Wei-Feng, 113
Tang, Hao, 73
Tang, Hao, 79
Tian, Jinwen, 28
Wan, Fang, 286
Wan, Hao-jiang, 278
Wan, Tao, 413
Wang, Chao-Bin, 63
Wang, Chao-Ming, 190
Wang, Chengrui, 443
Wang, Fei-Xue, 119
Wang, Guangwei, 28
Wang, Guo-Hua, 260
Wang, He, 310
Wang, Hong-Feng, 295
Wang, Hui-Qin, 310
Wang, Lian-Sheng, 270
Wang, Lin-lin, 94
Wang, Li-peng, 278
Wang, Ming-Chang, 340
Wang, Shuai, 197
Wang, Xiaojia, 278
Wang, Xu-Dong, 364
Wang, Xue-Wen, 190
Wang, Yuan, 270
Wei, Jun-Yu, 19, 225
Wei, Ming, 317
Wei, Zhang, 85
Weihua, Zhou, 429
Wen, Sheng-ke, 301
Wu, Huan, 435
Xiang, Tian-Chun, 364
Xiao, Fei, 28
Xu, Hao, 254
Xu, Si-Chen, 260
Yang, Han, 79
Yang, Ke-Li, 113
Yang, Qiu-Jie, 420
Yang, Yue-hui, 301
Yaoheng, Xie, 429
Ye, Ji-Hua, 180
Ye, Ya-Lin, 113
Yu, Chun-Lai, 286
Yu, Jinli, 239
Yuan, Quan, 113
Yun, Liu, 429
Yuzhu, Ding, 429
Zhang , Yan, 56
Zhang, Chao, 364
Zhang, Chao, 443
Zhang, Fan, 190
Zhang, Guang-Yong, 152, 401
Zhang, Hui, 295
Zhang, Shao-Fan, 35
Zhang, Wei, 204
Zhang, Wei-Shuo, 260
Zhang, Xiao-Qing, 152, 401
Zhang, Xiao-Yang, 435
Zhang, Yong-Dong, 171
Zhang, Yuan, 190
Zhang, Zhi-Yong, 190
Zhao, Mi-Zhuo, 152, 401
Zhao, Tian, 204
Zhao, Xin, 216

Zhao, Yu-Ting, 456
Zheng, Zhiming, 105
Zhi-wei, Chen, 210
Zhou, Jing, 443
Zhou, Jing, 443
Zhou, Zhou, 385
Zhou, Zhou, 413
Zhu, Jing-wei, 310
Zhu, Shu-Yang, 49
Zhu, Yi-Jun, 254
Zhu, Yi-Jun, 247
Zuo, Zhen, 19
Zuo, Zhen, 225